Lecture Notes in Computer Science 958

Edited by G. Goos, J. Hartmanis and J. van Leeuwen

Advisory Board: W. Brauer   D. Gries   J. Stoer

Springer

Berlin
Heidelberg
New York
Barcelona
Budapest
Hong Kong
London
Milan
Paris
Tokyo

Series Editors

Gerhard Goos, Universität Karlsruhe, Germany

Juris Hartmanis, Cornell University, NY, USA

Jan van Leeuwen, Utrecht University, The Netherlands

Volume Editors

Jacques Calmet
Fakultät für Informatik, Universität Karlsruhe
Postfach 69 80, D-76128 Karlsruhe, Germany

John A. Campbell
Dept. of Computer Science, University College London
Gower Street, London WC1E 6BT, United Kingdom

Cataloging-in-Publication data applied for

Die Deutsche Bibliothek - CIP-Einheitsaufnahme

**Integrating symbolic mathematical computation and artificial
intelligence** : selected papers / Second International Conference
AISMC 2, Cambridge, United Kingdom, August 3 - 5, 1994.
Jaques Calmet ; John A. Campbell (ed.). - Berlin ; Heidelberg ;
New York : Springer, 1995
  (Lecture notes in computer science ; Vol. 958)
  ISBN 3-540-60156-2
NE: Calmet, Jaques [Hrsg.]; International Conference AISMC <2, 1994,
  Cambridge>; GT

CR Subject Classification (1991): I.1-2, G.1-2

ISBN 3-540-60156-2 Springer-Verlag Berlin Heidelberg New York

© Springer-Verlag Berlin Heidelberg 1995
Printed in Germany

Typesetting: Camera-ready by author
SPIN 10486533   06/3142 – 5 4 3 2 1 0   Printed on acid-free paper

Jacques Calmet John A. Campbell (Eds.)

Integrating Symbolic Mathematical Computation and Artificial Intelligence

Second International Conference, AISMC-2
Cambridge, United Kingdom, August 3-5, 1994
Selected Papers

Springer

Foreword

This volume is based on the papers that were accepted for AISMC-2, the second conference on Artificial Intelligence and Symbolic Mathematical Computation. In many cases the material published here contains updated versions of the material that was presented at the conference, which was held at King's College, Cambridge, UK, on 3–5 August 1994. The number of accepted papers was intentionally limited, to give authors longer allocations of time than usual to present their results, and to promote the fruitful discussions that actually occurred after each presentation.

This was the second conference in a series that began in Karlsruhe almost exactly two years earlier. Following the success of these meetings, a third conference in the series is scheduled for an Austrian location in the late summer or the early Fall of 1996, organised from the Research Institute for Symbolic Computation at the Johannes-Kepler-Universität, Linz.

There was significant interest in the connections between mathematics and artificial intelligence in the early days of AI. The evidence is still present in knowledge of some of the early historical highlights of AI, particularly with the work of Gelernter and of Evans in (geometrical) theorem-proving and Slagle on indefinite integration. Now, after a quite long period when AI and mathematics appeared to have arranged an amicable separation, they are growing together again as indicated, for example, by a noticeable recent burst of activity in conferences on the interaction of mathematics and AI.

The work reported in this volume covers a more specific area than the whole of mathematics plus AI. The AISMC series of meetings is intended to focus on the relationships between symbolic mathematical computing (the activity in which the historical work mentioned above belonged, and which has given rise to a number of now well-known software systems such as Reduce and Mathematica, as well as substantial efforts by mathematicians on the algorithms and their properties supporting the operations that such systems provide). This should encourage AI specialists to test their ideas and heuristics on mathematical applications, which have the virtue of being more clear-cut indicators of success or failure than applications in most other areas, and also encourage mathematicians to make use of AI approaches to heuristics and the representation of mathematical heuristic knowledge in their own research.

The material that we publish here comes from diverse origins, demonstrating that the range of interest in the intersection of AI and symbolic mathematical computation is very wide. Exploring the intersection on this wide front is helpful in clearing the ground for future concentration on topics that can unify future work in the field. We have requested authors to include clear demonstrations of the relevance of their work to this intersection, in either the papers presented at AISMC-2 or in updated versions. (This explains the absence of the presented paper "A Generic Computer Algebra Library in Oberon", by D. Gruntz and W. Weck of ETH in Zürich, from this volume).

As was the case after the first conference, there has been a unanimous demand from the attendees to continue the AISMC series on a regular basis. In the projected 1996 conference, it is planned that the engineering aspect of the domain will be stressed, to promote a facet of the relevant research that certainly exists but has not been present to a significant extent in AISMC-1 and AISMC-2.

We thank King's College heartily for offering a perfect setting for this conference. In addition, we are grateful to the local organisers for their efforts, and to the International Science Foundation, which has made it possible for Russian and Ukrainian participants to attend, in a time of substantial currency and travel problems, and to present their papers. This last expansion of the geographical horizons of AISMC has added to the value of the activities of AISMC-2, and we hope to maintain those horizons in future meetings.

June 1995   Jacques Calmet and John A. Campbell

Conference Committee: Jacques Calmet (Germany)
John A. Campbell (UK)

Program Committee: J. Calmet (Germany)
J.A. Campbell (UK)
I. Cohen (Sweden)
Y. Demazeau (France)
J. Fitch (UK)
R. Garigliano (UK)
A. Guttmann (Australia)
P. Ladkin (USA)
A. Miola (Italy)
A. Norman (UK)
J. Perram (Denmark)
Z. Ras (USA)
A. Smaill (UK)
I.A. Tjandra (Canada)
D. Weld (USA)
D. Yun (USA)

Contents

Interactive Theorem Proving
and
Computer Algebra

Johannes Ueberberg

Mathematisches Institut
Arndtstrasse 2, D-35392 Giessen
e-mail: Johannes.Ueberberg@math.uni-giessen.d400.de

1 Introduction

Interactive Theorem Proving, ITP for short, is a new approach for the use of current computer algebra systems to support mathematicians in proving theorems. ITP grew out of a more general project - called Symbolic Incidence Geometry - which is concerned with the problem of the systematic use of the computer in incidence geometry (for a detailed description of the project Symbolic Incidence Geometry the interested reader is referred to [3]).

ITP is a project to develop program packages to support mathematicians in proving theorems in incidence geometry. At the moment we concentrate on finite linear spaces which are a particular class of geometries with finitely many points and finitely many lines (for a definition see 2.1. The programs of ITP are intended to be some sort of "intelligent paper and pencil". They are conceived as a dialog between the user and the system. Typical inputs of the user are as follows:

(1) Let p be a point.
(2) Let g be a line incident with p.
(3) Suppose that there are exactly n points incident with p.

With the first two inputs the user introduces some geometric configurations. The third input is an introduction of some combinatorial assumptions about the class of geometries under consideration. Following each input of the user the systems applies various proof techniques in order to deduce additional information.

In this paper we give a rather short survey about the concepts of ITP. A comprehensive study is in preparation [13] containing the algorithms and the necessary information about the implementation.

This article is organized as follows: In Section 2 we introduce the notion of a finite linear space and we present some of its properties. Section 3 is devoted to the discussion of the so-called partitions. They permit to represent an abstract geometry defined by a set of axioms on a computer, and to deal with imputs of the form:

Let g be a line incident with p.

In Section 4 the knowledge base is introduced. It contains the combinatorial information about the class of linear spaces under consideration. The application of the proof techniques by the system is the topic of Sections 5 - 7. Finally in Section 8 we discuss how use ITP to prove a classical theorem about finite linear spaces, namely the celebrated theorem of de Bruijn and Erdös.

Acknowledgement. I wish to thank Albrecht Beutelspacher for many valuable discussions.

2 Linear Spaces

Definition 2.1 A **linear space** is a geometry consisting of a set $\mathcal{P}$ of points and a set $\mathcal{L}$ of lines such that the following conditions are fulfilled:

(i) Any two points are joined by a unique line.
(ii) On every line there are at least two points. There are at least three non-collinear points, that is, three points that are not on a common line.

Important examples of linear spaces are affine and projective spaces. A **finite linear space** is a linear space with finitely many points and finitely many lines.

Let **L** be a finite linear space, and let x be a point of **L**. The number r_x of lines of **L** through x is called the **degree** of x. Dually, the **degree** of a line l of **L** is the number k_l of points on l.

The following proposition on finite linear spaces is elementary but very useful.

Proposition 2.1 *Let **L** be a finite linear space.*
*a) Let x be a point and let l be a line of **L** such that x and l are not incident. Then $r_x \geq k_l$.*
b) We have

$$\sum_{x \in \mathcal{P}} r_x = \sum_{l \in \mathcal{L}} k_l.$$

Proof. a) The set of lines through the point x consists of k_l lines of **L** joining x and a point of l and of $c = c(x, l)$ lines through x that are disjoint to l. Since $c \geq 0$, it follows that $r_x \geq k_l$.

b) follows by counting the incident point-line-pairs (x, l).

In order to illustrate the concepts of ITP we shall use the following elementary theorem on linear spaces. For an easier reference we enumerate the steps of the proof. In Section 8 we demonstrate how to use ITP to prove the Theorem of de Bruijn and Erdös.

Theorem 2.1 *Let **L** be a finite linear space fulfilling the following conditions.*

(i) *There is a constant $c \geq 0$ such that for each non-incident point-line-pair (x, l) the number of lines through x parallel to l equals c, that is, $r_x = k_l + c$.*
(ii) *Given two lines l and m of **L** there is always a point of **L** that is neither on l nor on m.*

Then every line of **L** *is incident with the same number n of points.*

Proof. The steps of a traditional proof are as follows:
1. Let g be a line of **L** of degree n.
2. In view of condition (i) it follows that $r_x = n + c$ for all points x not on g.
3. Let h be a second line of **L**.
4. By condition (ii) there is a point y on **L** that is neither on g nor on h.
5. Therefore we get $r_y = n + c$ and $r_y = k_h + c$.
6. It follows that $k_h = n$.
7. So we have seen that every line of **L** is of degree n. □

Readers interested in the theory of linear spaces are referred to [1]

3 Partitions

In the following we consider the question how these steps can be performed by a computer.

The concept of the partitions is of central importance. On the one hand it reflects the axiomatic structure of the geometry under consideration. In the case of linear spaces the fundamental axiom

Any two points are joined by a line.

is translated to the system via partitions. On the other hand it allows to apply parameter relations such as

$$r_x \geq k_l \text{ or } r_x = k_l + c.$$

In the proof of Theorem 2.1 we first introduce a line g of degree n (Step 1). To apply the equation $r_x = k_l + c$ for non-incident point-line-pairs (x, l) we partition the point set of **L** in the points on g and the points off g. For every point x not on g we obtain the equation $r_x = n + c$ (Step 4).

In the dialog between the user and the system the user introduces the new points and lines whereas the system computes the resulting partitions for the point and the line set. The points and lines introduced by the user are called **singular points** and **singular lines**. If two or more singular lines intersect in a non-singular point x, then x is called a **special point**.

Definition 3.1 Let **L** be a linear space.

a) If p is a singular point of **L**, then the set $\{p\}$ is called a **singular point component**.

b) Let G be a set of at least two singular lines. Then the set

$$\{x \in \mathcal{P} \mid x \text{ is non-singular and } x \in g \text{ for all } g \in G\}$$

is called a **special component**.

c) Let g be a singular line. Then the set of all non-singular and non-special points on g is called a **linear component**.

d) The set of all non-singular points of **L** that are not incident with any singular line is called the **free point component**.

e) The family of all point components is called a **point partition** of **L**.

Let g and h be two intersecting singular lines, and suppose that there is a singular point p incident with g but not with h.

Then the point partition consists of five components; the singular component $\{p\}$, the special component $\{g \cap h\}$, the linear components $\{x \in g \mid x \neq p$ and $x \notin h\}$ and $\{x \in h \mid x \notin g\}$, and finally the free component $\{x \in \mathcal{P} \mid x \notin g$ and $x \notin h\}$.

The line partitions are defined in a similar way. For the exact definition and the algorithms for the computation of the partitions cf. [13].

4 The Knowledge Base

In the proof of Theorem 2.1 we collected some information about the various parameters of the linear space: In Step 1 we made the assumption that the line g is of degree n. Then we deduced the relation $r_x = n + c$ for all points x not incident with g.

Following the terminology of logic programming languages like PROLOG these parameter relations are called *facts*. The system will store these facts in a program component called *knowledge base*.

The fact $r_x = n + c$ has been obtained by applying the equation

$$r_x = k_l + c \text{ for } x \notin l$$

(condition (i)) on the fact $k_g = n$. The expression $r_x = k_l + c$ for $x \notin l$ is called a rule. More formally, a *rule* consists of a parameter relation and of the description of the incidence relation that has to be fulfilled for the correct application of the rule. The parameter relation of the above rule is the equation $r_x = k_l + c$, the incidence relation of this rule is the condition that the point x and the line l are not incident. Further examples of rules are given by Proposition 2.1.

5 Formulation of Parameter Relations

One way to obtain new relations between the various parameters is to apply the rules on the facts and on the actual partitions. In Step 1 of the proof of Therorem 2.1 we introduced a line g of degree n. The system will partition the point set into two components C_1 and C_2

$$C_1 := \{x \in \mathcal{P} \mid x \in g\},$$
$$C_2 := \{x \in \mathcal{P} \mid x \notin g\}$$

and the line set into three components D_1, D_2, and D_3:

$$D_1 := \{g\},$$
$$D_2 := \{l \in \mathcal{L} \mid l \neq g, l \cap g \neq \emptyset\},$$
$$D_3 := \{l \in \mathcal{L} \mid l \cap g = \emptyset\}.$$

In order to apply the rule

$$r_x = k_l + c \text{ for } x \notin l,$$

the system has to search for the pairs (C, D) of a point component C and a line component D such that for each point x of C there is a line of D not incident with x.

6 Analysis of Parameter Relations

Once the parameter relations are established, the resulting system of equations and inequalities has to be solved. In Step 5 of the proof of Theorem 2.1 we obtained the following two equations:

$$r_y = n + c; \; r_y = k_h + c.$$

The system of these two equations has to be solved in order to obtain the final result $k_h = n$ (Step 6). There are several methods for solving systems of polynomial equations and inequalities. The most important are based on the theory of Gröbner bases [4], [2] and the quantifier elimination [5], [9]. We use the program packages for the computation of Gröbner bases by Melenk, Möller and Neun [8] and for linear quantifier elimination by Sturm. The solutions of these systems of equations and inequalities are facts on the linear space under consideration. The system has to insert these facts in the knowledge base and of course it has to return them to the user.

However there are many problems that cannot be solved by these programs. In this case the user has to solve these equations and inequalities himself and to communicate the solutions to the system.

7 Abstract Conclusions

In general, mathematicians deal with sets containing infinitely many elements or with possibly infinite classes of objects, for example the class of all finite groups or the class of all linear spaces. For our purpose it is important to know *how* mathematicians prove theorems on infinite sets or on classes of objects, since we want to simulate this capability on a computer.

Let us consider the following elementary proposition.

Proposition 7.1 *The square of an odd number is a again an odd number.*

Proof. 1. Let a be an odd number.
2. Then a is of the form $2n + 1$ for some integer n.
3. It follows that $a^2 = (2n + 1)^2 = 4n^2 + 4n + 1 = 2\left(2n^2 + 2n\right)^2 + 1$.
4. Hence a^2 is odd.
5. The square of an odd number is again an odd number.

Proposition 7.1 is an assertion on the infinite set of all odd numbers. Since we cannot check this assertion for each odd number individually, we need a more abstract method. This method is to choose a *representative* element of the set under consideration. The choice of such a representative element is done in Step 1 of Proposition 7.1. Of course this principle is well known to every mathematician, and it is used in virtually every proof.

In Step 2 of Proposition 7.1 the properties of the set of all odd numbers is reformulated for the representative element a. With this reformulation, it is shown in Steps 3 and 4 that a^2 is odd. Finally in Step 5 the conclusion that a^2 is odd is transfered (*shifted*) to the original set of all odd numbers. In general this last step is not written up explicitly, since any mathematician draws this conclusion automatically. We call this principle to transfer the properties of a representative element to the original set the *Shifting Principle*. It is of central importance for the development of a system for ITP. In Logic the shifting principle is formulated in the so-called closure theorem and the theorem on constants (cf. [10]).

In Theorem 2.1 we applied the shifting principle in Steps 3 and 7: In Step 3 the representative element h of the set $\mathcal{L}'$ of all lines of **L** different from g has been chosen, and in Step 7 the conclusions about the line h have been shifted to $\mathcal{L}'$.

The system applies the shifting principle automatically.

The architecture of the system is as follows:

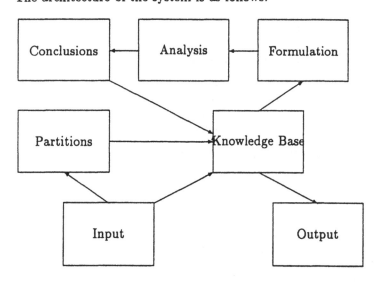

8   Case Study - The theorem of de Bruijn and Erdös

Theorem 8.1 (de Bruijn, Erdös) *Let L be a linear space, and let v and b be the number of points and the number of lines, respectively. Then $b \geq v$.*

In the following we explain how the system for ITP and finite linear spaces can be used to prove this theorem. We suppose that the knowledge base contains the rules

$$r_x \geq k_l \text{ for } x \notin l \text{ and} \tag{1}$$

$$\sum_{x \in \mathcal{P}} r_x = \sum_{l \in \mathcal{L}} k_l. \tag{2}$$

Furthermore we suppose that the first rule is applied automatically by the system whereas the second rule is only applied on request of the user.

1. Input. *Let $k_l \leq n$ for all lines l.* Since no singular points or lines have been introduced the actual point partition consists of only one component C containing the whole point set. Similarily the line partition consists of one component D containing all lines of **L**. The system inserts the fact

$$k_l \leq n \text{ for all } l \in D$$

in the knowledge base. We use the notation $k_l(D) \leq n$ to express that $k_l \leq n$ for all $l \in D$.

2. Input. *Let g be a line of degree n.* The system computes the actual point and line partitions; the point partition consists of the following two components:

$$C_1 := \{x \in \mathcal{P} \mid x \in g\} \text{ (linear component)},$$
$$C_2 := \{x \in \mathcal{P} \mid x \notin g\} \text{ (free component)},$$

the line partition consist of three components, namely

$$D_1 := \{g\} \text{ (singular component)},$$
$$D_2 := \{l \in \mathcal{L} \mid l \cap g \neq \emptyset\},$$
$$D_3 := \{l \in \mathcal{L} \mid l \cap g = \emptyset\}.$$

The knowledge base is extended by the fact

$$k_g = n \text{ or, equivalently, } k_l(D_1) = n.$$

Applying the rule (1) on the pair (C_2, D_1) yields the fact

$$r_x(C_2) \geq n.$$

At this point the proof splits into two cases depending on whether there is a second line of degree n or whether all lines different from g are of a degree strictly smaller than n. We restrict ourselves to the first case.

3. Input. *Let h be a second line of degree n.*
Case 1. Suppose that g and h are disjoint. Then the point partition consists of three components, namely

$$C_3 := \{x \in \mathcal{P} \mid x \in g\} \text{ (linear component)},$$
$$C_4 := \{x \in \mathcal{P} \mid x \in h\} \text{ (linear component)},$$
$$C_5 := \{x \in \mathcal{P} \mid x \notin g, x \notin h\} \text{ (free component)}.$$

Applying rule (1) on the pair $(\{h\}, C_3)$ the system obtains the fact

$$r_x(C_3) \geq n,$$

and it follows that $r_x \geq n$ for all $x \in \mathcal{P}$.

4. **Input.** *Apply rule (2).* The actual point partition consists of the three components C_3, C_4, and C_5. The line partition consists of the following six components:

$$D_4 := \{g\},$$
$$D_5 := \{h\},$$
$$D_6 := \{l \in \mathcal{L} \mid l \cap g \neq \emptyset, l \cap h \neq \emptyset\},$$
$$D_7 := \{l \in \mathcal{L} \mid l \cap g \neq \emptyset, l \cap h = \emptyset\},$$
$$D_8 := \{l \in \mathcal{L} \mid l \cap g = \emptyset, l \cap h \neq \emptyset\},$$
$$D_9 := \{l \in \mathcal{L} \mid l \cap g = \emptyset, l \cap h = \emptyset\}.$$

For the system the equation

$$\sum_{x \in \mathcal{P}} r_x = \sum_{l \in \mathcal{L}} k_l$$

takes the following shape:

$$|C_3| \cdot r_x(C_3) + |C_4| \cdot r_x(C_4) + |C_5| \cdot r_x(C_5) = |D_4| \cdot k_l(D_4) + \cdots + |D_9| \cdot k_l(D_9).$$

Using the facts $r_x \geq n$ for all $x \in \mathcal{P}$ and $k_l \leq n$ for all $l \in \mathcal{L}$ the system gets

$$(|C_3| + |C_4| + |C_5|)\, n \leq (|D_4| + \cdots + |D_9|)\, n.$$

The system is aware of the fact that the union of all point components is the whole point set, or, in other words, that $|C_3| + |C_4| + |C_5| = v$. Of course the same holds for the line partition, and the system obtains the following inequality.

$$vn \leq bn.$$

Hence $v \leq b$. □

The case that g and h intersect goes along the same lines even though it is a bit more complicated. The complete dialog between the user and the system to prove the theorem of de Bruijn and Erdös will be contained in [13].

References

1. L. M. BATTEN, A. BEUTELSPACHER: *The Theory of finite Linear Spaces. Combinatorics of Points and Lines*, Cambridge University Press, Cambridge (1993).
2. T. BECKER, V. WEISPFENNING: *Groebner Bases*, Springer Graduate Texts 141, Springer Verlag Berlin Heidelberg, New York (1993).
3. A. BEUTELSPACHER, J. UEBERBERG: *Symbolic Incidence Geometry. Proposal for doing geometry with a computer*, SIGSAM Bull. **27**, No. 2 (1993), 19 - 29 and No. 3 (1993), 9 - 24.

4. B. BUCHBERGER: *Gröbner bases*, in preparation.

5. G. E. COLLINS: *Quantifier elimination for the elementary theory of real closed fields by cylindrical algebraic decomposition*, Lecture Notes in Computer Science **33**, Springer-Verlag, Berlin, Heidelberg, New York (1975) 134 - 183.

6. N. DE BRUIJN, P. ERDÖS: *On a combinatorial problem*, Indag. Math. **10** (1948), 421 - 423.

7. P. DEMBOWSKI: *Semiaffine Ebenen* Arch. Math. **13** (1962), 120 - 131.

8. H. MELENK, H. M. MÖLLER, W. NEUN: *Groebner - A package for calculating Groebner bases*, part of the documentation for REDUCE 3.4.

9. R. LOOS, V. WEISPFENNING: *Applying linear quantifier elimination*, to appear.

10. J. SHOENFIELD: *Mathematical Logic*, Addison Wesley, Massachusetts (1967).

11. J. UEBERBERG: *Einführung in die Computeralgebra mit REDUCE*, Bibliographisches Institut Mannheim, Leipzig, Wien (1992).

12. J. UEBERBERG: *Symbolic Incidence Geometry and Finite Linear Spaces*, Discrete Math. **129** (1994), 205 - 217.

13. J. UEBERBERG: *Interactive Theorem Proving and Symbolic Incidence Geometry*, in preparation.

A Practical Algorithm for Geometric Theorem Proving

Ashutosh Rege *   John Canny*

Computer Science Division, University of California, Berkeley, CA, 94720

Abstract. This paper describes a practical algorithm for the problem of geometric theorem proving. Our work is motivated by several recent improvements in algorithms for sign determination and symbolic-numeric computation. Based on these, we provide an algorithm for solving triangular systems efficiently using straight-line program arithmetic. The geometric theorem prover so obtained works over both real closed and algebraic closed fields and handles the problem of degeneracy via the use of randomisation. The report concludes with a description of an implementation and provides preliminary benchmarks from the same.

1 Introduction

In this paper, we describe a practical algorithm (and its implementation) for automatic geometric theorem proving and related problems. Earlier methods for geometric theorem proving were based on a synthetic geometric reasoning approach using natural deduction, forward and backward chaining and the like [6], [5]. Following Wu's seminal work ([12], [13]), a significant amount of the recent work has focussed on an algebraic approach involving determining the feasibility of systems of polynomial equations [3], [9], [7]. Our work is based on this latter methodology and is motivated by recent improvements in algorithms for symbolic-numeric computation with polynomials [2].

The papers by Wu and others showed how a subclass of geometric theorems can be proved by expressing the hypotheses and conclusions as polynomial equations. A geometric theorem can then be proved by essentially determining the variety corresponding to the hypotheses and checking whether the varieties of the conclusions contain it. Different methods have been proposed for doing this - Wu's method involves reduction of the hypotheses system of polynomials to a triangular form (i.e. one in which each successive polynomial has at most one extra variable) and then using successive pseudo-division to determine the feasibility of the conclusion. Other work has focussed on using Gröbner bases to determine whether a system of polynomials (obtained from the original polynomials) has a common zero [9], [7].

---
* E-mail: {rege,jfc}@cs.Berkeley.edu. Fax: +1-510-642-5775. Supported by a David and Lucile Packard Foundation Fellowship and by NSF P.Y.I. Grant IRI-8958577

Wu's original method is incomplete in the sense that the feasibility of the conclusions is determined over an algebraically closed field whereas geometric theorems are often assertions about real numbers. It is not possible to refute a conjecture over the reals using this approach and in general, confirmation of a theorem can not be obtained automatically from a proof over an algebraically closed field. Kutzler and Stifter [9] and Ko [8] discuss two different approaches that can be used to alleviate this problem.

Another important consideration in geometric theorem proving is the notion of *genericity*. It is necessary to rule out *degenerate* cases such as non-distinct points or zero-radius circles. Wu's method and others generate a set of subsidiary polynomials representing such cases so that the conclusion polynomial vanishes at the common zeros of the hypotheses which are not zeros of the subsidiary polynomials.

From the above discussion, it seems important that a geometric theorem prover should satisfy the following criteria

1. It should work over the reals
2. It should handle degenerate cases
3. It should be practicable

In this paper we discuss a method for geometric theorem proving which satisfies the first two criteria and based on an initial implementation, seems to be rather efficient in practice.

The basic problem we consider is the following : *given a system of hypotheses polynomial equations in triangular form, determine whether the set of their common zeros over the reals is contained in the set of real zeros of the conclusion polynomial.* (One can use any standard algorithm for triangulating if the hypotheses are not triangular to begin with, see e.g. [4]). The observation here is that, in order to prove a theorem or refute it, one is only interested in the *sign* (+, - or 0) of the conclusion polynomial at the real zeros of the hypotheses polynomials. We provide an efficient method to encode the roots of a triangular system using the Sylvester resultant. To determine the signs, we make use of Sturm sequences which enable us to determine the number of real roots of the hypotheses where the conclusion is non-zero. Thus our algorithm can determine the truth or falsehood of an assertion over the reals.

The problem of genericity is handled by instantiating the independent variables (or parameters) of the given theorem with random values. Probabilistically, this takes care of degenerate cases without affecting the validity of the conclusion.

Computation in the prover is done over the field of reals (or rationals) extended by variables. The numbers in this extension field are then rational functions in these variables. It is, however, very expensive to compute with explicit rational functions. We circumvent this difficulty by using *straight-line programs* or SLP's to represent all intermediate calculations. SLP's are extremely useful for various types of computations with polynomials over field extensions. For example, the cost of multiplying two SLP's is a constant whereas that of explicitly multiplying the two rational functions they represent would be quadratic in the degree. In the algorithms we need for theorem-proving, one is primarily interested in the signs of the coefficients of polynomials over the extension field. These can be determined very easily and efficiently when the computation is done over the field of SLP's.

In order to make our prover more efficient, we use "mixed" arithmetic for algebraic computations : the idea is to represent a number as a structure with two values, the first is the integer value of the number modulo a prime p and the second is a floating-point approximation to that number. We use the finite field part for equality tests and the floating point part for relative ordering of two numbers and sign-determination.

Geometric theorem proving falls in the more general class of problems defined by the first-order theory of the reals. The work described in the subsequent sections is part of ongoing research at Berkeley aimed at developing an efficient practicable toolkit for solving more general problems. We plan on applying the tools and techniques described in this paper to a broader scope of problems including solving systems of polynomial equations, point-location in varieties, existential theory of the reals etc. (See [2] for an overview.) Further, one can use these techniques to extend our prover to work with polynomial inequalities and also to solve more general problems in constraint-based reasoning.

The paper is organized as follows : Section 2 gives an overview of the theorem prover. Section 3 sketches the algorithm for solving triangular systems. We conclude with Section 5 which describes the implementation and gives preliminary benchmarks on some problems.

2　Overview of the Theorem Prover

We will give a schematic overview of our theorem prover in this section. The notation used in this and subsequent sections is the same as introduced earlier. The prover takes as input a set of hypotheses polynomials $h_1, \ldots, h_n$ and a conjecture or conclusion polynomial g in $\mathbb{Q}[u_1, \ldots, u_m, x_1, \ldots, x_n]$. Here, as before, the x_i's represent the dependent variables and the u_j's, the independent variables. In what follows, we will abuse notation and write the hypotheses as polynomials over the x_i's only. It returns "true" or "false" according to whether the theorem is true or not over the reals. The prover can also be used to work over $\mathbb{C}$. The modules of the prover can be summarized as follows :

- **Triangulation :** The hypotheses are first triangulated using the triangulation algorithm described earlier to obtain the polynomials

$$f_1(x_1), f_2(x_1, x_2), \ldots, f_n(x_1, \ldots, x_n)$$

- **Root representation for triangular systems :** This module takes as input the triangular system

$$f_1(x_1) = f_2(x_1, x_2) = \ldots = f_n(x_1, \ldots, x_n) = 0$$

and returns a *symbolic* representation of the roots of the system : the output is a univariate polynomial $p(s)$ and rational functions $r_1(s), \ldots, r_n(s)$. If the roots of $p(s) = 0$ are $\alpha^{(i)} \in \mathbb{C}$, for $i = 1, \ldots, N$, then

$$\xi^{(i)} = r(\alpha^{(i)})$$

One can think of $r(s)$ as a parametric curve in $\mathbb{C}^n$ which passes through all the roots $\xi^{(i)}$, $i = 1, \ldots, n$. The values of s at which it passes through a solution of $f_1 = f2 = \ldots = f_n$ are precisely the roots of $p(s) = 0$. The fact that we have a symbolic representation of the roots allows us to determine correctly the feasibility of the conjecture polynomial given the hypotheses polynomials.

- **Substitution** : We put $x_i = r_i(s)$, $1 \le i \le n$, in the conjecture polynomial g. This reduces g to a univariate polynomial $g(s)$.
- **Sign Determination** : This module determines the sign, i.e. +, - or 0, of the instantiated conjecture $g(s)$ at the real roots of $p(s) = 0$ using Sturm sequences.
- **Theorem proving or refutation** : The conjecture should vanish at all of the common real roots of the hypotheses polynomials. Thus the sign determination module should return 0 for the sign of $g(s)$ at every root. If not, there exists a root where the conjecture does not hold and is therefore false OR the triangulation process introduced extra roots. To verify that the latter is not the case, we instantiate the original hypotheses polynomials h_j with the rational functions $r_i(s)$ as we did for g. We can now do sign-determination for the entire system $h_1(s), \ldots, h_n(s), g(s)$ at the roots of $p(s) = 0$. This procedure yields for each root of $p(s)$ an ordered sign sequence in $\{+, -, 0\}^{n+1}$ where the ith element in the sign sequence corresponds to the sign of the ith polynomial in $h_1(s), \ldots, h_n(s), g(s)$ at that root of $p(s)$. It is then trivial to check which of the roots of $p(s)$ actually correspond to roots of the original system of hypotheses and based on the signs of g at these correct roots, the prover returns true or false.

3 An Algorithm for Solving Triangular Systems

This section sketches the basic algorithm for solving triangular systems used in our geometric theorem prover. Proof of correctness, complexity analysis etc. can be found in the complete paper [10].

The input to the algorithm is a system of triangular polynomial equations in $\mathbb{Q}(x_1, \ldots, x_n)$,

$$f_1(x_1) = f_2(x_1, x_2) = \cdots = f_n(x_1, \ldots, x_n) = 0$$

Let $\xi^{(i)} \in \mathbb{C}^n$ denote the solutions to the given system.

As described earlier, the algorithm returns a *symbolic* representation of the roots of the system i.e. the output is a univariate polynomial $p(s)$ and rational functions $r_1(s), \ldots, r_n(s)$. If the roots of $p(s) = 0$ are $\alpha^{(i)} \in \mathbb{C}$, for $i = 1, \ldots, n$, then

$$\xi^{(i)} = r(\alpha^{(i)})$$

Our algorithm is based on an approach for computing (p, r) for more general systems due to Renegar [11]. We use his basic method but take advantage of certain properties of triangular systems.

The algorithm proceeds iteratively; at the ith step it adds the polynomial f_i and eliminates the extra variable x_i introduced by the polynomial to get a rational univariate representation $r_1(s_i), \ldots, r_n(s_i)$ of the roots of the first i polynomials. More precisely, at the start of the ith step, the algorithm has computed

$$r_1(s_{i-1}), \ldots, r_{i-1}(s_{i-1}) \text{ and } p_{i-1}(s_{i-1})$$

The algorithm now introduces $f_i(x_1, \ldots, x_i)$. It sets $x_1 = r_1, \ldots, x_{i-1} = r_{i-1}$ in f_i. For the new variable, x_i, the algorithm sets,

$$x_i = \frac{s_i - l(x)}{l_i} = \frac{s_i - l_0 - l_1 x_1 - \cdots - l_{i-1} x_{i-1}}{l_i}$$

where $x_1, \ldots, x_{i-1}$ are instantiated to $r_1, \ldots, r_{i-1}$ respectively. Thus we obtain from f_i a rational function g_i in s_{i-1} and s_i. We denote by g_{i_n} the numerator polynomial of g_i and by g_{i_d} the denominator.

The algorithm now computes the (univariate) Sylvester resultants $R(g_{i_n}, p_{i-1})$ and $R(g_{i_d}, p_{i-1})$ with respect to the variable s_{i-1}. The observation here (the proof being given in the complete version of the paper) is that the latter resultant divides the former exactly. Thus we can set

$$R = \frac{R(g_{i_n}, p_{i-1})}{R(g_{i_d}, p_{i-1})}$$

If the polynomial $l(x) = l_0 + l_1 x_1 + \cdots l_{i-1} x_{i-1}$ is specialized to a random linear polynomial L, then with probability one, the values $l(\xi^{(j)})$ will be distinct for distinct $\xi^{(j)}$. The roots of the resultant R will also be distinct. The algorithm sets $p_i(s_i) = R$.

We obtain the functions r_i by differentiating R. We define r_i as

$$r_i(s_i) = \frac{\left(\frac{dR}{dl_i}\right)}{\left(\frac{dR}{dl_0}\right)}\Bigg|_{l=L}$$

It can be verified that the r_i's have the correct values at $s_i = \alpha^{(j)}$.

At termination the algorithm eliminates all the variables $x_1, \ldots, x_n$ and returns rational functions $r_j(s)$ in the variable $s = s_n$ and the polynomial $p(s) = p_n(s_n)$. The complexity of this algorithm is singly exponential in n.

4 Sign Determination and Theorem Proving

Once we have a symbolic representation (p, r) of the roots of the triangular system of hypotheses, it is relatively straightforward to determine whether the theorem is true or not. The first step is to reduce the conjecture polynomial c to a univariate one by substituting $x_i = r_i(s)$, $1 \leq i \leq n$. This reduces c to a univariate polynomial $g(s)$. We now need to compute the sign of $g(s)$ at the real roots of $p(s)$. We can use Sturm sequences to do that as follows :

If $f(s)$ and $g(s)$ are polynomials, let the Sturm sequence of f and g be denoted $r_0, r_1, \ldots, r_k$, where $r_0(s) = f(s)$, $r_1(s) = g(s)$, and the intermediate remainders are computed via

$$r_{i+1} = q_i r_i - r_{i-1} \tag{1}$$

where q_i is the pseudo-quotient of the polynomial division of r_{i-1} by r_i. For a real value v, let $\mathrm{SA}(f, g, v)$ denote the number of changes in sign in the sequence $r_0(v), r_1(v), \ldots, r_k(v)$ and $\mathrm{SC}(f, g)$ denote $\mathrm{SA}(f, g, +\infty) - \mathrm{SA}(f, g, -\infty)$. The classical Sturm theorem states that $\mathrm{SC}(f, f')$ equals the number of real roots of $f(s)$. The most general form of the theorem states that $\mathrm{SC}(f, f'g)$ is the number of real roots of $f(s)$ where $g(s) > 0$, minus the number where $g(s) < 0$, common roots making no difference to the count. For our purpose, we consider $\mathrm{SC}(f, f'g^2)$ which counts real roots of f where $g \neq 0$. Thus we compute $\mathrm{SC}(p, p'g^2)$; if this number is 0 the theorem is true. If not, we could have the problem discussed earlier of having introduced extra roots in the triangulation procedure. To check for this eventuality, we instantiate the original hypotheses h_i with the rational functions $r_j(s)$ as described in the overview of the theorem prover. We now have a collection of polynomials $h_1(s), \ldots, h_n(s), g(s)$ whose signs we wish to determine at the roots of $p(s) = 0$. This can be done using the sign-determination algorithm of [1]. We then check the signs of g at the actual common roots of the h_i as described in the overview.

5 Implementation

In this section, we briefly describe a first implementation of the above algorithms. Most of the modules are part of a larger project involving the development of a toolkit for solving problems in non-linear algebra. The basic modules perform polynomial arithmetic over different fields such as finite fields, rationals, infinitesimal extensions, "mixed" arithmetic and most importantly straight-line programs or SLP's.

5.1 Straight-line Programs

All our computations are done over the field of arithmetic straight-line programs. An SLP can be represented by a directed acyclic graph, with each node representing an operation such as addition, and a value. *All the polynomials are thus represented as polynomials over the field of SLP's, i.e. the coefficients are straight-line programs and not numbers.* As mentioned in the introduction, this allows us to represent coefficients which are rational functions in the parameters $u_1 \ldots u_m$. In other words the extension field $\mathbb{Q}(u_1, \ldots, u_m)$ has a 1-1 correspondence with SLP's defined over $u_1, \ldots, u_m$.

Arithmetic with SLP's is easy - for example to add two SLP's one creates a new node with the operation "+" and two edges directed from the new node to the two operand nodes. The value of an SLP node (over the base field such as $\mathbb{Q}$), can be obtained by simply adding the values of its children nodes. Thus evaluation involves a depth-first search of the SLP graph and can be done in time linear in the size of the SLP.

Computing derivatives as required by the algorithm in Section 3 is particularly straightforward, as is determination of the signs of the coefficients of various polynomials (required by the Sturm sequence algorithm). To compute the derivative SLP of an SLP node one simply creates a new SLP using the usual rules of differentiation. This has roughly double the size of the original SLP. For sign determination, one uses successive differentiation and evaluation of SLP's. The details of how this works can be found in the complete version of the paper.

5.2 Some Benchmarks

We have run our prover on some of the standard problems in geometric theorem proving. Unfortunately, at the time of writing we didn't have all of the geometric theorems in the format accepted by our code. We did run it on the following problems and obtained the times as shown in the following table : All the benchmarks were run on a Sun Sparc 10 machine. The code is written in C. Though very incomplete, the initial benchmarks seem to be rather promising and we hope to have a more complete set of benchmarks very soon. Further, the benchmarks were obtained from a first implementation and we expect to reduce the running time by using various optimizations. (The times given for Chou's prover are from [4] and were obtained on a Symbolics 3600. The geometric statements for these theorems are also from [4].)

Geometric Theorem	Chou's Prover	This Prover
Pascal Conic-1	14.43s	4.9s
Pappus	1.52s	1.3s
Pappus Dual	1.45s	1.6s
AMS Dual	4.05s	4.6s

6 Acknowledgments

We would like to thank Richard Fateman, Phil Liao and Will Evans for some useful discussions. Additional thanks are due to Phil Liao for providing us with the theorem statements in the proper format for our prover.

7 Bibliography

[1] J.F. Canny. An improved sign determination algorithm. In *AAECC-91*, 1991. New Orleans.

[2] John Canny. A toolkit for non-linear algebra. (manuscript), 1993.

[3] S.-C. Chou. Proving elementary geometry theorems using Wu's algorithm. In W.W. Bledsoe and D.W. Loveland, editors, *Theorem Proving : After 25 Years*. American Mathematical Society, 1984.

[4] S.-C. Chou. *Mechanical geometry theorem proving*. Mathematics and its applications. D. Reidel, Holland, 1988.

[5] H. Coelho and L.M. Pereira. Automated reasoning in geometry theorem proving with PROLOG. *J. Automated Reasoning*, 2:329–390, 1986.

[6] H. Gelernter. Realization of a geometry theorem proving machine. In E.A. Feigenbaum and J.E. Feldman, editors, *Computers and Thought*. McGraw-Hill, New York, 1963.

[7] D. Kapur. A refutational approach to geometry theorem proving. In D. Kapur and J.L. Mundy, editors, *Geometric Reasoning*. MIT Press, 1988.

[8] H.-P. Ko. Geometry theorem proving by decomposition of semi-algebraic sets - an application of Wu's structure theorem. In *International Workshop on Geometry Reasoning*, 1986. Oxford, England.

[9] B. Kutzler and S. Stifter. On the application of Buchberger's algorithm to automated geometry theorem proving. *J. Symbolic Comput.*, 2:409–420, 1986.

[10] A. Rege and J. Canny. A practical algorithm for geometric theorem proving. In preparation, 1994.

[11] J. Renegar. On the computational complexity and geometry of the first-order theory of the reals, parts I, II and III. Technical Report 852,855,856, Cornell University, Operations Research Dept., 1989.

[12] W. Wu. On the decision problem and mechanization of theorem proving in elementary geometry. *Sci. Sinica*, 21:150–172, 1978.

[13] W. Wu. Some recent advances in mechanical theorem proving of geometries. In W.W. Bledsoe and D.W. Loveland, editors, *Theorem Proving : After 25 Years*. American Mathematical Society, 1984.

Combining Theorem Proving and Symbolic Mathematical Computing

Karsten Homann and Jacques Calmet

Universität Karlsruhe
Institut für Algorithmen und Kognitive Systeme
Am Fasanengarten 5 · 76131 Karlsruhe · Germany
{homann,calmet}@ira.uka.de

Abstract. An intelligent mathematical environment must enable symbolic mathematical computation and sophisticated reasoning techniques on the underlying mathematical laws. This paper disscusses different possible levels of interaction between a symbolic calculator based on algebraic algorithms and a theorem prover. A high level of interaction requires a common knowledge representation of the mathematical knowledge of the two systems. We describe a model for such a knowledge base mainly consisting of type and algorithm schemata, algebraic algorithms and theorems.

Keywords: theorem proving, symbolic mathematics, knowledge representation

1 Introduction

The dream of "doing" mathematics on a computer is progressively becoming true. Ideally, an intelligent assistant for doing mathematics will be a user friendly interactive environment allowing to perform computations, to prove theorems and to support formal reasoning and advanced tutoring. Such an environment must thus rely on some sophisticated AI techniques, e.g. automated theorem proving, machine learning and planning.

At present, two clases of mathematical computations can be efficiently performed. On one side, computer algebra systems (CAS) usually offer a large collection of powerful algebraic algorithms and a straightforward programming language. In classical systems the mathematical knowledge, e.g. definitions of mathematical structures, properties of operators of a domain, domains of computation, range of algorithms and their mathematical specification, is hidden in the algebraic algorithms. AXIOM [JeSu92] allows the definition of abstract data types including operators and domains of computation. However, no AI methods (e.g. automated theorem proving, learning) are provided. CAS are very efficient in computing symbolic solutions through given algorithms but cannot derive new theorems or lemmas. On the other side, automated theorem provers (ATP) have achieved remarkable results in proving non-trivial mathematical theorems. But they lack embedded mathematical knowledge such as algebraic algorithms or intelligible representations of proofs and they are difficult to use. Moreover, they require huge search spaces.

It is thus natural to integrate theorem proving and symbolic mathematical computing in a common environment. We report on such an environment, $\lambda\epsilon\mu\mu\alpha^1$, which enables to compute with algebraic algorithms, to derive theorems, to deal with vertical or inclusion polymorphisms, and to learn and apply equation schemata. The explicit formalization of mathematical dependencies provides new possibilities to explain the solution steps.

This paper is structured as follows. Section 2 illustrates different levels of interaction between a symbolic calculator using algebraic algorithms and a theorem prover. A common mathematical knowledge base stores the domain specific problem solving knowledge and is described in section 3. An overview of $\lambda\epsilon\mu\mu\alpha$ in section 4 is followed by some concluding remarks in the last section.

2 Mathematical Problem Solving by Algorithms and Theorems

When solving problems, mathematicians follow a 'Mathcycle' [VeVe94]: conception, naive formulation, exploration, tentative proof, formulation, proof, publication, education, and use. Many packages which aid mathematicians in some of these steps have been developed, e.g. AM [LeBr84] for concept formulation, CAS for application of algorithms, ATP for verification and discovery of theorems, specification languages and knowledge representation. However, few mathematicians use these systems as everyday tools, because of some severe drawbacks which make them hard to use.

Classical CAS provide thousands of sophisticated algebraic algorithms which are difficult to handle by users. On the one hand, it is hard to select the appropriate algorithm from the amount of available algorithms, on the other hand, the interpretation of the solution needs deep mathematical understanding. The user doesn't receive any information about the solution steps from the system (Why is the output the solution to the given problem, or how to find the solution to a problem?). The mathematical knowledge is embedded implicitly in the algorithms and is inaccessible to the user, e.g. commutativity of polynomial addition, axioms of groups. However, the algorithmic encoding leads to high efficiency.

In traditional theorem provers it is difficult to specify axioms in the provers language, usually a first-order language and a normal form. Therefore, the representation of mathematical concepts (e.g. gcd, finite fields) is awkward and unnatural. Provers usually lack embedded algebraic algorithms. Although OTTER [McCu94] allows the declaration of user-defined functions together with their corresponding argument and result types, the extension of the system by new algorithms is very expensive, i.e. new implementation of the function and recompilation of the whole system. ATP compute huge search spaces and are inefficient. Additionaly, long and complex proofs are difficult to understand and

---

[1] $\mathcal{L}$earning $\mathcal{E}$nvironment for $\mathcal{M}$athematics and $\mathcal{M}$athematical $\mathcal{A}$pplications

should provide representations that point out the essential steps of the proof. However, their success in proving hard mathematical theorems is impressing. In contrast to algorithmic problem solvers, theorem provers provide proofs to explain their solutions.

We propose the integration of CAS and ATP. This integration can be achieved in different ways as illustrated in figure 1.

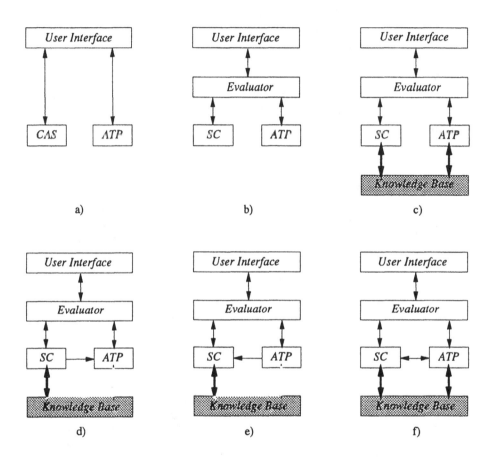

Fig. 1. Forms of interaction between algorithms and theorems

a) In the simplest case of interaction the user interface only provides a link to a CAS and an ATP, respectively. A user can access both systems and can apply algorithms or theorems to solve a given problem, depending on the class of the problem. Only this type of interaction allows the use of arbitrary CAS and ATP. However, the systems do not interact directly and a user must be familiar with both systems. Such an architecture combines the advantages, but also the drawbacks.

Example:

CAS: adding 0 to a polynomial in a polynomial ring is done by a function.
ATP: tries to prove that a left neutral element is also right neutral in a group.

b) Symbolic calculator (SC) and ATP can be extended by a common control unit (evaluator). This evaluator controls the selection of the modules by meta-knowledge on all functions and predicates. It also controls the application of algebraic algorithms in the SC and of theorems. The mathematical knowledge is represented separately in each module.

c) Algorithms often require type information about their arguments to be applicable. This information can be derived by theorems assuming that both modules share a common knowledge representation. This mathematical knowledge base also consists of meta-information on algorithms in form of algorithm schemata (cf. next section, e.g. figure 2).

Name	$\gcd(?a, ?b) = ?g$
Signature	$?A \times ?A \rightarrow ?A$
Constraints	isa $(?A, \text{Euclidean Ring})$
Definition	$(?g | ?a) \wedge (?g | ?b) \wedge (\forall c \in ?A : (c | ?a) \wedge (c | ?b) \Rightarrow (c | ?g))$
Subalgs	
Theorems	$\gcd(u, v) = \gcd(v, u)$
	$\gcd(u, v) = \gcd(v, u \bmod v)$
	$\gcd(u, 0) = u$
Function	

Fig. 2. Schema of algorithm gcd

Algebraic algorithms offer no capabilities to explain their solutions. These explanations can be generated using the theorem prover and the mathematical specification of the algorithms. The knowledge representation of both symbolic calculator and theorem prover must be adjusted to a common representation. With this form of interaction, theorems are not available within algorithms because the SC cannot access directly the ATP.

d) As mentioned in b), algorithms can be used for the efficient computation of predicates when proving theorems. However, the interaction in b) needs to transfer all necessary knowledge and parameters to the SC. This is avoided when a common knowledge base is used, and a direct link from SC to ATP allows the immediate call of an algorithm out of a proof. New versions of theorem provers (e.g. OTTER 3.0) allow the introduction of user-defined algorithms which must be identified by a special character (e.g. $GCD). The extension of the prover requires the recompilation of the whole system and each algorithm has to be implemented in C. CAS provide an extensive collection of very efficient mathematical algorithms, thus reimplementation is neither necessary nor meaningful. This kind of interaction would lead to a strong improvement of the efficiency of a theorem prover.

Example:

SC: various efficient algorithms for gcd calculation.
ATP: OTTER allows the definition of simple functions (e.g. gcd in figure 3). The performance can be increased strongly by calling instead the adequate gcd algorithm of the SC.

```
gcd(x,y) =        % greatest common divisor for nonnegative integers
          $IF($EQ(x,0),
              y,
              $IF($EQ(y,0),
                  x,
                  $IF($LT(x,y),
                      gcd(x,$DIFF(y,x)),
                      gcd(y,$DIFF(x,y)))))).
```

Fig. 3. Definition of function gcd in OTTER

e) The application of theorems is useful even when running algebraic algorithms (e.g. verification of conditions, properties of objects). This kind of integration (same is true for f)) requires to redesign new algorithms to use the prover. The advantage lies in using the powerful reasoning capabilities of the theorem prover in the SC.

Example:

SC: ... if #IsNormal(G,H) then ... [2]
ATP: tries to prove that all subgroups of index 2 are normal (figure 4).

f) A complete integration of algorithms and theorems is achieved by combining d) and e). At any step, arbitrary combinations of algorithms and theorems can be applied to solve a given problem. This combines the advantages of a) to e), but requires to fit SC and ATP to a common knowledge representation.

Example:

SC: in Berlekamp algorithm ... if #SquareFree(p) then ...
ATP: $\forall f \in Z_p[x] : SquareFree(f) \Leftrightarrow \$GCD(f, f') = 1$.
The SC can be used to compute the derivation of p and the gcd.

We have shown different levels of interaction between SC and ATP. The complete integration in f) requires the development of a new common semantics of SC and ATP, the reengineering of some algorithms, and the common explicit

---

[2] The special character # indicates a call to the theorem prover.

```
% existence of inverse
    4 P(x,g(x),e).
% closure
    5 P(x,y,f(x,y)).
% associative property
    6 -P(x,y,u) | -P(y,z,v) | -P(u,z,w) | P(x,v,w).
    7 -P(x,y,u) | -P(y,z,v) | -P(x,v,w) | P(u,z,w).
% the operation is well defined
   10 -P(z,y,x) | -P(w,y,x) | EQUAL(z,w).
% Denial of the theorem
   28 H(b).
   29 P(b,g(a),c).                  PROOF:
   30 P(a,c,d).
   31 -H(d).                        195 (29,7,4,5,37,33) P(c,a,b).
% demodulators                      213 (195,7,30,5) P(d,a,f(a,b)).
   33 EQUAL(f(x,e),x).              716 (213,6,5,4,38,39) P(d,g(b),e).
   37 EQUAL(g(g(x)),x).             721 (716,10,4) EQUAL(d,b).
   38 EQUAL(g(f(x,y)),f(g(x),g(y))). 722 (721) EQUAL(d,b).
   39 EQUAL(f(x,f(g(x),y)),y).      729 (31,722,28) .
```

Fig. 4. Proof using hyperresolution and standard p-formulation of the theorem: all subgroups of index 2 are normal.

representation of objects and mathematics in a mathematical knowledge base. The construction of this memory is described in the next section.

3 The Mathematical Knowledge Base

The mathematical knowledge base consists of type schemata, algorithm schemata, algebraic algorithms, theorems, symbol tables, and normal forms. In this paper, we will not discuss the representation of algebraic algorithms and theorems, because they are exclusively used by the prover or CA engine. Thus, a unique treatment, e.g. by defining theorem schemata, is desirable but does not improve the collaboration of both systems. However, it would be required to verify algorithms and generate theorems automatically.

The theory of algebraic specification provides a good framework to design the type system of a mathematical assistant. We adopt the specification language FORMAL-Σ [CaTj93] to represent the mathematical knowledge. It is well-suited to specify mathematical domains of computations, e.g. finite groups, polynomial rings, which are inherently modular. An algebraic specification introduces constants, operators and properties in their intended interpretation, and enables the reuse of subspecifications within a specification in accordance with the dependencies between particular specification modules of an abstract computational structure (ACS).

A type schema represents such a module and consists of:

- *Name*, a unique identifier
- *Based-on*, a list of inherited ACS
- *Parameters*, a list of ACS which are parameters
- *Sorts*, a list of new sorts
- *Operators*, declarations of new operators
- *InitialProps*, initial properties.

Figure 5 shows the schemata of some selected ACS (more details may be found in [CHT92]). These definitions build a based-on hierarchy of the mathematical domains of computation (figure 6).

Name	Monoid
Based-On	SemiGroup
Sorts	Mo $ne \in$ Elt
Operators	
InitialProps	$\forall x \in$ Elt: $ne\ f\ x = x$

Name	Group
Based-On	Monoid
Sorts	Gr
Operators	inv _ :: Elt $\to$ Elt
InitialProps	$\forall x \in$ Elt: $inv(x)\ f\ x = ne$

Name	Ring
Based-On	MultSemiGroup (*rename:* $(f, \times)$, $(ne, 1)$) AddAbelianGroup (*rename:* $(f, +)$, $(ne, 0)$, $(inv, -)$)
Sorts	Ri
Operators	
InitialProps	$\forall x, y, z \in$ Elt: $x \times (y + z) = (x \times y) + (x \times z)$ $\forall x, y, z \in$ Elt: $(y + z) \times x = (y \times x) + (z \times x)$

Fig. 5. Type schemata for Monoid, Group, and Ring.

The user doesn't receive any information about the solution steps from the system, e.g. why is the output the solution of the given problem, or how to find the solution of a problem. Therefore, algorithms are represented in terms of schemata. They allow the representation of meta-knowledge like:

- *Name*, a unique identifier of the schema with variable bindings
- *Signature*, describes the types of input and output
- *Constraints*, imposed on domain and range
- *Definition*, mathematical description of the output
- *Subalgs*, list of subalgorithms describing the embedded subtasks

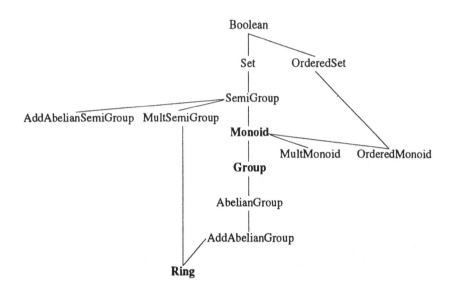

Fig. 6. Hierarchy of type schemata

- *Theorems*, describing properties of the algorithm
- *Function*, name of the corresponding executable algebraic function to compute the output.

Name	gcd-primitive(?a, ?b) =?g
Signature	?A × ?A → ?A
Constraints	isa (?A, UnivariatePolynomial(x, UFD))
Definition	
Subalgs	primitive-part pseudo-remainder content gcd multiply
Theorems	
Function	GcdPrimitive

Fig. 7. Schema of algorithm gcd-primitive

Similarly to type and equation schemata, algorithm schemata build a hierarchy of specialized versions, and specializations inherit definitions and theorems from more general algorithms. Examples of algorithm schemata are given in figures 2 & 7, figure 8 describes parts of the hierarchy of algorithm schemata. New properties of algorithms can be derived by the theorem prover.

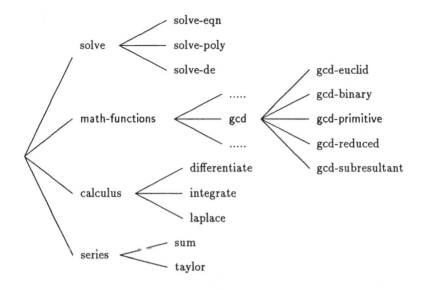

Fig. 8. Hierarchy of algorithm schemata

We introduced different kinds of interactions between SC and ATP. The mathematical knowledge is represented in a common knowledge base which consists of symbol tables, normal forms, theorems, algebraic algorithms, type and algorithm schemata. An environment corresponding to the interaction described in figure 1 f) is introduced in the next section.

4 An Intelligent Environment for Symbolic Mathematical Computing

An environment for solving mathematical problems which integrates theorem proving, symbolic computing, explanation-based learning and a knowledge representation system is given in figure 9. The schema-based representation of mathematical structures and algorithms enables the representation of meta-knowledge, e.g. constraints of parameters, dependencies of algorithms and theorems.

The user interface offers frames and graphs for handling schemata and displays the explanations about solutions of specific problems. An evaluator solves these problems by using a theorem prover and a symbolic calculator, and applying equation schemata (learning subsystem). The knowledge base consists of symbol tables, normal forms of the simplifier, algebraic algorithms of the symbolic calculator, algorithm schemata for the specification of algorithms, type schemata for abstract computational structures, as well as initial and derived equation schemata for simplifying expressions.

Equation schemata consist of mathematical rewriting rules which model domain knowledge, and user defined laws. New equation schemata can be learned by generalizing specialized solutions using explanation-based learning. Given

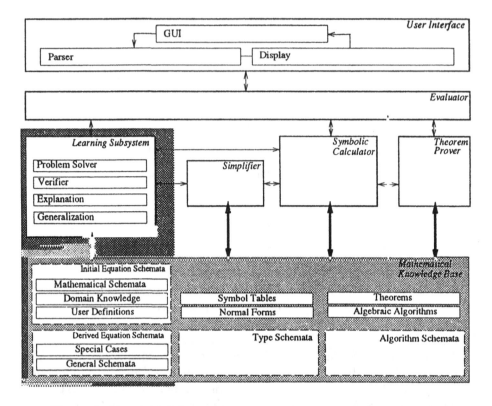

Fig. 9. Architecture of the intelligent environment for symbolic computing

problems are solved by applying schemata to eliminate obstacles [Shav90] in the calculation of unknown properties of a variable. An explanation why this is an appropriate solution to the problem is generated, the achieved schema is generalized to solve other problems, and finally, the knowledge base of equation schemata is updated with the new generalized schema.

5 An Example

For the purpose of having prototype systems, we created two ad hoc interfaces between theorem provers and CAS. These interfaces are controled by a common evaluator and implemented in CLISP and C respectively, however, our aim is to approach the integration at a higher level of interaction corresponding to figure 1 f).

The first prototype combines DTP [Gedd94], a simple first-order theorem prover, and MAGMA [BoCa94], a CAS for computations in algebra and particularly in group theory. The application of MAGMA algorithms is guided by DTP, which can solve new problems, e.g. finding elements with minimal index, and prove difficult properties by induction and applications of algorithms. Additionaly, the tremendous knowledge about groups represented in MAGMA is accessible by the prover.

Another prototype combines ISABELLE [Paul94], a generic theorem prover supporting set theory, type theory, higher order logic ..., and MAPLE [Heck93], a well known commercial CAS. Many domains of computation were defined in the provers language (e.g. numbers, polynomials) and new theorems were proven by the cooperation with MAPLE.

Example:

$$\text{Proof of } \forall n \geq 5 \; : \; n^5 \leq 5^n$$

by using ISABELLEs induction theorem:

```
[| n: Nat; a: Nat; P(a);
   !!x. [| x: Nat; a <= x; P(x) |]  ==>  P(x + 1);
   a <= n |]  ==>  P(n)
```

start: $5^5 \leq 5^5$, true by reflexivity of $\leq$
step and new goal: $n^5 \leq 5^n \Rightarrow (n+1)^5 \leq 5^{(n+1)}$

MAPLE is used to expand both sides:
$$n^5 + 5*n^4 + 10*n^3 + 10*n^2 + 5*n + 1 \leq 5*5^n$$
and the prover is trying to prove the upper bounds:
$$n^5 \leq 5^n \; (1), \quad 5*n^4 \leq n^5 \leq 5^n \; (2), \quad 10*n^3 \leq n^5 \leq 5^n \; (3),$$
$$10*n^2 \leq n^5 \leq 5^n \; (4), \quad 5*n+1 \leq n^5 \leq 5^n \; (5).$$
Again, these bounds are proven by an interaction of both systems, e.g. the second bound is derived by ISABELLEs rule

```
le_mult
[| a: Nat; b: Nat; c: Nat; d: Nat;
   a <= b; c <= d |]  ==>  a * c <= b * d
```

and $5 \leq n$ (given) and $n^4 \leq n^4$ (reflexivity). Finally, MAPLE transforms the conclusion $5*n^4 \leq n*n^4$ to $5*n^4 \leq n^5$.

6 Conclusion

We have outlined several advantages of combining theorem proving and symbolic mathematical computing. On the one hand, computer algebra systems profit from theorem provers, e.g. by explanations of the solution of algorithms and verification of properties of mathematical objects. On the other hand, they offer an extensive collection of efficient mathematical algorithms which can improve the efficiency of the theorem prover.

A high level of interaction requires a common representation of the mathematical knowledge of the two systems. Such a knowledge base mainly consists of type and algorithm schemata, algebraic algorithms and theorems. The adopted specification language for the specification of type schemata provides executability and offers a type system for both symbolic calculator and theorem prover.

Among the work in progress is the design of a "language" for the environment whose semantics allows a consistent treatment of algorithms and theorems, tools for the verification of algorithm schemata, extraction and learning of theorems out of algebraic algorithms, generation of algorithms from theorems, interaction of the learning component and the theorem prover and applications of the environment.

References

[BoCa94] W. BOSMA, J. CANNON, *Handbook of MAGMA Functions*, Sydney, 1994.

[CHT92] J. CALMET, K. HOMANN, I.A. TJANDRA, *Unified Domains and Abstract Computational Structures*, in J. Calmet, J.A. Campbell (eds.), International Conference on Artificial Intelligence and Symbolic Mathematical Computing, Karlsruhe, August 3–6, 1992, LNCS 737, pp. 166–177 , Springer, 1993.

[CaTj93] J. CALMET, I.A. TJANDRA, *A Unified-Algebra-Based Specification Language for Symbolic Computing*, in A. Miola (ed.), Design and Implementation of Symbolic Computation Systems, LNCS 722, pp. 122–133, Springer, 1993.

[Gedd94] D. GEDDIS, *The DTP Manual*, Stanford University, 1994.

[Heck93] A. HECK, *Introduction to MAPLE*, Springer, 1993.

[JeSu92] R.D. JENKS, R.S. SUTOR, *AXIOM*, Springer, 1992.

[LeBr84] D.B. LENAT, J.S. BROWN, *Why AM and EURISKO Appear to Work*, Artificial Intelligence 23, pp. 269–294, Elsevier, 1984.

[McCu94] W.W. MCCUNE, *OTTER 3.0 Reference Manual and Guide*, Technical Report ANL-94/6, Argonne National Laboratory, 1994.

[Paul94] L.C. PAULSON, *ISABELLE: A Generic Theorem Prover*, LNCS 828, Springer, 1994.

[Shav90] J.W. SHAVLIK, *Extending Explanation-Based Learning by Generalizing the Structure of Explanations*, Pitman, London, 1990.

[VeVe94] A. VELLA, C. VELLA, *Artificial Intelligence and the Mathcycle*, in J.H. Johnson, S. McKee, A. Vella (eds.), Artificial Intelligence in Mathematics, Oxford University Press, 1994.

Tools for solving problems in the scope of algebraic programming

Y.V. Kapitonova[1], A.A.Letichevsky[1], M.S. L'vov[2], V.A.Volkov[1]

[1] Glushkov Institute of Cybernetics, 40 Glushkov ave.,252207 Kiev, Ukraine
[2] Department of Informatics & Computer Technique, Kherson State Pedagogical Institute,40-let Oktyabrya str., N27, 325000 Kherson,Ukraine

Abstract. Algebraic programming system APS is considered as a tool for integrating computer algebra with artificial intelligence. The system is based on rewriting rule programming and algebraic program in APLAN, the source language of the system, in many cases may be considered as an executable specification of a problem. Two different kinds of solvers are specified in terms of rewriting rules. The first one is a universal solver that extends a pure PROLOG-like solver in different directions. One of the important property of this solver is the possibility for inclusion of special algorithms for solving equations in different algebras. Another solver is directed to solving problems on computational models (some kind of constraint networks). It searches for the solution of a problem in two stages - constructing the plan and solving equations. On the second stage the solver calls the universal one to get the solution of equations. The application of APS and its solvers to the development of system for mathematical education in secondary school is briefly described in the last section of the paper.

1 Introduction

Rewriting technique [1] is nowadays of a great interest as a reach field of investigations for theoretical computer science as well as a new information technology for practical use in many applications. Well known conferences such as [2], [3] consider different aspects of this field. There are a lot of systems which use rewriting as a basis for symbolic computation. The languages of OBJ family [4] and O'Donnell's languages [5] use rewriting as a basis for equational programming. ASF [6] and ASSPEGIQE [7] use rewriting as a basis for algebraic specifications. Rewriting is also used to extend the possibilities of systems based on other programming paradigms. Computer algebra systems such as Reduce [8] or MATEMATICA [9] are the examples.

Usually rewriting systems are considered as an extension of functional definitions and different restrictions such as regularity or confluence are supposed. For canonical term rewriting systems an arbitrary strategy of rewriting terminates and defines the unique result. The algebraic programming system APS [10], developed in Glushkov Institute of Cybernetics, is based on rewriting but, in distinction from traditional approach, it is possible in APS to combine arbitrary systems of rewriting rules with different strategies of rewriting which may be specified explicitly by means of procedural definitions. The separation of equational definitions in the form of rewriting rules from the strategy of rewriting essentially extends the possibilities of rewriting technique enlarging the flexibility and expressibility of it. The APS integrates four main programming paradigms (procedural, functional, algebraic and logic) by adjusted use of corresponding computational mechanisms as it was described in [10]. Procedures are used in the system to define rewriting strategies, functional definitions are special case of algebraic ones and logic programming is realized by defining strategies of solving problems on axiomatically defined subject domains in terms of rewriting.

The mentioned features of APS make it to be a good base for integrating computer algebra and artificial intelligence. There are many successful examples of such an integration nowadays. Scratchpad/AXIOM ([20], [17], [18]) based on the notions of domains and categories has been developed for description of algebraic structures. AXIOM includes reach hierarchy of classical mathematical domains and efficient algorithms for solving problems for them. Another example of integration of database and deductive facilities with computer algebra algorithms is system called Cayley/MAGMA [23] for computations in the group theory.

The hybrid knowledge representation system MANTRA [19] is a next step of such integration. It combines different formalisms for specification of mathematical domains and development the computational environment for solving problems combining the strong mathematical algorithms with heuristic search for solutions.

On the other hand, many famous large computer algebra system are developing for usage in the intelligent framework. The Praxis [21] system is implemented as intelligent user interface for the symbolic algebra system Macsyma, using a rule-based expert system. Reduce theory of computational domains [24] is attempt to design the formal means for description of interaction among Reduce algorithms.

Very important is the problem of programing language for efficient development of computation algorithms as deductive tools. This language must provide a semantics which is able to combine different programing paradigms. For example, the language LIFE [22] integrates the logic and functional programing, equations and inheritance. It allows combine database, reasoning, simplification and function evaluation. Many efforts was made also in the scope of constraint logic programming paradigm [27].

In spite of these examples the problem of efficient combination of computer algebra algorithms and searching methods from artificial intelligence has not yet satisfactory solution. The investigations continue and the development of programming tools for their support is actual.

In this paper two different kinds of solvers are specified in terms of rewriting rules. The first one is a universal solver that extends a pure PROLOG-like solver so, that arbitrary predicate formulas may be used in the right hand sides of clauses and left hand side may be the negation of an atomary formula. The queries are arbitrary predicate formulas (without quantifies). The semantics of the solver may be defined in terms of three-valued logic as it has been done in [13]. One of the important property of this solver is the possibility for inclusion of special algorithms for solving equations in different algebras. This possibility is illustrated by including linear equations over fields.

Another solver is directed for solving problems on computational Tyugu models. This concept close to constraint networks of today [16] has been developed in 70-th for specifications of engineering problems and used in a problem solver Utopist [15]. The computational model solver searches for the solution of a problem in two stages - constructing the plan and solving equations. On the second stage the solver calls the universal one to get the solution of equations.

The application of APS and its solvers to the development of system for mathematical education in secondary school [25] is briefly described in the last section of the paper [25]. This system is intended for supporting mathematical education process. It is able to solve algebraic and trigonometric problems as simplifications, proofs of identities and solutions of equations. APS was used for the development of the mathematical kernel of the system.

2 Introduction to APS

In this section some information about APS and its source language APLAN is presented. This information is necessary to understand next sections and some important features of algebraic programming technology used in APS. More complete description of APS may be found in [10].

Algebraic programs written in APLAN are located in algebraic modules (ap-modules). Each module contains description of names, description of operation symbols (marks) with corresponding syntactical information and initial assignments for names. Data structures are presented by terms, generated by operations starting with numbers, names and atoms (symbols which cannot have values). Special types of terms are equalities, which may represent rewriting rules, rewriting systems, procedures and statements which are parts of procedures. Any procedure in an ap-module may be called and performed by one of the system interpreters.

The following example of ap-module named pl_ac.ap will be used to explain these notions more precisely.

```
INCLUDE <rat.ap>

/*      Expanding polynomials represented in natural form      */
NAMES rdn, canpl;
NAMES pw,pow,bn;

/*                  Specification of canpl                     */
canpl:=proc(t)(can_ord(t,rdn,rdn));
```

```
rdn:=rs(q,x,y,z,u,k,n,a)(

    isnum(x) -> (x*y = y$x),
    isnum(y) -> (x*y = x$y),
    isnum(x) -> (x$y = y*x),

              x - y = x+(-1)*y,

              x $ 0 = 0,
              x $ 1 = x,
        (x + y) $ z = x $ z + y $ z,
        (x $ y) $ z = x $ (y * z),

          (x+y)*z = x*z+y*z,
          x*(y+z) = x*y+x*z,

          (x$y)*(z$u) = (x*z)$(y*u),
          (x$y)* z    = (x*z)$y,
           x   *(y$z) = (x*y)$z,

          (x$y)^n = x^n$y^n,
          (x*y)^n = x^n*y^n,
          (x^y)^n = x^(y*n),
          (x+y)^n = pw((x+y)^n)
);

pw:=proc(t)(
    can_ord(t,pow,rdn);
    return(t)
);

pow:=rs(x,y,z,n,k,q,a)(
    n>1 ->   ((x+y)^n = bn(1,x,y,1,n,n) +   y^n),
    n>1 -> (q*(x+y)^n = bn(q,x,y,1,n,n) + q*y^n)
);

bn:=rs(q,x,y,k,n,a)(
    (q,x,y,n,n,a) = q*x^n,
    (q,x,y,k,n,a) = (x^k*q$a)*y^(n-k) +
                    bn(q,x,y,k+1,n,(a*(n-k))/(k+1))
);

NAME T;

T:=((a+b)*(b+1/2)+(a-5*b)*(b+c))*(a+b+c)*(b+c-d);
```

```
task:=(
    canpl(T);
    prnpl(T)
);
```

The first sentence of the module includes to it previously designed module **rat.ap** which especially defines all operations used in **pl_ac.ap**, including arithmetical operations, separators ("," and ";"), and so on. It also defines operations over rational numbers and indicates that "+" and "*" are associative-commutative operations, inserting them to the data structure called **ac_list**. This information is used by some rewriting strategies.

The module specifies an algorithm which expands a polynomial with numerical coefficients reducing it to the natural canonical form - the sum of monomials. The operation $ is used to denote the multiplication of polynomial by number to distinguish it from the multiplication "*" of polynomials. Any symbol (atom or name) is considered as a variables of the polynomial.

Initial values of the names **rdn, pow** and **bn** are rewriting systems. Rewriting rules are represented by equalities and conditional equalities. The application of rewriting system to a term is made in the following way. Rules are matched with the top operation in the order they are written in the system. The first matched rule is applied to the top operation of a term. The matching is made in free algebra, but after substituting the right hand side of a rule and the values of variables (listed in the head of a system) which they got when matching, a new part of a term is reduced to the *basic canonical form*. This reduction is top-bottom application of interpreters for interpreted operations if there are any. The interpreters are built-in for some operations, or may be defined by user. The arithmetical operations and relations are interpreted on numbers in a usual way with some additional simplifications, such as $x + 0 = x$ or $x * 1 = 1$. Conditional rules are applied only if after matching the basic canonical form of a condition is 1.

To apply rewriting system to the term different strategies may be used. Each strategy defines some movement along the nodes of a tree in searching for the subterms to which the system is applicable, and some additional transformations defining implicitly applied rewriting rules or simplifications. Simple example of strategy is strategy **ntb(t,R)**. It applies the system R to the term t searching for mentioned subterms (redexes) top-bottom and left right. This is a lazy strategy. Dual to it is the bottom-up strategy **nbt(t,R)** which corresponds to call-by-value. In the considered case more complicated strategy **can_ord(t,R1,R2)** is used. This strategy works with two systems of rewriting rules. First system is applied top-bottom, second - bottom-up. At that when the strategy moves over the nodes bottom-up the subterms are ordered w.r.t. ac-operations by means of merging already ordered arguments of such operations.

Each strategy may be specified as a simple recursive procedure in APLAN in terms of two basic strategies **applr(t,R)** and **appls(t,R)** realized on the low level of APS (C language). The first one applies the system R to the top operation of **t** one time, the second repeats the application of **R** while possible. Here the special role of application operation must be mentioned. This operation is interpreted. If **x** is a function or the name of a function than this function is applied to its argument **y** when the expression **x(y)** is reduced to the basic canonical form. There are two types of functions in APLAN - rewriting systems and procedures. If the function is a procedure, the strategy **applr** is used to compute the value of a function. The name of a function may occur in the right hand side (as in the definition of system bn in the example above). In this case the function is computed recursively. Therefore **applr** is called recursive strategy and **appls** iterative or recursive-iterative one.

Now let us consider specification of the strategy **can_ord**:

```
NAME can_ord,can_up;

can_ord:=proc(t,R1,R2)loc(s,i)(
    t:=can(t);
    appls(t,R1);
        forall(s=arg(t,i),
                can_ord(s,R1,R2)
        );
    can_up(t,R2)
);

can_up:=proc(t,R)loc(s,i)(
    appls(t,R);
    while(yes,
        forall(s=arg(t,i),
            can_up(s,R)
        );
        appls(t,R)
    );
    t:=can(t);
    merge(t)
);
```

Function **can** calls the reduction to basic canonical form, **arg(t,i)** is i-th argument of **t**, in the loop **forall(s=arg(t,i),...)** i varies from 1 to the arity of **t**. Procedure **merge(t)** merges two arguments of **t** if the top operation of **t** is associative-commutative operation.

Data structures in APS really are graph terms although the initial representation of algebraic expressions, occurring in APLAN program are trees. Therefore APLAN has two different kind of assignments **x-->y** and **x:=y**. The first one sets the name **x**, considered as a pointer to the top node of the data structure, which is the value of **y**, the second one curries the copy of the top node of **y** with all its references to the place where the current value of the name **x** is placed. The exact definition of the loop **forall** may be expressed by means of simple for statement:

```
for(i:=1,i<=ART(t),i:=i+1,
    s-->arg(t,i);
    <body of the loop forall>
)
```

where `ART(t)` is the arity of `t` and the function `arg` returns not copy but the original top node of `i`-th argument of `t`.

The simplicity of `can_ord` makes it easy to prove the correctness of the procedure `canpl(t)` using induction on the size of the term `t` after exact definition of what is canonical form of a polynomial.

3 Universal solver

The solver described below searches for the solution of a problem on subject domain defined by the set of axioms, which are quantifierless formulas with variables assumed to be tied by universal quantifier. Each formula is supposed to be elementary (that is atomary formula or the negation of atomary formula) or represented as implication $P \Rightarrow Q$ where P is arbitrary formula, Q is elementary one. The solver will use PROLOG-like strategy for solving problems and each axiom in the form of implication will be used as PROLOG clause or inference rule that reduces problem to subproblems.

The signature of predicates may contains equality. The axioms define some equational theory for it. It consists with all equalities which are the sequences from the set of axioms. The solver may use some special algorithms for solving equations in this theory without referring to axioms. The same may be said about some other predicates.

An absolutely free algebra of terms of a given signature, generated by the set of constants A and the set of variables Z, will be denoted as $T(A, Z)$ (initial algebra with the set $A \cup Z$ of operations of arity 0). The variables of axioms are assumed to belong to the set $W = \{W(1), W(2), \ldots\}$.

An *elementary problem* is a pair (P, X), where P is an arbitrary predicate formula without quantifiers and with free variables from the set $V = \{V(1), V(2), \ldots\}$ and X is a V-*context* with values in $T(A, V)$ that is a substitution of a type $\{V(1) \leftarrow t_1, \ldots, V(n) \leftarrow t_n\}, t_1, \ldots, t_n \in T(A, V)$. A *solution* of an elementary problem (P, X) is a new V-context $Y = \{V(1) \leftarrow s_1, \ldots, V(n) \leftarrow s_m\}$ such, that $m \geq n$, s_i is an instance of t_i for $i = 1, \ldots, n$ and PY is a consequence of axioms. What means a consequence exactly is defined by operational (calculus) or denotational (set - theoretical) semantics of the language.

When the problem is being solved, the set of subproblems appears, and we may speak about the *complex problems*, that correspond to the sets of elementary problems and their solutions. Syntactically, the general notion of a problem is defined as follows.

1. Elementary problem is a problem;
2. Context is (a solved) problem or a solution;
3. `fail` is (unsolvable) problem;

4. If P and Q are problems, than $P \mid Q$ is a problem;
5. If P is a formula, Q is a problem, then (P, Q) is a problem.

Intuitively, solution of a problem $P \mid Q$ is a solution of P or a solution of Q, solution of (P, Q) is a solution of (P, X) where X is one of the solutions of Q. Therefore a notion of (undecidable) problem corresponds to a search AND-OR-tree. If the problem P is to be solved using special algorithms for solving equations or predicates, the name z of corresponding algebra must be joined to the problem. Problem $z(P)$ is called *specialized*.

Operational semantics of the language is defined by means of the partial function **solve** which maps problems to problems. The main property of this function is that $\text{solve}(P) = X$, where X is one of the solutions of the problem P (if there are any) or $\text{solve}(P) = X|Q$, where X is one of the solutions and one may obtain another solutions applying **solve** to Q. The best situation is when all solutions may be covered by this process. Function **solve** applies the system **solve_rs** to a problem with iterative strategy **appls**. Therefore the rewriting rules of this system may be considered as inference rules of corresponding calculus. The problem to be solved must be specialized. If there are no special algorithms for the problem it must be specialized by the name **fr** of free algebra. Following are the definitions in APLAN.

```
solve:=proc(p)(
    appls(p,solve_rs);
    return(p)
);

solve_rs:=rs(P,Q,R,X,Y,a,b,x,y,z)(
                    fail|Q =   Q,                      /*  1 */
                    (P|Q)|R = P|Q|R,                   /*  2 */
        is_not_sol(x)->(x|y = solve(x)|y),             /*  3 */

        z(    P,    fail) = fail,                       /*  4 */
        z(~(~(P))    ,X) = z(P,X),                      /*  5 */
        z(~(  P &  Q ),X) = z(~(P) || ~(Q),X) ,         /*  6 */
        z(~(  P || Q ),X) = z(~(P) &  ~(Q),X),          /*  7 */

        z(    P &  Q ,X) = z(Q,solve(z(P,X))) ,         /*  8 */
        z(    P || Q ,X) = z(P,X) | z(Q,X) ,            /*  9 */

            z(P,  Q|R ) = z(P,Q)| z(P,R),              /* 10 */
            z(P,y(Q,X)) = z(P,solve y(Q,X)),           /* 11 */

        z(P,(a,Y)|~(Q,X)) = z(P,solve((a,Y)|~(Q,X)) ), /* 12 */

        z ( (P = Q),   X) = (vl(z).solve_eq)((P = Q),X), /* 13 */
        z ( P,         X) = (vl(z).solve_pr)(P,X),       /* 14 */
```

```
        ((a, b),Y)|-(P,X) = (a,Y)|-(P,X) | (b,Y)|-(P,X),   /* 15 */
        ( R=>Q, Y)|-(P,X) = try (R,ART(X),unf(P=Q,X,Y)),   /* 16 */
        ( a,    Y)|-(P,X) = unf(P=a,X,Y)                    /* 17 */
);

is_not_sol:=proc(x)(return(~(mark(x)==mark_ar)));

try:=rs(R,n,X)(
    (R,n,fail) = fail,
    (R,n,X  ) = fr(rename(R,n,X),X)
);
```

First 11 rules deals with general problem and reduce it to elementary one. Sign || is disjunction. Rules 13 and 14 solve elementary problems referring to the algebra that specify it. The name of an algebra in specialized problem is the name of data structure called valuation. Components of a valuation are named by atoms and referred to by means of operation z.x which denotes the body of a component with the name x of a valuation z. Valuation for an algebra must contain at least two names: solve_eq and tt solve_pr for algorithms to solve equations and predicates, correspondingly. The description of free algebra is the following:

```
fr:=(
 solve_pr: rs(P,z,x,X)(
      (~(P(x)),X) = (vl(P).neg,make_nil(vl(P).head))|-(~(P(x)),X),
      ( P(x), X) = (vl(P).pos,make_nil(vl(P).head))|-( P(x), X)
 );
 solve_eq: rs(P,Q,X)(
      ((P = Q),X) = unify((P = Q),X)
 )
);
```

Function solve_eq for this algebra call unification procedure, realized on low level. It returns new context which represent the most general unifier of P and Q or symbol fail if unification of the terms is impossible. For solving predicates (elementary formulas) function solve_pr refers to definition of a predicate. Non-interpreted predicate symbol is the name of a valuation where all axioms related to this predicate are differed on two groups: positive and negative. Following is the example of predicate definition:

```
P1:=ax(x,y,z)(
    pos: (
        P1(A,B),
        P1(x,A+x),
        P1(x,y)|| ~(P2(y,z) => P1(x+z,y)
    ),
```

```
    neg:(
        ~(P1(C,D)),
         P3(y,y)=> ~(P1(x,y,z))
    )
);
```

This is the external representation. The head of this definition contains the list of variables which must be translated to W(1),W(2),.... The head is also used for creating the empty context (function make_nil). Function solve_pr returns new type of a problem: *inference problem*. It has a pattern $A\|B$ where A is a list of axioms and B is an elementary problem. This kind of a problem is considered by rules 15-17. Function un is the modification of unification for two contexts, the solution (new context) may contain two kind of variables - axiom variables and problem ones. The function rename renames axiom variables if any to new problem ones.

The contexts are represented as one-dimensional arrays and the condition is_not_sol(x) (is not a solution) in the rule 3 checks if the type of x is array. This rule shows that the search for the solution is depth-first-search. To get breadth-first- search or parallel strategy, this rule must be modified to is_not_sol(x)->(x|y = solve(y)|x) and a function that make only some steps of solving problem must be called from solve_rs instead of solve.

Another example of an algebra is the algebra lin_alg that contains an algorithm for solving linear equations over field:

```
lin_alg:=(
    solve_pr:rs(P,X)((P,X) = (vl(fr).solve_pr)(P,X));
    solve_eq: proc(p)(
        canpl(p);
        yes:=1;
        appls(p,solve_lin_rs);
        ntb(p,del_mlt);
        return(p)
    )
);
```

To solve predicate the algebra refers to free case. The procedure canpl is modified so that symbols different from unknowns V(i) would be considered as constasnts and be included to coefficients. Rewriting systems used in solve_eq are the following.

```
solve_lin_rs:=rs(A,B,E,X,i)(

    (V(i)$A+B = 0,V(i)=nil,X) = starg(X,i,canplf((-1)*(1/A)*B)),
    (V(i)  +B = 0,V(i)=nil,X) = starg(X,i,canplf((-1)*     B)),
    (V(i)$A   = 0,V(i)=nil,X) = starg(X,i,0),
    (V(i)     = 0,V(i)=nil,X) = starg(X,i,0),
```

```
(E,                 V(i)=A,  X) = (canplf(sub(V(i)=A,E)),X),

    (0=0,X) = X,
    (A=0,X) = make_sys(A=0,unknown(A),X),
    (A=B,X) = (canplf(A+(-1)*B=0),X)
);

make_sys:=rs(E,i,X)(
    (E,nil,X) = fail,
    (E,  i,X) = (E,V(i)=arg(X,i),X)
);

unknown:=rs(A,B,i)(
    A+B = unknown(A),
    A$B = unknown(A),
    V(i)= i,
    A   = nil
);
```

Some technical explanations. Function `canplf` is functional modification of `canpl`. Function `starg(X,i,z)` updates the array X setting its i-th argument to z.

Semantics. Semantics of the solver, may be described in the terms of three-valued logic of Klinee on the base of the paper [12] as it was done in [13]. This logic is the logic of partially defined predicates, and its truth values $\perp, 0, 1$ correspond to undefined, false and true values. This set of values is assumed to be partially ordered so that $\perp \sqsubset 0, 1$ and $0, 1$ are not comparable. Let us fix some signatures of operations and predicates (it does not matter if these signatures are multisorted or one-sorted), and consider algebras with partially defined predicates of the given signature. Gomomorphism is defined as a mapping which preserves the operations and is monotonous w.r.t. logical values of predicates. Equality is also considered as partial predicate which relates with real equality so that $(d = d\prime) = 1 \leftrightarrow d = d\prime$ (therefore if $d \neq d\prime$, then $(d = d\prime) = 0, \perp$).

Algebra D is called to be *labelled* by a set of constant symbols A, if the labelling mapping $v : A \to D$ is defined. If $d = v(a)$, the element d is called to be labelled by symbol a. Labelled algebra is supposed to be generated by all labelled elements.

Now let $\Phi(A, W)$ be a language of arbitrary predicate formulas without quantifiers with the set of constants A and variables from the set W, constructed by means of logical connectives $\vee, \wedge, \neg$. Formula $P(x_1, \ldots, x_n) \in \Phi(A, W)$ is called to be true on the algebra D, labelled by A, $(D \models P(x_1, \ldots, x_n))$ if it is true on D for all ground values of therms $x_1, \ldots, x_n$. Let F be a subset of $\Phi(A, W)$. An algebra D labelled by A is called to be a *model* of a set F if each formula from F is true on D. A set F of formulas is *consistent* if it has a model.

Let D and $D\prime$ be two algebras, labelled by the sets of constants A and $A\prime$, correspondingly. D is called to be an *approximation* of $D\prime$ if $A \subset A\prime$ and there exists a gomomorphism $\gamma\colon D \to D\prime$ such, that $\gamma(v(a)) = v\prime(a)$ for all $a \in A$, where v and $v\prime$ are the labelling functions for D and $D\prime$, respectively. Two algebras are called to be isomorphic if each approximates other. The approximation relation is a partial order on the class of all labelled algebras (of the same type) considered up to isomorphism. The minimal element in the class of all algebras, labelled by the same set A, is absolutely free algebra $T(A)$ of ground terms with nowhere defined predicates.

Algebra D labelled by a set A approximates a set F of formulas if it approximates every model of that set, labelled by a set $A\prime \subset A$.

Theorem 1. *The class of all approximations of a consistent set of formulas F, labelled by the same set A, has a maximal element, defined uniquely up to isomorphism.*

The maximal approximation may be constructed as a factor-algebra $T(A)/F = T(A)/\rho(F)$, where $\rho(F)$ is the congruence, defined by relation: $t = t'(rho(F)) \Leftrightarrow F \models t = t'$. For predicates define $p(t_1, \ldots, t_n) = 1 \iff F \models p(t_1, \ldots, t_n)$ and $p(t_1, \ldots, t_n) = 0 \iff F \models \neg p(t_1, \ldots, t_n)$. Therefore, if no one of two alternative is true, $p(t_1, \ldots, t_n) = \bot$. For detailed proof of this theorem and theorems mentioned below see [13]. Maximal approximation is not generally a model of F, but it contains a complete information about all models of F: formula $P \in \Phi(A, X)$ is true on all models of F if it is true on $T(A)/F$. The consistency of F for three-valued logic is the same as the classical consistency. Really, let us denote $(F)^=$ the extension of F by the axioms of equality ($x = x, x = y \Rightarrow y = x, x = y \wedge P(x) \Rightarrow$). Than F is consistent if $(F)^=$ is classically consistent (that is has two-valued model). This statement is the consequence from the theorem below. The model of F is called to be complete if every ground atomary formula has the value different from $\bot$.

Theorem 2. *The set F is consistent if it has a complete model.*

The theorem and its corollary provides the possibility to use the ordinary resolution calculus with paramodulation as a complete deductive system for inference of all semantical consequences from the given set F of formulas. Indeed, every logical formula in Kleene logic may be reduced to conjunctive normal form and F may be assumed as a set of disjuncts. Then for any atomary formula P, $F \models P \Leftrightarrow (F)^= \cup \{\neg P\}$ is classically inconsistent, but it means that P is inferred from F in the calculus of resolution with paramodulation. Now to set the connection of solver, described above, with this deductive system is the matter of well known technique.

The solver is obviously consistent, that is every solution provided by it is the solution on a maximal approximation. But it is not complete (even if there are no specialized predicates and equations) for several reasons. The first is that solver uses SLD-resolution inference which is known to be incomplete in the case of non-Horn clauses. So the full resolution calculus must be modeled by solver. The simplest (but not the best) way is to extend the set of statements of evaluated program by intermediate results at each step of inference. Another reason is that the solver realizes the depth-first-search. It may be transformed to the breadth-first search device (less efficient) and than it will be complete if there are no axioms, containing equality (the maximal approximation in this case is absolutely free algebra). To make it complete in any case, the rules for paramodulation must be added.

4 Problem solving on computational models

Computational model is a set of variables (elements of a model) and the set of relations or constraints, binding possible values of these variables. Special type of constraints is algebraic equations. A problem on computational model is a question of a type: "find the values of $y_1, \ldots, y_m$ assuming the values of $x_1, \ldots, x_n$ to be known". Known values may be given as a constants of subject domain, defined by a model, or symbolically. The last case may be recognized as a special type of a problem of program synthesis. Example of computational model (in terms of APLAN data structures):

```
triangle:=comp_model(

    elements:(a,b,c,alpha,beta,gamma);

    equations:(

        alpha+beta+gamma=pi,

        b^2=a^2+c^2+2*a*c*cos(beta),
        c^2=a^2+b^2+2*a*b*cos(gamma),

        a/sin(alpha)=b/sin(beta),
        b/sin(beta) =c/sin(gamma),
        c/sin(gamma)=a/sin(alpha)
    );

    solutions:........;
    solve_eq:.........;
    solve_pr:.........;
    ..................
}
```

Each relation may be considered as a sours for finding solution of *elementary problem*. For instance, the equation b2=a2+c2+2*a*c*cos(beta) may be used for finding a if b,c,beta are known, or for finding beta if a,b,c are known. Let us suppose that for each element of a model the set of all possible elementary problems are given, say, in the form (for example above):

```
solutions:(
      a:(
         b c alpha        compute(a: a^2=b^2+c^2+2*b*c*cos(alpha))||
         b c beta         compute(a: b^2=a^2+b^2+2*a*b*cos(beta))||
         b c gamma        compute(a: c^2=a^2+b^2+2*a*b*cos(gamma))||
         b alpha beta     compute(a: a/sin(alpha)=b/sin(beta))||
         c alpha gamma    compute(a: c/sin(gamma)=a/sin(alpha))
      );

      b:(
         a c alpha        compute(b: a^2=b^2+c^2+2*b*c*cos(alpha))||
         a c beta         compute(b: b^2=a^2+c^2+2*a*c*cos(beta))||
         a c gamma        compute(b: c^2=a^2+b^2+2*a*b*cos(gamma))||
         a alpha beta     compute(b: a/sin(alpha)=b/sin(beta))||
         c beta gamma     compute(b: b/sin(beta) =c/sin(gamma))
      );
         . . . . . . . . . . . . . . . . . . . . . . .
      )
```

These solutions may appear as a result of preliminary analysis of equations of a model, or a procedure which generates possible solutions for each element of a model may be put as a value of the name **solutions** to be used in run time. Following is the rewriting system for making the plan for solving a problem on computational model. It is used with the iterative strategy and transforms a problem to the plan of its solution. A problem is given in the form **problem x => y** where **x** is the list of known elements **y** the list of unknowns. Elements in a list are separated by applications (blanks).

```
make_plan:=rs(P,Q,R,Q1,Q2,F,x,y,z,u)(

                   problem(x => y) = (e,x => y e, e) || no_solutions,

                   (Q1 || Q2) || Q  =  Q1 || Q2 || Q,
                   no_solutions || R  = R || no_solutions,
        (P,x => e,           e) || R  =   simp(P),
        (P,x => (Q1   Q2)Q, z) || R  =   (P,x => Q1 Q2 Q, z) || R,
        (P,x => (Q1 || Q2)Q, z) || R  =   (P,x => Q1 Q, z) ||
                                   (P,x => Q2 Q, z) || R,

        (P,x => compute F Q, y z) || R = (P F, y x => Q, z) || R,

   is_in(y,x) ->( (P,x => y Q, z)  || R = (P y,    x => Q, z) || R ),
   is_in(y,z) ->( (P,x => y Q, z)  || R =  R ),

        (P,x => y Q, z) || R  =  R || (P,    x =>  (get_sol y) Q, y z)
);
```

```
simp:=rs(x,y,z)(
        e x    =    simp(x),
       (x y) z =    simp(x y z),
        x y    = x simp(y)
);
```

Function `get_sol(y)` calls the solutions for the element y. It may return, for instance, `vl(model_name).solutions).y` where `model_name` is a global name which contains the name of a model. Intermediate data structure `(P,x => Q,z)` which appears in the left hand sides of rewriting rules, has the following meaning. P is already made part of a plan, x - list of known elements, list of unknowns with first element possibly changed to the disjunction of possible solutions, z is the list of elements for which it were calls to `get_sol` function in right-left order. If generalize the notion of problem to the form which appears in the rewriting rules, it may be proved that the rules preserve equivalence of problems, and each elementary problem may appear no more than once in the problem. Therefore the iterative strategy terminates and gives correct plan. It has linear time complexity w.r.t. to the number of elementary problems, and a plan cannot be simplified, there is no redundant computations.

In the example, considered above the problem `problem a b gamma => beta c alpha` will be rewritten to the plan

```
a b a b gamma (c : c ^ 2 = a ^ 2 + b ^ 2 + 2 * a * b * cos gamma)
(alpha : a ^ 2 = b ^ 2 + c ^ 2 + 2 * b * c * cos alpha)
gamma (beta : alpha + beta + gamma = pi) c alpha
```

To get the symbolic or numeric solution, the plan must be (automatically) converted to the problem for universal solver:

```
trn(
    V(2) ^ 2 = a ^ 2 + b ^ 2 + 2 * a * b * cos gamma &
    a ^ 2 = b ^ 2 + V(2) ^ 2 + 2 * b * c * cos V(3) &
    V(3) + V(1) + gamma = pi, array(nil,nil,nil)
)
```

and solved by special solver for computational model `trn`.

5   Application of APS to school mathematical education

A system for computer support of mathematical training in secondary school on the base of APS is now under development. The first version of this system, called AIST has been discussed in [25]. Second version of this system, called TerM (Terra Mathematica) has more powerful algorithms of solving equations and better theoretical background, based on well-defined algebraic hierrachy. The solvers, described in this paper, are included to the system and used for organization of computational processes on a base of different special algorithms.

The main task of the system TerM is support for algebraic and trigonometric problems on simplifications, proofs of identities and solutions of equations (further on the term "algebraic problem" is used namely for this type of problems).

The mathematical activity of a student consists of recognizing properties of mathematical objects and its transformations according to the rules, strictly defined in a corresponding mathematical theory. System either verifies transformation made by a student (Short-Step mode) or automatically executes transformation according to the students instruction (Long-Step mode).

Thus, the solution process of an algebraic problem is a sequence of steps. Every successive step is the result of some algebraic transformation of the previous one. The sequence steps from the setting up of the task and terminates with its solution.

Like other pedagogical computer systems, TerM provides the user by the mathematical Reference Book.

Windows shell Copybook is designed as a computer model of the student's copybook where student can solve a current problem, save and look through the solving of the previous problems. The natural notation of algebraic expressions in a schooltype syntax is provided with specialized formula editor.

TerM contains the subsystem Solver aimed at automatic solving of trigonometric problems. The algorithm is based on the equation-type classification (this approach is similar to the conception of the PRESS [26] system). The classification algorithms use canonical forms and hierarchy of algebras. This hierarchy especially the following algebras:

Trig - The field of rational trigonometric expressions whose arguments belong to Arg and coefficients are from Coef (sin, cos, tan, cot is a signature)

Coef - The Coefficients field

Arg - Vector's space of arguments

Sol - Algebra of Sets of arithmetic progression-solutions of a trigonometric equations

BField - The basic field of zero - characteristic. We use field of rational numbers

BIntDom - (Commutative) Integral Domain. Ring of integers

These algebras constitute the basis. Other algebras may be defined by recoursive typing with constructive definitions of gomomorphisms and inclusions.

6   Conclusion remarks

We have presented the basic ideas of problems solving in the algebraic programming environment and its application. This technique is effective for integrating computations and logic in the common framework. We are going to use these methods to develop automatic means for specializations of general algorithms on a base of partial evaluation with respect to rewriting rule systems. The general theory of computing invariants of programs will be used to develop the algorithms.

References

1. N. Dershovitz, J.-P. Jouannaud, Rewrite systems. In Jan van Leeuwen (Ed.), *Handbook of theoretical computer science, v.B* Elsevier, 1990.
2. C. Kirchner (Ed.), Rewriting techniques and Applications. Proceedings, *LNCS* vol. 690, 488, Springer, 1993.
3. M. Rusinovich, J.L. Remy (Eds.), Conditional term rewriting systems. Proceedings, 1992, *LNCS* vol. 656, 501, Springer, 1993.
4. J. Gogen, C. Kirchner, H. Kirchner, A. Megrelis, and T. Wincler, An introduction to OBJ-3. In Jouannaud and Kaplan (Eds.), *Proc. 1st Intern. Workshop on Conditional Term Rewriting Systems*, Springer, 1988.
5. M. J. O'Donnell, Term rewriting implementation of equational logic programming. In P. Lescanne (Ed.), *Rewriting Techniques and Applications*, volume 256 of *LNCS*, 1–12, Springer, 1987.
6. J. A. Bergstra, J. Hearing, and P. Klint (Eds.), *Algebraic Specification*. ACM Press and Addison Wesley, 1989.
7. M. Bidoit and C. Choppy, ASSPEGIQUE: an integrated environement for algebraic specifications. In *Proc. Intern. Joint Conf. on Theory and Practice of Software Development*, 246–260, Springer, 1985.
8. G. Rayna, REDUCE. Software for Algebraic Computation. Springer, 1989.
9. S. Wolfram, *Mathematica*^{TM}. *A System for Doing Mathematics by Computer*. Addison-Wesley, 1988.
10. A.A. Letichevsky, J.V. Kapitonova, S.V. Konozenko, Computations in APS, *Theoretical Computer Science* 119(1993), 145–171, Elsevier,1993.
11. A.A. Letichevsky, J.V. Kapitonova, S.V. Konozenko, Algebraic Programs Optimization. In *Proc. of the Int. Symp. on Symbolic and Algebraic Computation* (ISSAC'91), July 15-17 1991, Bonn, Germany, ACM Press, New York 1991.
12. M. Fitting, A Kripke-Kleene semantics for logic programs, *J.Logic Programming*, 4, 295–312, 1985.
13. J.V. Kapitonova, A.A. Letichevsky, On constructive mathematical descriptions of subject domains, *Kibernetica*, 4, 1988.
14. J.M. Hullot, Canonical forms and Unification, In Proc.of 5-th Conference on Automated Deduction, LNCS v. 87, 318-334, Springer, 1980.
15. Enn Tyugu, Solving problems on computational models. J. Computational Mathematics and Math. Phys., 10:716–33, 1970.
16. Ugo Montanary, Networks of constraints: Fundamental properties and application to picture processing. *Information Sciences*, 7(2):95–132, 1974.
17. J.H. Davenport, B.M. Trager, Scratchpad's View of Algebra I: Basic Commutative Algebra. *LNCS* 429, 40–55, Springer, 1990.
18. J.H. Davenport, P. Gianni, B.M. Trager, Scratchpad's View of Algebra II: A Categorial view of factorization. In *Proc. of the Int. Symp. on Symbolic and Algebraic Computation* (ISSAC'91), July 15-17, 1991, Bonn, Germany, 32–39, ACM, New York 1991.
19. J. Calmet, I.A. Tjandra, A Unified-Algebra-based Specification Language for symbolic Computing. In *Design and Implementation of Symbolic Computation System*, 14–27, Springer, 1993.
20. R.S. Sutor (Ed.), Axiom. *User's Guide.*, The Numerical Algorithm Group Limited, 1991.
21. M. Clarkson, Praxis: rule-based expert system for Macsyma, *LNCS* 429, 264–265, Springer, 1990.

22. H. Ait-Kaci, A. Podelski, An overview of LIFE. *LNCS* 504, 42–58, Springer, 1991.
23. G. Butler, J.J. Cannon, Cayley, version 4: the user language, *LNCS* 358, 456–466, Springer, 1989.
24. B.J. Bradfort, A.C. Hearn, J.A. Padget, E. Schrufer, Enlarging the REDUCE domain of computation. In *SYMSAC 1986*, 100–106, ACM, New York, 1986.
25. M.S. L'vov, A.B. Kuprienko, V.A. Volkov, Applied Computer Algebra System AIST: Computer Support of Mathematical Training, In *Proc. Int. Workshop on the Computer Algebra Application,*, July 9, 1993, Kiev, Ukraine.
26. B. Silver, *Meta-level Inference*. Elsevier Science, Amsterdam, Netherlands, 1986.
27. Jacques Cohen, Constraint logic programming languages. Communications of the ACM, 33(7):52–68, 1990.

Planning A Proof Of The Intermediate Value Theorem

Myles Chippendale

Rutherford Appleton Laboratory, Chilton, Didcot, Oxon. Email - mjc@inf.rl.ac.uk

Abstract. This paper descibes work done in The Mathematical Reasoning Group (MRG)[1] in the Department of Artificial Intelligence at Edinburgh to allow the automatic proof of theorems in constructive analysis, with particular emphasis on the intermediate value theorem. This work included formalising theories of the rational and real numbers in a constructive type-theory and the development of meta-level tools to allow a proof of the theorem to be planned and constructed.

1 Introduction

The Edinburgh MRG has developed an approach to automatic theorem proving based on the idea of proof planning using meta-level reasoning. This approach involves work at two levels.

The object level is incorporated in an interactive proof development system called *Oyster*. Oyster is a rational reconstruction of Nuprl [4] in Prolog. It is based on Martin-Löf's type theory. This is a constructive logic which means that we can extract a program from a proof which will contain its algorithmic content.

The meta-level is incorporated in a proof planner called *Clam*. When Clam has planned a proof, this plan can be applied in Oyster to automatically prove the theorem. The emphasis of Clam's abilities is on proofs by induction, in which a method of proof called rippling is especially important.

The purpose of the work presented in this paper was to verify that the ideas of proof planning and rippling could be successfully applied to the domain of real analysis. This involved constructing theories of the rationals and reals as well as developing meta-level tools.

2 The Object Level

In this section we present some background behind the Oyster interactive proof development system and the theories that were built using it.

---

[1] The work described in this paper was undertaken by the author as part of his MSc thesis at the Department of Artificial Intelligence at Edinburgh.

2.1 Constructive Type Theory

Oyster uses a constructive type theory which consists of a form of λ-calculus with a type theory built in. There is a small set of primitive types (*e.g.* the Peano natural numbers, the integers) and a number of *type constructors* that can be used to build complex types out of simpler ones. For instance, for any types A and B, we have that $A \wedge B$ is a type, usually called the product type. Its elements are of the form $\langle a, b \rangle$ where $a \in A$ and $b \in B$. All the usual connectives of first-order logic are type constructors. Some properties of these types are presented in table 2.1. Some other types that are available and are relevant are:

- *The subset type*, $\{x : A|B\}$, is inhabited by objects a such that $a \in A$ and such that $B[a/x]$ is true, *i.e.* by objects of type A that have the property B. Strictly speaking, the objects of this type should be pairs (a, b) where a is an object of A and b is a proof that a satisfies $B(x)$. This, however, leads to difficulty in practice since objects of this type are not objects of the base type (and so we cannot use functions over the base type on our subset type). The approach taken is to regard elements of a subset type as members of the base type, with the restriction that if, for any member of a subset, the proof that the defining property of the subset holds for it is used in a constructive way, more work must be done in supplying this proof.
- *The quotient type*, A/E, is to be interpreted as a partition of the base type A using the equivalence relation E. A problem with this type is that it is possible to define functions over the base type that do not respect the equivalence relation, in the sense that $f(a) \neq f(b)$ for $a, b \in A$ such that aEb. A solution to this is to explicitly supply definitions of functions over quotient types, and prove that they respect the equivalence relation.
- *The Acc type*, $Acc(A, \prec)$, is a type that characterises a well-founded relation $\prec$ over the type A. This type will be discussed in more detail below.

Type	Object form	where ...	Name
$A \to B$	$\lambda x \cdot b$	for every $a \in A$, $b[a/x] \in B$	Function type
$A \wedge B$	$\langle a, b \rangle$	$a \in A,\ b \in B$	Cartesian Product
$A \vee B$	$\text{inl}(a), \text{inr}(b)$	$a \in A,\ b \in B$	Disjoint Union
$(\forall a \in A)B(a)$	$\lambda x \cdot b$	for every $a \in A$, $b[a/x] \in B$	Dependent Function type
$(\exists a \in A)B(a)$	$\langle a, b \rangle$	$a \in A,\ b \in B[a/x]$	Dependent Product type

Table 1. Summary of type information

These types are collected together into *universes*, $U_1, U_2, \ldots$. Each U_i is formed by closing the primitive types and universes U_j for $j < i$ under the type constructors.

There are four judements that we can make about types and their members:

- $T \in U_i$. This states that T is a type. This can be considered as meaning that the type is well-formed
- $t \in T$. t is an element of the type T.
- $T_1 = T_2 \in U_i$. T_1 is equal to T_2 as a type, *i.e.* objects of type T_1 are also objects of type T_2.
- $t_1 = t_2 \in T$. t_1 and t_2 are considered equal as objects of type T.

To use this type theory as a logic, we use the principle of *propositions-as-types*, due to Curry and Howard. The main idea is that a proposition P is repesented by a type T_P such that P is true iff T_P has a member. Conversely, given any type, we can regard it as a proposition that is true iff the type is inhabited. The members of a type can be regarded as proofs of the validity of (the proposition coresponding to) a type. Further, since the members of a type are functional programs that construct, destruct and otherwise manipulate objects of types, we find that these proof objects encapsulate the algorithmic content of the proof procedure (outlined later) that deduced the truth of the proposition.

For example, consider the proposition $A \rightarrow B$. This corresponds to the type $A \rightarrow B$ (where the first $\rightarrow$ is the logical connective "implies" and the second one is the constructor for the function type). This proposition is true if the type is non-empty, *i.e.* if there is an object of the type $\lambda x \cdot b$ where for every $a \in A$, $b[a/x] \in B$. So, we can prove that the proposition is true if we find a function that maps every element of A to an element of B.

The standard constructive interpretation of what constitutes a proof of $A \rightarrow B$ is "a construction that takes a proof of A and transforms it into a proof of B". By using the idea of proof as a member of a type and interpreting "a construction that ... transforms" as a function in our language, we see that our idea of proof corresponds with the constructive idea of proof, at least in the case of implication. In fact the two ideas of proof are coincident for each of the logical connectives and quantifiers.

We note here that negation is not a primitive of our type theory but is defined by

$$\neg P := P \rightarrow \bot$$

where $\bot$ is a type that contains no objects. This is the standard constructive treatment of negation.

2.2 Inference Rules

The mechanism for inference used in Oyster is based upon intuitionistic sequent calculus. An *intuitionistic sequent* has the form

$$H_1, \ldots, H_n \vdash G$$

where the H_is are hypotheses and G is a conclusion. Informally this represents the formula

$$H_1 \wedge \ldots \wedge H_n \to G$$

In Oyster the form of the sequents is

$$a_1 : A_1, \ldots, a_n : A_n \vdash G \text{ ext } G^{ext}$$

where, for each i, A_i is a type and a_i is an object of that type. G^{ext} is an object of type G, called the *extract term*. We often write this as $\Gamma \vdash G$ ext G^{ext}.

The inference rules of this calculus define which sequents are consequences of which other sequents. For example, The $\wedge$-introduction rule has the form

$$\frac{\Gamma \vdash A \text{ ext } A^{ext} \qquad \Gamma \vdash B \text{ ext } B^{ext}}{\Gamma \vdash A \wedge B \text{ ext } \langle A^{ext}, B^{ext} \rangle}$$

These rules are applied in a backwards chaining manner to a *goal* (a sequent that we wish to prove) to produce a set of 0 or more subgoals that must be proved. A goal is proved when we apply an inference rule that results in 0 subgoals. As we apply these rules, the extract term is, in general, *refined*. For example, to prove the goal

$$\Gamma \vdash A \wedge B \text{ ext } G^{ext}$$

we can apply the above inference rule to obtain the subgoals

$$\Gamma \vdash A \text{ ext } A^{ext} \quad \text{and} \quad \Gamma \vdash B \text{ ext } B^{ext}$$

and refine the extract term from G^{ext} to $\langle A^{ext}, B^{ext} \rangle$. A^{ext} and B^{ext} may need further refinement until they, and hence G^{ext}, are fully instantiated.

The above rule is an *introduction rule*. These introduce the relevent type constructor into the consequent of the sequent which is the conclusion of the rule and build up the extract term from objects of types. Other rules are *elimination rules*. These introduce the relevent type constructor into the hypothesis list of the conluded sequent. The extract terms of these rules are generally functions that "break apart" or otherwise manipulate objects of types.

From the above we can see that a proof of a sequent is a tree of goals, the links from a goal to its children being an application of an inference rule. Furthermore, the method of refinement and the relation between introduction and elimination rules ensure that the extract term that is built up during the process of proving a sequent contains the algorithmic content of that proof. In particular, this extract term will be a functional program that can be applied to objects of the type $H_1, \ldots, H_n$, the hypotheses of the top level goal, to produce an object of type G, the consequent of the top level goal.

2.3 Tactics

A *tactic* is an arbitary piece of Prolog code for applying rules of inference. The effect of applying a tactic to a goal $\Gamma \vdash G$ is a (possibly empty) list of subgoals, $[G_1, \ldots, G_n]$ which need to be solved, assuming that the tactic succeeds (i.e. is actually applicable).

Tactics are usually created by combining rules of inference using a restricted control language. This means that the level of inferencing can be raised from the low-level inference rules of our logic to a level that is, for instance, closer to the level at which a human mathematician might reason, but without losing soundness.

Tactics that have been developed for Oyster include ones that apply induction, generalise theorems, evaluate definitions *etc.*

2.4 The Acc Type

One of the most important tactics in Oyster is the *induction* tactic. It can be applied to a sequent to produce subgoals corresponding to the base case(s) and step case(s) of the particular induction scheme being used in a proof. Oyster has induction schemes for Peano natural numbers, integers, lists *etc* but does not have an induction scheme for real numbers.

We shall see later that the proof of the intermediate value theorem relies on induction; in particular it relies on well-founded induction. The mechanism that was used to implement this scheme is due to Nordström [5] and is called the Acc type.

The judgement $a \in Acc(A, \prec)$ is interpreted as meaning a is a member of the well-founded part of A with respect to the relation $\prec$. Classically, an element is part of the well-founded part of a relation if it is not a member of an infinite descending chain. Using the Acc type, if we wish to prove a goal of the form

$$\Gamma \vdash x : Acc(A, \prec)$$

then we need to prove the subgoals

$$\Gamma \vdash x \in A$$

$$\Gamma, y : A, y \prec x \vdash y : Acc(A, \prec)$$

i.e. to prove x is in the well-founded part of the relation, we must show that all the elements that precede x are also in the well-founded part of the relation. This is a simplified version of the introduction rule.

To use the Acc type to prove a goal of the form

$$\Gamma \vdash \phi(x)$$

we need to prove the subgoal

$$\Gamma, w : A, u : (\forall v : \{y : A | y \prec w\} \rightarrow \phi(v)) \vdash \phi(w)$$

where Γ contains the hypothesis $x : Acc(A. \prec)$. This states that we must prove $\phi(w)$ for an arbitrary w in $Acc(A, \prec)$ given that $\phi(v)$ holds for all v that preceed w. This is a simplified version of the elimination rule. We can see that it corresponds to the principle of well-founded induction.

The extract term of the goal $\Gamma \vdash \phi(x)$ is $wo\_ind(x, [w, u, T])$ where T is the extract term for the subgoal. This evaluates to T with w replaced by $x*$ (the result of evaluating x) and u replaced by $\lambda y \cdot (wo\_ind(y, [w, u, T]))$.

For a full treatment see [5].

3 The Rationals And Reals

In this section we describe some of the work done in constructing theories of the rational and real numbers.

3.1 The Rationals

The theory of the rationals was constructed as in most classical treatments, *i.e.* by identifying an equivalence class of pairs of integers with a rational number, and by defining operations and orderings in the usual way. Some theorems of rational arithmetic were proved, including, for instance, a theorem stating the decidability of equality over the rationals.

3.2 The Reals

The development of the reals outlined here follows the approach of Errett Bishop [1]. His work demonstrated the fact that much of higher mathematics could be built in a constructive framework.

The first step in building a theory of the reals is to construct the real numbers. There are two common methods in classical mathematics for this: *Dedekind cuts* and *Regular sequences*. Bishop uses the later formulation. We say that a sequence of rationals $x := (x_n)$ is *regular* iff

$$\forall m, n \in \mathcal{N}^+ \; |x_m - x_n| \leq m^{-1} + n^{-1}$$

where $\mathcal{N}^+$ is the set of positive natural numbers. We also define that two sequences of rationals, x and y are equal ($x =_{\mathcal{R}} y$) iff

$$\forall n \in \mathcal{N}^+ \; |x_n - y_n| \leq 2n^{-1}$$

We then define the real numbers to be the equivalence class of regular sequences of rationals using the equivalence relation $\lambda x, y \cdot (x =_{\mathcal{R}} y)$.

The order relations are defined by

$$x <_{\mathcal{R}} y \Leftrightarrow \exists n \in \mathcal{N}^+ n^{-1} < y_{2n} - x_{2n}$$

$$x \leq_{\mathcal{R}} y \Leftrightarrow \forall n \in \mathcal{N}^+ \; -n^{-1} < y_{2n} - x_{2n}$$

where the ordering relations on the right hand side of the above definitions are over the rationals. From now on we will drop the subscript $\mathcal{R}$ and consider all equalities and orderings to be over the real numbers.

These are, respectively, a strict and a non-strict ordering that have many of the properties of the usual ordering relations on the reals. In particular, they behave like the classical relations "up to double negation", *i.e.* given a propostion α involving either of the two ordering relations, if α is classically valid, then $\neg\neg\alpha$ is contructively valid.

One of the main difficulties in using a constructive theory of the reals is that the ordering relations and equality are, in general, undecidable. In particular the usual form of dichotomy only holds in the sense that at most one of $x \leq y$ and $y \leq x$ is true. The formula $\forall x, y \in \mathcal{R} \; x \leq y \vee x \geq y$ is not constructively valid, and hence cannot be used as a decision procedure. There is. however, a weaker form of dichotomy that is constructively valid, namely

$$\forall x, y, z \in \mathcal{R} \; y < z \rightarrow (x < z \vee y < x)$$

This theorem was proved interactively in Oyster. By the properties of $A \vee B$ in constructive maths, the above formula constitutes a decision procedure. In particular it was used in the proof of the following theorem

$$\forall x, c \in \mathcal{R}, k \in \mathcal{N} \; (x < c + 2^{-k}) \vee (c - 2^{-k} < x)$$

This theorem is used extensively in the constructive proof of the intermediate value theorem that is presented later.

The arithmetic operations on the reals are defined according to Bishop's work, *e.g.*

$$x + y := (x_{2n} + y_{2n})$$

$$x \cdot y := x_n \cdot y_n$$

We also have the very rudiments of some real analysis set up. We have the definitions

$$cts(f, a, b, n, k) \Leftrightarrow \forall x, y \in [a, b] \; (|x - y| \leq (b - a) \cdot 2^{-n} \rightarrow |f(x) - f(y)| < 2^{-(k+1)})$$

and

$$continuous(f, a, b) \Leftrightarrow \forall e \in \mathcal{N} \; \exists d \in \mathcal{N}^{+} \; cts(f, a, b, d, e)$$

where a and b are real numbers and $f : \mathcal{R} \rightarrow \mathcal{R}$. These definitions are motivated by the proof of the intermediate value theorem. The second one is a definition of continuity that is equivalent to the standard definition of continuity.

4 The Meta-Level

In this section we present some background to the Edinburgh MRG's proof planner, Clam. This provides the facility to automate the proof of theorems in Oyster. It is based on the idea of using *meta-level inferencing* to construct *proof plans*.

A proof plan is a tree of *methods* which specifies a way in which a sequent can be proved in Oyster. A method is a meta-level specification of a tactic and has four main constituents:

- *Input Slot* : a specification of the object-level formula to which the method is applicable.
- *Preconditions Slot* : a specification of the meta-level conditions that must hold if the method is to be applicable.
- *Postcondition Slot* : a specification of meta-level conditions that will hold after the successful application of the method. The contents of this slot are calculated within Prolog by a procedure that simulates the effect of the associated tactic.
- *Ouput Slot* : a specification of the object-level formulae that will be produced as subgoals when the method is applied successfully.

When a plan has been constructed, the tactics associated with each method of a plan are applied to sequents in Oyster. It is not guaranteed, however, that a particular tactic will be applicable, and in that sense methods are actually partial specifications of tactics.

For a more detailed account of proof plans, see [3]. For more details about the Clam theorem prover, the reader should consult [6].

4.1 Rippling And Fertilisation

After applying induction to a sequent, we must prove the base case and the step case. In the step case, we are trying to prove the induction conclusion given that the induction hypothesis is true. One mechanism for doing this is that of *rewrite rules*. General rewriting, however, suffers from the problem of combinatorial explosion and non-termination.

The solution to these problems used in Clam is the heuristic method of *rippling*. This method searches for a sequence of *wave-rules* which will rewrite the induction conclusion until the induction hypothesis can be used to prove the sequent (remembering that inference rules are applied in a backwards chaining manner). The process of applying the induction hypothesis to the rewritten induction conclusion is called *fertilisation*.

Wave-rules are rewrite rules which contain meta-annotations that must match similar annotations on the induction conclusion for the rule to be applicable. These are designed such that each application of a wave-rule makes the induction conclusion more like the induction hypothesis.

As an example, let us consider the proof of the associativity of plus:

$$\forall x, y, z \in \mathcal{N} \; x + (y + z) = (x + y) + z$$

If we prove this using successor induction on x, we have that the inductive hypothesis is

$$x + (y + z) = (x + y) + z$$

and the induction conclusion is

$$\boxed{s(\underline{x})}^{\uparrow} + (y + z) = \left(\boxed{s(\underline{x})}^{\uparrow} + y\right) + z$$

The extra annotations on the induction conclusion are inserted automatically by Clam and correspond to the meta-level information that is exploited in rippling. The boxed term, $s(x)$, is the *induction term*, the underlined term, x, is the *wave hole*, and the constructor function, $s(\ldots)$ is known as the *wave front*.

To do this rippling we need wave-rules. The basic form of a wave-rule is

$$F\left(\boxed{S(\underline{U})}^{\uparrow}\right) :\Rightarrow \boxed{T(\underline{F(U)})}^{\uparrow}$$

where F, S and T are terms with one distinguished argument, and $:\Rightarrow$ is to be read as "rewrites to". T may be empty, but S and F must not be. S and T are respectively the old and new wavefronts, and U and $F(U)$ are the old and new wave holes. Note that the effect of rewriting using a wave rule is that the the wave front S is rippled past F.

In our example above, we can use a wave rule formed from the step case of the recursive definition of $+$, namely :

$$\boxed{s(\underline{u})}^{\uparrow} + v \Rightarrow \boxed{s(\underline{u + y})}^{\uparrow}$$

where the unary function $\lambda x \cdot x + v$ corresponds to F, S and T are the successor function, s, and U is the variable u.

If we apply this wave rule twice to the induction conclusion we get the equations

$$\boxed{s(\underline{x})}^{\uparrow} + (y + z) = \left(\boxed{s(\underline{x})}^{\uparrow} + y\right) + z$$

$$\boxed{s(\underline{x + (y + z)})}^{\uparrow} = \boxed{s(\underline{x + y})}^{\uparrow} + z$$

$$\boxed{s(\underline{x + (y + z)})}^{\uparrow} = \boxed{s(\underline{(x + y) + z})}^{\uparrow}$$

Notice that we now have a copy of the induction hypothesis within the wave holes of the last equation, and that our wave-rule no longer applies. At this point we can cancel the s on each side using a wave-rule derived from the substitutivity axiom for s, namely

$$\boxed{s(\underline{u})}^{\uparrow} = \boxed{s(\underline{v})}^{\uparrow} \Rightarrow u = v$$

leaving a copy of the inductive hypothesis. Fertilisation applies and the proof is complete.

The simple definition of wave-rules given above can be generalised to allow more complex forms of wave-rules. For our purposes there are two points in particular to be noted:

1. We can have wave-rules with multiple wave holes. We shall see later that we use wave-rules with two wave holes. These have the form

$$F(\boxed{S(\underline{U_1, U_2})}^{\uparrow}) :\Rightarrow \boxed{T(F(\underline{U_1}), F(\underline{U_2}))}^{\uparrow}$$

2. We also allow *conditional* wave-rules. These have the form

$$P \Rightarrow F(\boxed{S(\underline{U_1, U_2})}^{\uparrow}) :\Rightarrow \boxed{T(F(\underline{U_1}), F(\underline{U_2}))}^{\uparrow}$$

where P is a precondition that must hold before we can apply the wave rule.

Clam has a wave rule parser built into it so that it can automatically parse equations and theorems as wave-rules. Given a theorem of the form $P \rightarrow T_1 = T_2$ it will try to parse it as wave-rules of the form $P \Rightarrow T_1 :\Rightarrow T_2$ or $P \Rightarrow T_2 :\Rightarrow T_1$. It can also parse theorems of the form $P \rightarrow T_1 \rightarrow T_2$ as wave-rules of the form $P \Rightarrow T_2 :\Rightarrow T_1$. Note that the direction of rewriting is the opposite direction to the implication in the theorem. This because we are backwards chaining when using Oyster/Clam.

For more information about rippling see [2].

5 The Intermediate Value Theorem

In this section we descibe in detail the automatic proof of the intermediate value theorem in constructive mathematics. We first of all present a classical proof which will serve as a motivation for the work that needs to be done in Oyster and Clam to prove the construtive version.

5.1 The Classical Proof

Theorem : The Intermediate Value Theorem Given two real numbers a and b such that $a < b$ and a function $f : \mathcal{R} \rightarrow \mathcal{R}$ such that f is continuous on $[a, b]$, then for all real numbers c such that $f(a) \leq c \leq f(b)$ there exists a real number, $x \in [a, b]$, such that $f(x) = c$.

Proof The main idea of the proof is to construct two sequences, (a_n) and (b_n) which both converge to the same real number. This is the x that we are trying to find. These sequences are defined as follows:

We define $I_0 = [a_0, b_0] = [a, b]$.

Given $I_n = [a_n, b_n]$, if $f(\frac{1}{2}(a_n + b_n)) \leq c$ then we define $a_{n+1} = \frac{1}{2}(a_n + b_n)$ and $b_{n+1} = b_n$. Otherwise we define $a_{n+1} = a_n$ and $b_{n+1} = \frac{1}{2}(a_n + b_n)$.

This construction means that I_{n+1} is either the left or right-hand half interval of I_n. Note that with this definition every interval $I_n = [a_n, b_n]$ has the property that $f(a_n) \leq c \leq f(b_n)$.

All that is left is to check that the two sequences converge (both are monotonic and bounded) to the same number $(b_n - a_n = (b - a) \cdot 2^{n-1})$ and that this is the number we want (using $f(a_n) \leq c \leq f(b_n)$).

Comments On The Classical Proof

1. The statement of the theorem given above is too strong to prove in constructive mathematics.

    We must weaken the conclusion from $\exists x \in [a, b] \, (f(x) = c)$ to

    $$\forall k \in \mathcal{N} \, \exists x \in [a, b] \, (|f(x) - c| < 2^{-k})$$

    i.e. we can find an x such that $f(x)$ is arbitarily close to c.

2. The (classical) decidability of the ordering relations is used when deciding whether $f(a_n + b_n)$ is less than c or not.

    To overcome this problem we must use the weak form of dichotomy discussed above.

3. The process of constructing the two sequences is infinite (since the sequences are infinite). If we were to follow the same approach in Oyster, we would never finish building the objects we want.

    However, since we only need to find an approximate root, we can build two finite sequences to a predetermined length that ensures our solution is as accurate as needed.

    But how can this be done in practice? We can see that building such sequences is a *recursive* process. This immediately suggests that we should use some form of *induction* to create these sequences, or some analogue of them that can be used to both prove the theorem and also give us a recursive function that can calculate the real x that we are looking for.

Our approach, therefore, will be to construct an induction scheme mimicking bisection of the interval.

5.2   An Induction Scheme

As stated above, we will use the Acc type to construct an induction scheme that can be used in the proof of the theorem. For this we must construct a base type and a predecessor relation.

The induction scheme that was eventually used was based on the idea of bisecting the interval that is used in the classical proof of the IVT. The idea is that we have several "layers" of intervals. The top most layer consists of one interval, the second layer is composed of the two half intervals of the top one. This is repeated such that layer n consists of the two half intervals of every interval in layer $n - 1$. See figure 1.

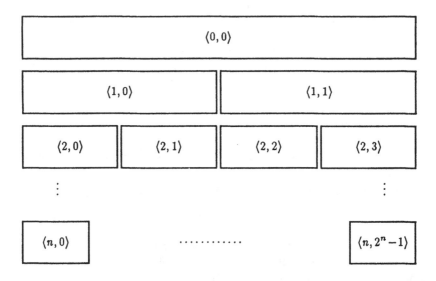

Fig. 1. The hierachy of intervals

Formally we define a set

$$\Omega_N := \{\langle n, m\rangle | 0 \leq n \leq N, 0 \leq M \leq 2^n - 1\}$$

where n and m are natural numbers. $\langle n, m\rangle$ will represent the mth interval at depth n. We have a bound N on the number of layers to ensure that there are a finite number of intervals in our base type, so that the predecessor relation $\prec$ will be well-founded over Ω_N.

When defining $\prec$, the idea is that an interval is preceded by its two half intervals. In terms of Ω_N, the relation we require is

$$\langle n, m\rangle \prec \langle n', m'\rangle \Leftrightarrow (n - 1 = n' \wedge \frac{m}{2} = m')$$

Note that this relies on the fact that integer division always results in an integer (so $\frac{2m}{2} = m$ and $\frac{2m+1}{2} = m$).

We can now plug our base type and predecessor relation into the Acc type and automatically be rewarded with an induction scheme. This, however, is the not the induction scheme that is used in Oyster. An induction scheme based on a predecessor relation takes the form of destructor induction, *i.e.* given an object

x we *find* its predecessors and assume that the proposition ϕ we are trying to prove holds for them. This is as opposed to constructor induction which proves that ϕ holds for an object x which is constructed from *previously given objects* for which ϕ holds.

Although Clam can perform destructor induction, the overall form of a proof and the program extracted from it, and the rippling process are all more natural and easier to understand if we perform constructor induction. To this end we define a constructor function for our base type Ω_N and use a constructor version of induction. The constructor function $up(i_1, i_2)$ takes two intervals and returns the interval that they precede. This is defined by

$$up(\langle n_1, m_1 \rangle, \langle n_2, m_2 \rangle) := \begin{cases} \langle n_1 - 1, \frac{m_1}{2} \rangle & n_1 \neq 0 \\ \langle 0, 0 \rangle & \text{otherwise} \end{cases}$$

or

$$up(\langle n_1, m_1 \rangle, \langle n_2, m_2 \rangle) := \begin{cases} \langle n_2 - 1, \frac{m_2}{2} \rangle & n_2 \neq 0 \\ \langle 0, 0 \rangle & \text{otherwise} \end{cases}$$

where we again rely on the properties of integer divison.

We take the position that $up(i_1, i_2)$ is defined by the first equation. This only relies on i_1. If we make use of i_2 then we must ensure that i_1 and i_2 are in fact the left and right-hand half intervals of the same interval. This is done by defining the relation

$$downpair(\langle n_1, m_1 \rangle, \langle n_2, m_2 \rangle) \Leftrightarrow (n_1 = n_2 \wedge \frac{m_1}{2} = \frac{m_2}{2} \wedge m_1 + 1 = m_2)$$

With these definitions in place, we find that the constructor version of induction that we use in Oyster is

$$\frac{\Gamma, \langle n, m \rangle : \Omega_N \vdash n = N \rightarrow \phi(\langle n, m \rangle)}{\Gamma, i_1 : \Omega_N, i_2 : \Omega_N, v0 : downpair(i_1, i_2), v1 : \phi(i_1), v2 : \phi(i_2) \vdash \phi(up(i_1, i_2))}{\Gamma, i : \Omega_N \vdash \phi(i)}$$

where N is in the type $\mathcal{N}$ and ϕ is in the type $(\Omega_N \rightarrow U_2)$.

It should be noted that these "intervals" are not sets of real numbers, as in the usual meaning of the term. We can give them some of the properties of normal intervals using the definitions

$$el(\langle n, m \rangle, a, b) := 2^{-n}((2^n - m)a + mb)$$

$$er(\langle n, m \rangle, a, b) := 2^{-n}((2^n - (m+1))a + (m+1)b)$$

$$inside(x, i, a, b) \Leftrightarrow x \geq el(i, a, b) \wedge x \leq er(i, a, b)$$

where a and b are respectively the left and right endpoints of the top most interval $\langle 0, 0 \rangle$. $el(\langle n, m \rangle, a, b)$ returns the left endpoint of the interval $\langle n, m \rangle$, er returns the corresponding right endpoint. Note that in particular the expression $x \in [a, b]$ is represented in Oyster as $inside(x, \langle 0, 0 \rangle, a, b)$.

For the rest of the paper we will use the term "interval" to refer either to a member of the type Ω_N or to the real interval that it represents (given a left and right endpoint of the top most interval). It should be clear form the context which usage is intended.

5.3 The Generalisation

The induction scheme presented above has the conclusion that for all intervals $i \in \Omega_N$, we have $\phi(i)$. If we take $\phi(i)$ as being the conclusion of the IVT, i.e.

$$\exists x \in i \, (|f(x) - c| < 2^{-k})$$

we can see that it will not necessarily be the case that $\phi(i)$ holds for *all* $i \in \Omega_N$. We would certainly expect it to hold for $i \in \Omega_0$, i.e. the top interval $[a, b]$ (since this is what we are trying to prove) but we cannot guarantee that the proposition holds for all intervals in Ω_N where $N > 0$.

In the classical case, we find that intervals $[a_i, b_i]$ such that

$$(f(a_i) < c \wedge f(b_i) < c) \vee (c < f(a_i) \wedge c < f(b_i))$$

(*i.e.* the f of the endpoints are on the same side of c) cannot be guaranteed to have a root. The constructive version of this condition in the formalism that we have developed is

$$((f(el(i, a, b)) < c + 2^{-k} \wedge f(er(i, a, b)) < c + 2^{-k}) \vee$$
$$(c - 2^{-k} < f(el(i, a, b)) \wedge c - 2^{-k} < f(er(i, a, b))))$$

Let us call the above proposition $\psi(i)$.

So, either we are guaranteed to find a root, or the above condition holds. Hence we take as our generalisation of the theorem

$$\forall i \in \Omega_n \, a < b \rightarrow cts(f, a, b, n, k) \rightarrow (\phi(i) \vee \psi(i))$$

which holds for all intervals $i \in \Omega_N$.

5.4 Planning A Proof Of The Theorem

In this section we descibe the information that Clam needed to plan a proof of the generalised version of the intermediate value theorem.

Clam's induction method was extended so that it could use bisection induction presented above. This method was used by Clam as its first step in planning the proof. This left two subgoals to be planned:

The Base Case:

$$\forall i \in \Omega_N \, a < b \rightarrow cts(f, a, b, n, k) \rightarrow \pi_1(i) = N \rightarrow \phi(i) \vee \psi(i)$$

where $\pi_1(\langle n, m \rangle) := n$.

This lemma had been proved interactively in Oyster, so Clam could immediately apply this lemma to prove the subgoal. We note that our definition of $cts(f, a, b, n, k)$, i.e.

$$cts(f, a, b, n, k) \Leftrightarrow \forall x, y \in [a, b] \, (|x - y| \le (b - a) \cdot 2^{-n} \rightarrow |f(x) - f(y)| < 2^{-(k+1)})$$

was designed so that given the bound k on the accuracy of the root, we could choose a depth N for our type of intervals Ω_N that would allow the proof to go through.

The Step Case: The goal is, given the induction hypotheses

$$(\phi(i_1) \vee \psi(i_1)) \text{ and } (\phi(i_2) \vee \psi(i_2))$$

prove the induction conclusion

$$\phi(\boxed{up(\underline{i_1}, i_2)}^{\uparrow}) \vee \psi(\boxed{up(\underline{i_1}, i_2)}^{\uparrow})$$

where i_1 and i_2 are intervals and we are given that $downpair(i_1, i_2)$ holds.

This was proved by rippling.

There are two theorems that had been interactively proven in Oyster that were used in the rippling process. These are

$$downpair(i_1, i_2) \rightarrow \phi(i_1) \vee \phi(i_2) \rightarrow \phi(up(i_1, i_2))$$

and

$$downpair(i_1, i_2) \rightarrow \psi(i_1) \wedge \psi(i_2) \rightarrow \psi(up(i_1, i_2))$$

The first expresses the fact that if there is an x such that $|f(x) - c| < 2^{-k}$ in either i_1 or i_2, then there is also such an x in $up(i_1, i_2)$ (in fact, it is the same x).

The second says that if f of the endpoints of i_1 are on the same side of c, and if f of the endpoints of i_2 are on the same side of c then this property holds for $up(i_1, i_2)$. This is because the right hand endpoint of i_1 is the left hand endpoint of i_2.

These were parsed by Clam as the wave-rules:

$$downpair(i_1, i_2) \Rightarrow \phi(\boxed{up(\underline{i_1}, i_2)}^{\uparrow}) :\Rightarrow \boxed{\underline{\phi(i_1)} \vee \underline{\phi(i_2)}}^{\uparrow}$$

$$downpair(i_1, i_2) \Rightarrow \psi(\boxed{up(\underline{i_1}, i_2)}^{\uparrow}) :\Rightarrow \boxed{\underline{\psi(i_1)} \wedge \underline{\psi(i_2)}}^{\uparrow}$$

Both of these wave-rules are applicable to our induction conclusion. They were applied by Clam leaving the goal

$$\boxed{\underline{\phi(i_1)} \vee \underline{\phi(i_2)}}^{\uparrow} \vee \boxed{\underline{\psi(i_1)} \wedge \underline{\psi(i_2)}}^{\uparrow}$$

Clam automatically planned a wave-rule of the form

$$\boxed{\underline{A} \vee \underline{C}}^{\uparrow} \vee \boxed{\underline{B} \wedge \underline{D}}^{\uparrow} :\Rightarrow \boxed{\underline{A} \vee B \wedge \underline{C} \vee D}^{\uparrow}$$

and applied it to our goal leaving

$$\boxed{\underline{\phi(i_1)} \vee \psi(i_1) \wedge \underline{\phi(i_2)} \vee \psi(i_2)}^{\uparrow}$$

At this point each of our induction hypotheses matches one part of the wave hole in our goal. Clam selects a form of fertilisation to finish the proof by using the hypotheses to prove each conjunct.

6 Remarks

The plan that Clam constructed was applied to the theorem, thus proving it in Oyster.

We note here that some of the lemmas used in proving the theorem still had open proof obligations. These were generally either type obligations (showing that certain types were well formed) or basic arithmetic facts that were not proved due to time constraints.

The aim of this work was to investigate the applicability of the proof planning approach to the domain of real analysis. We found that the only extension that was needed to Clam to allow it to plan a proof of the intermediate value theorem was to construct a new induction scheme. In particular rippling was used in the proof of the step case and needed no modification.

References

1. Bishop, E. and Bridges, D.: Constructive Analysis. Springer-Verlag, 1985.
2. Bundy, A. and van Harmelen, F. and Smaill, A.: Extensions to the Rippling-Out Tactic for Guiding Inductive Proof. Proceedings of CADE-10.
3. Bundy, A.: The Use of Explicit Plans to Guide Inductive Proofs. Dept. of Artificial Intelligence, Edinburgh, Reaearch Paper 349.
4. Constable, R.L. and Allen, S.F. and Bromley, H.M. and others.: Implementing Mathematics with the Nuprl Proof Development System. Prentice Hall, 1986.
5. Nordström, B.S.: Terminating General Recursion. 1987.
6. van Harmelen, F. and Ireland, A. and Stevens, A. and Negrete, S.: The Clam Proof Planner, User Manual and Programmer Manual (version 1.5) Dept. of Artificial Intelligence, Edinburgh, 1992

A General Technique for Automatically Optimizing Programs Through the Use of Proof Plans

Peter Madden[1] and Ian Green[2]

[1] Max-Planck-Institut für Informatik,
Im Stadtwald, W-6600 Saarbruecken, Germany
E-mail: madden@mpi-sb.mpg.de

[2] Department of AI, University of Edinburgh,
80 South Bridge, Edinburgh EH1 1HN, Scotland
E-mail: img@aisb.edinburgh.ac.uk

Abstract. The use of *proof plans* — formal patterns of reasoning for theorem proving – to control the (automatic) synthesis of efficient programs from standard definitional equations is described. Proof plans are used to control the (automatic) synthesis of functional programs, specified in a standard equational form, by using the proofs as programs principle. Thus the theorem proving process is a form of program optimization allowing for the construction of an efficient, *target*, program from the equational definition of an inefficient, *source*, program.

A general framework for synthesizing efficient programs, using tools such as higher-order unification, has been developed and holds promise for encapsulating an otherwise diverse, and often ad hoc, range of transformation techniques. A prototype system has been implemented which has the desirable properties of automatability, correctness and restricted search within a small (meta-level) search space.

Different optimizations are achieved by placing different characterizing restrictions on the form of this new sub-goal and hence on the subsequent proof. Meta-variables and higher-order unification are used in a technique called *middle-out reasoning* to circumvent eureka steps concerning, amongst other things, the identification of recursive data-types, and unknown constraint functions. Such problems typically require user intervention.

We illustrate the methodology by a novel means of affecting *constraint-based* program optimization through the use of proof plans for mathematical induction.

1 Synopsis

In this paper we investigate how *proof plans* – formal patterns of reasoning for theorem proving – can be used for controlling the synthesis of *efficient* functional programs from standard sets of equational definitions. By exploiting meta-level control strategies, a general framework for *automatically* synthesizing efficient programs has been developed. A key meta-level strategy is called *middle-out reasoning*, henceforth MOR, which involves the controlled use of higher-order meta-variables at the meta-level planning phase. This allows the planning to proceed even though certain object-level objects are (partially) unknown. Subsequent planning provides the necessary information which, together with the original definitional equations, will allow us to instantiate such meta-variables through higher-order unification (HOU) procedures. MOR allows for the circumvention of eureka steps during the optimization process concerning, amongst other things, the identification of recursive data-types, and unknown constraint functions. Such steps have typically required user-intervention in more traditional ("pure") transformational systems such as unfold/fold [3]. The control provided by proof planning allows us to view such syntheses as verification together with MOR.

The proof planning approach to controlling the synthesis of efficient programs was originally investigated within the context of synthesizing tail-recursive programs from naive definitions by using a tail-recursive generalization strategy [7]. In this paper we present a *general* framework for automatically synthesizing efficient programs through the use of proof planning and MOR. We illustrate the methodology by describing a novel form of generalization strategy which, together with an induction strategy and MOR, automatically affects *constraint-based* optimizations: the *constraint-based generalization* proof plan is used for generating families of efficient programs from definitions which include expensive expressions.

In previous generalization proof plans, such as that described in [7], MOR has been limited to introducing (higher-order) meta-variables into goal statements according to the pre-conditions of the generalization proof plan. We refere to this kind of MOR as *generalization*-MOR, or simply gen-MOR. We significantly extend the mechanism by which MOR operates by allowing for the use of higher-order meta-variables in rewrite rules *in addition* to those introduced via the proof plan application. Such meta-variables are introduced via the exploitation of higher-order recursive definition schemas. These can be viewed as higher-order schematic rule templates. This significantly increases the scope for delaying proof commitments until subsequent theorem proving provides the requisite information to identify the relevant data structures. We shall refer to this usage of MOR as template MOR

Different characterizations of proofs can be formalized as proof plan pre-conditions that subsequently effect the kind of optimization exhibited by the synthesized functions. In particular, the way in which meta-variables are introduced, via gen-MOR, into the proof of the goal statement(s) specifying the program being synthesized. Constraint-based optimizations are, for example, characterized differently from tail-recursive optimizations. This basically accounts for the new form of generalization systemized in the constraint-based generalization. However, so as to illustrate many features of the general framework we shall, in this paper, choose a running example which consists of synthesizing an efficient program, from standard equational definitions, which is *both* tail-recursive *and* constraint-based. The example will illustrate both the usage of gen-MOR and the new template MOR

We believe that a large class of otherwise diverse, and often ad hoc, transformation strategies can be encompassed within this uniform proof plan framework. We show how the proof planning framework provides the necessary meta-level control over HOU and proof structure. Furthermore, the underlying logic we use guarantees the total correctness of the synthesized function with respect to the specification. Toward the end of this paper we compare the proof planning approach to synthesizing efficient programs with existing optimization strategies and discuss its advantages.

Contents In the remainder of this section we explain what precisely constraint-based optimization is with the introduction of our running example. In §2 we briefly describe the proofs as programs paradigm and illustrate an (interactive) synthesis of the example in §1.1. We also describe, respectively, the object level OYSTER proof refinement system and the meta-level CLAM proof planner. We provide an outline of the planning strategies employed and present the notation used to illustrate the rewriting process. In §3 we describe the *general* framework for optimization by proof planning. §4 addresses one kind of optimization encapsulated by the general framework described in §3: the use of proof plans for the purposes of constraint-based optimization. We provide pre- and post-conditions for the application of a constraint-based transformation proof plan, and we describe the use of higher-order schematic rule templates. In §4.3 we revisit the example of §1.1 and show how, using proof planning together with MOR, the optimization process is automated. In §5 we address the benefits of our approach as compared with standard transformational approaches such as unfold/fold.

1.1 An Example Optimization

Hesketh et al. considers the automatic synthesis of *tail-recursive* programs from in-efficient non-tail-recursive programs using standard sets of equational definitions [7]. However, tail-recursive programs may, in turn, present scope for further optimization. Consider the following two *tail-recursive* definitions of procedures for simultaneously producing both the sum and reverse of the input list:[3]

$$rev\_sm(nil, w) = (sum(w), w); \tag{1}$$
$$rev\_sm(hd :: tl, w) = rev\_sm(tl, hd :: w) \tag{2}$$

* and where:
$$sum(nil) = 0; \tag{3}$$
$$sum(a :: x) = a + sum(x) \tag{4}$$

$$rev\_sm_{C\_t}(l, x) = rev\_sm_C(l, x, 0) \tag{5}$$
$$rev\_sm_C(nil, w, sum(w)) = (sum(w), w); \tag{6}$$
$$rev\_sm_C(hd :: tl, w, sum(w)) = rev\_sm(tl, hd :: w, hd + sum(w)). \tag{7}$$

The rev_sm procedure has an expensive expression, $sum(w)$, in its base-case equation (1). The program's inefficiency stems from the double traversal of its input, which builds up a reversed list in the second parameter, w, and then finally performs a summation of this reversed list. The inefficiency may be removed by introducing the new constrained tail-recursive function definition, rev_sm_C, which uses only a single traversal of its input list in order to obtain both the summation together with its reversed list.

The removal of such expensive expressions has been investigated within the context of program transformation and is called *constraint-based transformation* [4] (or

---

[3] Tail recursive definitions have the feature that recursive calls occur as the outermost function of the procedure body, and an *accumulator*, w in the examples, is used to construct the output as the recursion is entered.

sometimes *finite differencing* [11]). In general, it involves the replacement of expensive expressions in program loops, or recursion, by equivalent expressions which are incrementally maintained. As indicated by the above example, this is done through the introduction of new parameters – *constraint parameters* – whose values are specified using constraints (thus, regarding rev_sm_C, the constraint parameter is $sum(w)$). We shall call functions such as rev_sm *unconstrained*, and their optimal counterparts, such as rev_sm_C, *constrained*.

[4] outlines how functional programs containing expensive expressions can be optimized using standard *unfold/fold* type rewrites to obtain constrained function definitions such as that for rev_sm_C. The unfold/fold strategy was pioneered by Darlington, and its most influential implementation has been within the NPL program transformation system [3, 6]. It has since been extended and reconstructed in various guises for various applications. The state of the art unfold/fold system design, in terms of performance and control, is described in [4]. Unfold/fold transformation typically involves the sequential application of rewrites which use definitions (specifically using instantiated definitions to replace terms) and known properties of functions in order to derived a target program which is independent of the source definition. However, the identification of the new constraint parameters, as $sum(w)$ in the rev_sm_C example, constitute *eureka steps*, thus presenting obstacles for providing a general automatic procedure for constraint-base optimization. Regarding program optimization in general, such *eureka steps* correspond to the problems of identifying *explicit definitions* for target programs: that is, new definitions where the target program is defined explicitly in terms of the source. Indeed, within the context of unfold/fold transformations, providing explicit definitions is the key to the optimization process: by subsequently folding the explicit definition (or derivations thereof) with the original source equations recursion is introduced into the target program. Further difficulties include the search involved with identifying and applying rewrites to explicit definitions in order to derive the recursive target definitions. For example, unfold/fold transformations are motivated by the desire to find a successful fold. This involves extensive search and the somewhat arbitrary application of laws thus presents difficulties regarding automation. We discuss these difficulties in more depth in §5.

In this paper we consider a general technique for circumventing the aforementioned eureka steps, and for reducing the search control problems in the rewriting process, by exploiting proof plans.

2 Proofs as Programs

Constructive logic allows us to correlate computation with logical inference. This is because proofs of propositions in such a logic require us to construct objects, such as functions and sets, in a similar way that programs require that actual objects are constructed in the course of computing a procedure.[4] Historically, this duality is accounted

---

[4] Thus we can not, for example, compute (or constructively prove) that there are an infinity of prime numbers by assuming the converse and deriving a contradiction, rather we must produce a program that computes them (or a proof that we can always construct another one greater than the ones known so far).

for by the *Curry-Howard isomorphism* which draws a duality between the inference rules and the functional terms of the λ-calculus [5, 8].

Such considerations allow us to correlate each proof of a proposition with a specific λ-term, λ-terms with programs, and the proposition with a specification of the program. Hence the task of generating a program is treated as the task of proving a theorem: by performing a proof of a formal specification expressed in constructive logic, stating the *input-output* conditions of the desired program, an algorithm can be routinely extracted from the proof. A program specification can be schematically represented thus:

$$\forall inputs, \exists output. \; spec(inputs, output)$$

Proofs of such specifications must establish (constructively) how, for any input vector, an output can be constructed that satisfies the specification.[5] Thus any synthesized program is guaranteed correct with respect to the specification. Different constructive proofs of the same proposition correspond to different ways of computing that output. By placing certain restrictions on the nature of a synthesis proof we are able to control the efficiency of the target procedure. Thus by controlling the form of the proof we can control the efficiency with which the constructed program computes the specified goal. Here in lies the key to synthesizing efficient programs. For example, we can synthesize constrained functions from unconstrained ones by placing the restriction that the new constraint parameter is some function on the non-recursive (non-inductive) parameter (we illustrate this in §2, and in more detail in §4.1). We can also guarantee that a synthesized program is tail recursive by ensuring that the witnesses of the two existential quantifiers, one in the induction hypothesis and one in the induction conclusion, are identical [14].[6] By making these witnesses identical we ensure that the function does not change value as the recursion is exited. Alternatively, we can use special schematic rules to affect the nature of the recursion exhibited by the program under construction. In §4.2 we illustrate how tail-recursive behaviour can be ensured through the application of such rules.

The OYSTER System The OYSTER system is an implementation of a constructive type theory which is based on Martin-Löf type theory, [9].[7] OYSTER is written in Quintus Prolog, and run at the Prolog prompt level, so it is controlled by using Prolog predicates as commands. Proof tactics can be built as Prolog programs, incorporating OYSTER commands. The language uniformity of the logic programming environment allows for the construction of *meta-theorems* which express more general principles, concerning the object level theorem proving. So, for example, we are able to construct *tactics* which combine the object-level rules of the system in various ways and apply them to proof (sub)goals.

At any stage during the development of a proof it is possible to access the *extract term* of the proof constructed so far. Each construct in the extract term corresponds to a

---

[5] Thus constructive logic *excludes* pure existence proofs where the existence of *output* is proved but not identified.

[6] A witness constitutes an instantiation of an existential quantifier thus providing evidence of the existence asserted.

[7] OYSTER is the Edinburgh Prolog implementation of NuPRL; version "nu" of the *Proof Refinement Logic* system originally developed at Cornell [Horn 88, Constable *et al* 86].

proof construct. As such, the extract term reflects the computational content of the proof of the theorem. The extract programs consist of λ-calculus function terms, $\lambda(x, f_x)$ where f is the computed function and f_x the output when f is applied to input x.

For the purposes of illustrating our methodology we do not need to make the type information contained in the proofs explicit. Indeed, it is adequate and aids clarity to present, in this paper, our proofs in a classical framework.

Example: Synthesis of Constraint Function rev_sm_C To illustrate the synthesis process we shall outline the synthesis of rev_sm_C using the definition of the unconstrained function rev_sm. We indicate those proof steps which, regarding traditional program transformation re-writing systems, correspond to *eureka* steps.

A key feature of the proof plan approach to automatically synthesizing efficient programs is to use the inefficient program definition to specify the required procedure. We specify the output for the constrained function, rev_sm_C, in terms of the unconstrained rev_sm thus:

$$\forall x, \forall w, \exists z.\; z = rev\_sm(x, w) \tag{8}$$

We then introduce a new sub-goal, providing an explicit definition for rev_sm_C which includes a constraint parameter instantiated to $sum(w)$:

$$\forall x, \forall w.rev\_sm(x, w) = rev\_sm_C(x, w, sum(w)) \tag{9}$$

The identification of $sum(w)$ as the constraint parameter (required for the full identification of the explicit definition) is the first eureka step. Note that the proof satisfies the *constraint-based restriction* that the constraint parameter is some function on the non-recursive (non-inductive) parameter.

Also note that since we have provided the identity of the constraint parameter that the proof will in fact be a verification proof. In §4.3 we illustrate how (higher-order) meta-variables are used to partially identify such constraint parameters and there by avoid such eurekas. That is, synthesis without eureka steps can be affected through higher-order verification proofs.

In §4.3 we show how such eureka steps can be automated through the use of MOR. In general terms, MOR allows us to delay choice commitments by introducing (higher-order) meta-variables at the meta-level application of rules of inference. Subsequent planning provides the requisite information to instantiate the meta-variables by (higher-order) unification. Thus, the main difference between verification and synthesis proofs is precisely the identification of those structures for which MOR is used. This idea is captured by our slogan that *synthesis is equivalent to verification plus meta-variables*.

The introduction of the new goal (9) also leaves us with the trivial proof obligation that the new goal entails the original one (i.e. (9) ⊢ (8)):

$$\forall x, \forall w.rev\_sm(x, w) = rev\_sm_C(x, w, sum(w)) \vdash \forall x, \forall w, \exists z.\; z = rev\_sm(x, w)$$

To prove (9) standard stepwise induction on x is used:

Induction Base. The base case is as follows:

$$\vdash \forall w.rev\_sm(nil, w) = rev\_sm_C(nil, w, sum(w)) \tag{10}$$

Using the definition of rev_sm, the left hand side of (10) rewrites to $(sum(w), w)$:

$$\vdash (sum(w), w) = rev\_sm_C(nil, w, sum(w))$$

Induction Step. The step case of the induction is:

$$\forall x, \forall w.rev\_sm(tl, w) = rev\_sm_C(tl, w, sum(w))$$
$$\vdash \forall x, \forall w.rev\_sm(hd :: tl, w) = rev\_sm_C(hd :: tl, w, sum(w)) \tag{11}$$

We can use the definition of rev_sm to rewrite the left hand side of (11) to $rev\_sm(tl, hd :: w)$. However the re-writing process is blocked on the right-hand side: the available recursive definitions provide no suitable re-writes to unfold the rev_sm_C term any further. Hence to arrive at the following equation:

$$... \vdash \forall x, \forall w.rev\_sm(tl, hd :: w) = rev\_sm_C(tl, hd :: w, hd + sum(w))$$

a second eureka step is required to enable the re-writing of $rev\_sm_C(hd :: tl, w, sum(w))$ to $rev\_sm_C(tl, hd :: w, hd + sum(w))$. In §4 we illustrate how a further new form of MOR allows us to avoid the eureka step.

The step case is completed by stripping of the universal quantifiers and instantiating the w in the induction hypothesis to $hd :: w$ in the induction conclusion (reducing the induction step to true since both hypothesis and conclusion are identical). Analyses of the base and step cases of the proof provide the base and recursive branch for the rev_sm_C procedure. In the next section we discuss the program extraction process.

2.1 Proof Plans – Automating the Proof Process

The induction strategy, together with other commonly used proof tactics, has been systematized in a metalogic in the automatic plan formation program CLAM [2]. This consists of formal *proof plans* where each tactic is specified by a *method* which includes pre- and post-conditions. By using *meta-level reasoning*, CLAM executes the individual proof plans to obtain a combination of tactics customized to the particular theorem at hand. Execution of this tactic combination, at the object level, will then produce a proof of that theorem.

A key strategy of the CLAM proof planner is *rippling*. Our explanation of rippling, and the corresponding notation, will be necessarily simplified. For a fully comprehensive account the reader should consult [1]. In an inductive proof, the goal of the rippling proof plan is to reduce the induction step case to terms which can be unified with those in the induction hypothesis. This unification is called *fertilization*, and is facilitated by the fact that the induction conclusion is structurally very similar to the induction hypothesis except for those function symbols which surround the induction variable in the conclusion. These points of difference are called *wave-fronts*. Thus, the remainder of the induction conclusion – the *skeleton* – is an exact copy of the hypothesis. Wave fronts consist of expressions with holes – *wave holes* – in them corresponding to sub-terms in the skeleton. Wave-fronts are indicated by placing them in boxes, and the wave-holes are underlined. For example, the induction conclusion (11) becomes annotated thus:

$$\vdash \forall x, \forall w.\, rev\_sm(\boxed{hd :: \underline{tl}}^{\uparrow}, \lfloor w \rfloor) = rev\_sm_C(\boxed{hd :: \underline{tl}}^{\uparrow}, \lfloor w \rfloor, sum(\lfloor w \rfloor))$$

To understand the additional notation, the arrows and terms surrounded by $\lfloor \ \rfloor$, we must explain the function of the structural rewrite rules, or *wave-rules*. Rippling applies wave-rules so as to remove the difference (wave-fronts) from the conclusion, thus leaving behind the skeleton and allowing fertilization to take place. For the purposes of this paper we need to identify three kinds of wave-rule, each distinguished by the direction in which they move wave-fronts in the conclusion. Wave-fronts may be moved outwards, *rippling-out*, such that they surround the entire induction conclusion thus allowing a match between everything in the wave-front with the induction hypothesis. Wave-rules for rippling-out are called *longitudinal* wave-rules and an upward arrow signals the outward direction of movement. Examples of longitudinal wave rules are:

$$\boxed{s(\underline{U})}^{\uparrow} + V \Rightarrow \boxed{s(\underline{U + V})}^{\uparrow}$$

$$sum(\boxed{Hd :: \underline{Tl}}^{\uparrow}) \Rightarrow \boxed{Hd + \underline{sum(Tl)}}^{\uparrow}$$

$$U + (\boxed{\underline{V} + W}^{\uparrow}) \Rightarrow \boxed{\underline{(U + V)} + W}^{\uparrow} \qquad (12)$$

Throughout this paper upper-case variables denote *meta-variables* such that the above rules are best understood as *rule schemata*. Note that we include (12) to show that wave rules need not only be formed from the step cases of inductive definitions ((12) is formed from the associative law of +). Wave-rules may also move wave-fronts sideways, *rippling sideways*, such that they surround non-induction universal quantified variables (such as accumulators). The sub-terms which sideways rippled wave-fronts surround are called *sinks* and are demarcated by $\lfloor \ \rfloor$. This allows the wave fronts, and the universally quantified variable that they surround, to be identified with the corresponding universally quantified variable in the hypothesis. Thus again fertilization can take place. Such a wave-rule is called a *transverse* wave-rule, e.g.

$$\boxed{s(\underline{X})}^{\uparrow} + Y \Rightarrow X + \boxed{s(\underline{Y})}^{\downarrow}$$

$$rev\_sm(\boxed{Hd :: \underline{Tl}}^{\uparrow}, W) \Rightarrow rev\_sm(Tl, \boxed{Hd :: \underline{W}}^{\downarrow})$$

Rippling into sinks typically involves an application of a longitudinal wave-rule followed by a transverse wave rule. It may also, however, require *rippling-in*: a *reverse* application of a longitudinal wave-rule. A downward arrow signals this inward direction of movement. In §4.3 we provide a worked example that illustrates both kinds of wave rule, and all three directions of rippling.

Rippling has numerous desirable properties. A high degree of control is achieved for applying the rewrites since the wave-fronts in the rule schemas must correspond to those in the instance. This leads to a very low search branching rate. Rippling is guaranteed to terminate since wave-front movement is always propagated in a desired direction toward some end state (a formal proof of this property is presented in [1]).

3 General Technique for Optimization by Proof Plans

Constraint-based optimization is representative of only one of the kinds of optimization possible by using our general technique. The synthesis of tail-recursive programs from naive definitional equations has also been implemented as a *tail-recursive* proof plan [7]. Other kinds of optimization that we are now investigating include deforestation transformations, fusion transformations and tupling transformations [13, 4, 10]. In this section we describe the *general* technique for controlling the syntheses of efficient programs from the definitions of inefficient programs. The technique encapsulates all the aforementioned kinds of optimizations. We intend to continue expanding this range of optimizations.

Proof plans are used to control the (automatic) synthesis of functional programs, specified in a standard equational form, $\mathcal{E}$, by using the proofs as programs principle. The goal is that the program extracted from a constructive proof of the specification is an optimization of that defined solely by $\mathcal{E}$. Thus the theorem proving process is a form of program optimization allowing for the construction of an efficient, *target*, program from the definition of an inefficient, *source*, program.

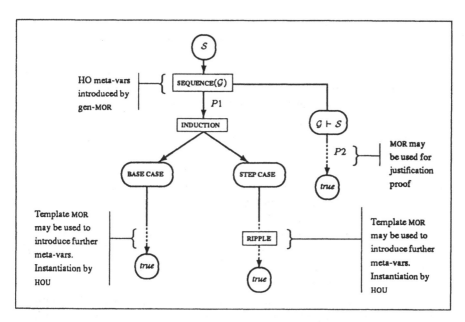

Figure 1: Schematic Representation of Generalization Synthesis Proof

The proof planning approach to optimization is depicted above by fig.1 where we show the general form of the inductive generalization proof. The strategy can involve four main steps. Firstly, a target program specification, S, is formed from the source program's equational definitions $\mathcal{E}$. S is then set up as the conjecture to prove. Secondly, the technique involves *sequencing* into the proof of S a new sub-goal, $\mathcal{G}$. The sub-goal $\mathcal{G}$ is produced as an output of the constraint-based generalization proof-plan. $\mathcal{G}$ is partially identified by the use of higher-order (HO) meta-variables, thus initiating the gen-MOR process. The application of the sequencing rule produces two subgoals: the first being the original goal S with G as an additional hypothesis (the so-called *justification* goal), and the second being G itself.[8] The inductive proof, $P1$, of (sub)goal G is then responsible for synthesizing the more efficient computation of the input-output relation specified in S. G will be some form of generalization on S, such as in tail-recursive generalization [7], or, as in our example, it may place additional constraints on S so as to affect constraint based synthesis. Whatever the relation between S and G, the main requirements are that a more efficient procedure can be synthesized through proving G than through proving S *and* that G entails S.

The application of the constraint-based generalization proof plan, the induction proof plan and any subsequent proof plans (such as symbolic evaluation) is automatically co-ordinated by the CLAM proof planner according to the which proof plan has it's pre-conditions satisfied by the current goal statement. So, for example, the pre-conditions of a generalization proof plan will be satisfied by S, the proof plan will be applied producing the corresponding post-conditions. These post-conditions will satisfy the pre-conditions of the induction proof plan which will then be subsequently applied. Proof plans may also be applied as sub-plans. For example the ripple proof plan is called within the application of the induction proof plan.

The third step of the strategy consists of using gen-MOR to fully identify G through higher-order unification (HOU). Recall that this is the case if only the form, but not the precise content of G, is known at the sequencing step. HOU instantiates the meta-variables by matching subsequent proof derivations with the available rewrite rules (which will always include those formed from $\mathcal{E}$).[9] In this way, the need to treat the identification of G as a eureka step is removed by higher-order unification using rewrite rules formed from the available definitional equations.

In traditional program transformation, eureka steps may also occur during the iden-tification of the recursive branch(es) of the target program definition. Within the context of proof synthesis, this would correspond to providing the requisite recursive date-types at the induction step case of the proof. The application of template MOR, the fourth step of the strategy, allows us to circumvent such eureka steps by introducing meta-variables during the rippling stages of the induction step. Template MOR applies higher-order schematic rewrite-rules which have the effect of introducing new HO meta-variables. These allow for the delaying of proof commitments concerning the identity of recursive data-types until further proof development enables the meta-variables to be instantiated (more detail is provided in §4.2).

The precise nature of G, the induction rule applied to G, the kind of higher-order rewrite-rule employed by template MOR, and any restrictions on how the meta-variables are instantiated, fully determine the type of recursion constructed in the target proof (and thereby the efficiency of the extract algorithm). The process is entirely mechanical, and the resulting program is guaranteed to satisfy the program specification (S).

---

[8] In some case, the proof of the justification goal may also require MOR. This is not the case with the example we shall use where the justification goal is trivial.

[9] Some efficient functions can be synthesized without the use of higher-order meta-variables. In such cases MOR is not required. and fig.1 would simply resemble a standard existential inductive proof (i.e. we would omit the sequencing step and the justification proof $P2$).

4 Proof Plans for Controlling Constrained (Tail) Recursion

Recall that the main objective of constraint based transformations is the replacement of expensive expressions in programs. Constraint-based transformations cover a wide variety of function types. Three kinds of constraint-based transformation may be distinguished according to the properties of the expensive expression(s): the expensive expression(s) may occur either at the repetitive places of recursive definitions, or in the terminating branches of a procedure, or the expensive expression is based on the fixed, or constant, parameter of a recursive function. We shall for now only consider an example of the latter kind – e.g. the rev_sm example – although what is said here concerning proof restrictions and pre- and post-conditions for the constraint based generalization proof plan applies to the other kinds of constraint-based transformation.

4.1 Pre- and Post-Conditions For Constraint-Based Generalization

Recall that the identification of the new constraint parameter in rev_sm_C, as $sum(w)$ is a *eureka step*. The *eureka step* can be circumvented by MOR: introducing a new parameter identified in terms of meta-variables (which will then be subsequently instantiated through HOU). To ensure that we obtain the efficient constrained function definition we place the restriction that the new constraint parameter will be some function, represented by a meta-variable M, on the non-recursive (non-inductive) parameter, i.e. w, of the original rev_sm program.

In general, this *constraint based restriction* ensures the removal of the expensive expression, and is implemented as a CLAM proof plan method. The input goal for the constraint based proof plan is of the following form:

$$\forall x, \forall y, \exists z. \ z = f_n(x, y) \tag{13}$$

where y is a vector that denotes 0 or more additional parameters. The effects will consist of a synthesis and a justification goal where the former is of the following general form (where the constraint parameter is partially identified by $M(y)$, and where M is a meta-variable):

$$\forall x, \forall y. \ f_n(x, y) = f_m(x, y, M(y)) \tag{14}$$

and where the justification goal requires a trivial proof that (14) $\vdash$ (13), viz:

$$\forall x, \forall y. f_n(x, y) = f_m(x, y, M(y)) \vdash \forall x, \forall y, \exists z. z = f_n(x, y) \tag{15}$$

We also require that the target of the constraint-based synthesis is tail-recursive. Unlike the tail-recursive syntheses reported in [7], proofs of goals such as (14) are *equality proofs* as opposed to existential proofs (a direct proof of (13) would constitute an existential proof, but we require the introduction of the meta-variable in the manner described). Thus, we cannot use the tail-recursive restrictions on the identity of existential quantifiers described in [7].[10] Instead, for such equality proofs, we achieve the

---

[10] That is, that the witnesses of the two existential quantifiers, one in the induction hypothesis and one in the induction conclusion, should be identical. This would ensure that the value of the function before the recursion is entered (determined by the induction hypothesis) to be the same as the value as the recursive call is exited (determined by the induction conclusion).

desired tail-recursive form through the application of schematic, or higher-order, rule templates.

4.2 Higher-Order Rule Templates

The constraint-based proof plan has access to higher-order rule templates. These are higher-order rule schemas which, upon application, provide partially identified recursive definitions. The templates are designed to provide definitions of the desired form from which new *higher-order* wave-rules may be formed. These wave-rules facilitate the rippling process and the higher-order meta-variables introduced into the proof, as a result of their application, become instantiated through subsequent theorem proving.

In our example we desire, in addition to containing constraint parameters, a tail-recursive program which takes three arguments. Thus, the corresponding template is as follows:

$$F(nil, W, D) \Rightarrow G_3(W, D) \tag{16}$$

$$F(\boxed{Hd :: Tl}^{\uparrow}, W, D) \Rightarrow F(Tl, \boxed{G_1(Hd, \underline{W})}^{\downarrow}, \boxed{G_2(Hd, \underline{D})}^{\downarrow}) \tag{17}$$

where G_1, G_2 and G_3 are second-order meta-variables. In general, for a function of n arguments there will be n meta-variables $A_1, ..., A_n$:

$$F_n(nil, A_1, ..., A_n) \Rightarrow G_n(A_1, ..., A_n)$$

$$F_n(\boxed{Hd :: Tl}^{\uparrow}, A_1, ..., A_n) \Rightarrow F_n(Tl, \boxed{G_1(Hd, \underline{A_1})}^{\downarrow}, ..., \boxed{G_{n-1}(Hd, \underline{A_n})}^{\downarrow})$$

Such a use of higher-order variables within rule schemas is a new kind of MOR. It is not the same as the MOR which introduces meta-variables in the constraint-based proof plan preconditions (nor as MOR is identified in previous publications such as [7]). Although both uses employ the meta-variables to stand in for "unknown constructs", the pre-condition usage delays proof commitments by partially identifying goal statements, whereas the above usage constructs partially identified (wave) rule templates. The above example corresponds to a tail-recursive template. This allows us to introduce further meta-variables during re-writing (i.e. to delay further proof commitments), in addition to those introduced by the proof plan preconditions, whilst ensuring that the function being constructed adheres to a tail-recursive form.

4.3 The Synthesis of rev_sm_C Revisited

We now repeat the rev_sm_C example except this time we include the rippling annotations so as to illustrate how the *eureka steps* are circumvented, and how search is tamed, through the use MOR.[11] HO meta-variables are used to postpone the commitments to existential witnesses and the identification of the new constrained goal. As in §2, the specification goal is,

$$\forall x, \forall w, \exists z. \, z = rev\_sm(x, w) \tag{18}$$

---

[11] Note that the example is an instance of the general framework represented by fig.1.

This forms the input to the constraint-based proof plan (the pre-conditions of which ensure that this is the first successfully applied proof plan). The output consists of the synthesis goal (19):

$$\vdash \forall x, \forall w. rev\_sm(x, w) = rev\_sm_C(x, w, M(w)) \tag{19}$$

and the trivial justification goal:

$$\forall x, \forall w. rev\_sm(x, w) = rev\_sm_C(x, w, M(w)) \vdash \forall x, \forall w, \exists d. d = rev\_sm(x, w)$$

where M is a (higher-order) meta-variable, and $M(w)$ the partially identified constraint parameter. (19) is set up as the conjecture to prove (the specification goal).

Amongst the rewrite rules are those formed from the available recursive definitions given in §1.1. Rewrites (21) and (23) are wave rules formed from the recursive branches (2) and (4). The non-wave rules (20) and (22) are formed from the corresponding terminating branches (1) and (3).

$$rev\_sm(nil, w) \Rightarrow (sum(w), w); \tag{20}$$

$$rev\_sm(\boxed{Hd :: \underline{Tl}}^{\uparrow}, W) \Rightarrow rev\_sm(Tl, \boxed{Hd :: \underline{W}}^{\downarrow}) \tag{21}$$

$$sum(nil) \Rightarrow 0; \tag{22}$$

$$sum(\boxed{Hd :: \underline{Tl}}^{\uparrow}) \Rightarrow \boxed{Hd + \underline{sum(Tl)}}^{\uparrow} \tag{23}$$

As yet, rev_sm_C, is undefined, but this is precisely where the use of the transformation templates comes into play: by instantiating F in the higher-order rewrites, (16) and (17), we introduce second-order meta-variables, G_1, G_2 and G_3, and provide a partially identified *tail recursive* definition for rev_sm_C which yields the following schematic rewrites:

$$rev\_sm_C(nil, W, D) \Rightarrow G_3(W, D) \tag{24}$$

$$rev\_sm_C(\boxed{Hd :: \underline{Tl}}^{\uparrow}, W, D) \Rightarrow rev\_sm_C(Tl, \boxed{G_1(Hd, \underline{W})}^{\downarrow}, \boxed{G_2(Hd, \underline{D})}^{\downarrow}) \tag{25}$$

Note that (25) can now be employed as a *higher-order wave-rule*. All we have assumed here is that rev_sm_C is a tail-recursive program with three arguments, we have not begged the question concerning the identity of the recursive data-types for the rev_sm_C definition.

Following $\forall$-introduction, induction is performed on (19) yielding the following induction cases:[12]

Induction Base. The base case is as follows:

$$\vdash \forall w. rev\_sm(nil, w) = rev\_sm_C(nil, w, M(w)) \tag{26}$$

By symbolic evaluation (using the rewrite yielded by (1) and (24):

$$\vdash \forall w. (sum(w), w) = G_3(w, M(w)) \tag{27}$$

(27) reduces to true by tautology, and HOU instantiates $G_3(w, d)$ to (d, w) and M to $\lambda x. sum(x)$.

---

[12] A feature of the goal-directed proofs is that introduction rules have the effect of eliminating existential quantifiers in the consequents of sequents. Conversely, elimination rules have the effect of introducing an existential instantiation in the hypotheses.

Induction Step. The identification of $M(w)$, the constraint-parameter, as $sum(w)$ during the base case proof enables us to instantiate M to sum in the step case of the proof. Thus, at the induction step we have the following sequent to prove (where (28) is the induction hypothesis, and (29) the induction conclusion):

$$\forall w.\ rev\_sm(tl, w) = rev\_sm_C(tl, w, sum(w)) \qquad (28)$$

$$\vdash\ \forall w.\ rev\_sm(\boxed{hd :: \underline{tl}}^\uparrow, \lfloor w \rfloor) = rev\_sm_C(\boxed{hd :: \underline{tl}}^\uparrow, \lfloor w \rfloor, sum(\lfloor w \rfloor)) \quad (29)$$

Rippling sideways using, on the l.h.s., (21) and, on the r.h.s., (25):

$$\vdash\ \forall w.\ rev\_sm(tl, \boxed{\boxed{hd :: w}^\uparrow}^\downarrow) = rev\_sm_C(tl, \boxed{\boxed{G_1(hd, \underline{w})}^\uparrow}^\downarrow, \boxed{G_2(hd, \underline{sum(\lfloor w \rfloor)})}^\uparrow)$$

Rippling-in using (23) instantiates, in the process, G_2 to $\lambda x, y.x + y$ through HOU:

$$\vdash\ \forall w.\ rev\_sm(tl, \boxed{\boxed{hd :: w}^\uparrow}^\downarrow) = rev\_sm_C(tl, \boxed{\boxed{G_1(hd, \underline{w})}^\uparrow}^\downarrow, \boxed{sum(\boxed{hd :: \underline{w}}^\uparrow)}^\downarrow)$$

Fertilization with the induction hypothesis, (28), now applies: w in the induction hypothesis is instantiated to $hd :: w$ from the induction conclusion. In the process, G_1 is instantiated to $\lambda x, y.x :: y$.

Analysis of the proof yields the desired tail-recursive program with the constraint-parameter $sum(w)$:

$$rev\_sm_C([\,], w, sum(w)) = (sum(w), w);$$
$$rev\_sm_C(h :: t, w, sum(w)) = rev\_sm_C(t, h :: w, h + sum(w))$$

4.4 Fixed Expensive Expressions

In the case of functions where the expensive expression is based on a fixed, or constant, parameter no MOR is required. For example, consider the following function:

$$g(nil, \epsilon) = nil;$$
$$g(h :: t, \epsilon) = sqr(\epsilon) \times h :: g(t, \epsilon).$$

Here the second parameter, ϵ, remains constant throughout the recursion (and hence so will the expensive expression, based on the constant, $sqr(\epsilon)$).

Again, the procedure we follow to optimize this function, by lifting the expensive expression out of the recursion, is to introduce a new (generalized) function with an additional parameter, d, which will be a function on the non-recursive parameter, ϵ. However, in the case of fixed parameters, there is no need for meta-variables since in such cases we know that d is to be identified with the constant expensive expression $sqr(\epsilon)$. Hence the new (generalized) function, g_new, is introduced as follows:

$$\forall x, \exists d.g(x, \epsilon) = g\_new(x, \epsilon, d) \ where \ d = sqr(\epsilon)$$

After applying induction, followed by rippling, an analysis of the resulting proof yields the following optimization:

$$g\_new(nil, \epsilon, d) = nil;$$
$$g\_new(h :: t, \epsilon, d) = d \times h :: g\_new(t, \epsilon, d).$$

5 Benefits and Comparisons

Using proof plans to synthesize efficient algorithms presents search and control advantages over unfold/fold style program development [3]. Unfold/fold transformations are motivated by the desire to find recursive terms which can be used for *folding* with definitional equations. This can involve extensive search in order to find, if at all, a successful fold. The proof plan analysis, on the other hand, is motivated by the desire to find witnesses at the induction step of a synthesis *proof*. Once this has been achieved, through rippling, which may include MOR, then the proof is completed in much the same way as any inductive synthesis proof: by a process of *unfolding* until all terms in the conclusion match terms in the proof hypotheses. The fact that in rippling, the wave-fronts in the proof must correspond to those in the wave-rule schemas provides considerable control with a low branching rate. The unfold/fold transformations on the other hand require numerous applications of laws for which any overall strategy is difficult to characterize. Thus rippling is far easier to automate. The most persuasive empirical evidence for this being within the context of automatic proof plan application through the automation of the rippling out process (*cf*. [1]). A further benefit of the rippling strategy is that it is guaranteed to terminate: wave-fronts are always propagated in a direction toward achieving a match between conclusion and hypothesis (also *cf*. [1]).

Rippling, incorporated with MOR, allows us to circumvent eureka steps required by sequential rewriting transformation strategies such as unfold/fold. This clearly aids automation since key decisions regarding the identity of recursive terms in the target program can be delayed, rather than user-supplied, until subsequent planning provides the requisite information.

Since any of the synthesis proofs approach *must* satisfy the specification formed from the standard equational form, $\mathcal{E}$, of the function being synthesized then the proof extract program is guaranteed to satisfy the specification, and hence to compute the function defined by $\mathcal{E}$. Furthermore, by providing the characteristics of the various generalizations, in the form of restrictions on the proof, we can ensure that the desired optimization is built into the algorithm being synthesized.

By proving that the synthesized program satisfies the original specification, we avoid the need to establish that any rewrite rules used are in themselves correctness (equivalence) preserving. This will, as a general rule, require as much effort as providing an explicit proof of correctness for the source to target transformations. For example, many of the systems that employ the *unfold/fold* strategy rewrite the recursive step(s) of a source program through the application of various *equality* lemmas, each of which needs to be proved (by induction) if the source to target transformation is to preserve equivalence [12, 6]

6 Conclusion

We described a general technique for controlling the synthesis of efficient programs using automatic proof planning. The technique encapsulates a diverse range of program optimizations, and benefits from the principled search and control strategies of proof plans. In particular, the syntactic pattern matching properties of rippling mean that we avoid the control problems encountered by the arbitrary application of rules and laws in program transformation systems. The optimization process is automatic and correctness is guaranteed. The incorporation of the MOR technique enables the circumvention of eureka steps which have prevented total automation in program transformation systems.

References

1. A. Bundy, A. Stevens, F. van Harmelen, A. Ireland, and A. Smaill. Rippling: A heuristic for guiding inductive proofs. *Artificial Intelligence*, 62:185–253, 1993. Also available from Edinburgh as DAI Research Paper No. 567.

2. A. Bundy, F. van Harmelen, C. Horn, and A. Smaill. The Oyster-Clam system. In M.E. Stickel, editor, *10th International Conference on Automated Deduction*, pages 647–648. Springer-Verlag, 1990. Lecture Notes in Artificial Intelligence No. 449. Also available from Edinburgh as DAI Research Paper 507.

3. R.M. Burstall and J. Darlington. A transformation system for developing recursive programs. *Journal of the Association for Computing Machinery*, 24(1):44–67, 1977.

4. W. N. Chin. *Automatic Methods for Program Transformation*. PhD thesis, University of London (Imperial College), 1990.

5. H.B. Curry and R. Feys. *Combinatory Logic*. North-Holland, 1958.

6. J. Darlington. A functional programming environment supporting execution, partial evaluation and transformation. In *PARLE 1989*, pages 286–305, Eindhoven, Netherlands, 1989.

7. J. Hesketh, A. Bundy, and A. Smaill. Using middle-out reasoning to control the synthesis of tail-recursive programs. In D. Kapur, editor, *11th Conference on Automated Deduction*, pages 310–324, Saratoga Springs, NY, USA, June 1992. Published as Springer Lecture Notes in Artificial Intelligence, No 607.

8. W.A. Howard. The formulae-as-types notion of construction. In J.P. Seldin and J.R. Hindley, editors, *To H.B. Curry; Essays on Combinatory Logic, Lambda Calculus and Formalism*, pages 479–490. Academic Press, 1980.

9. Per Martin-Löf. Constructive mathematics and computer programming. In *6th International Congress for Logic, Methodology and Philosophy of Science*, pages 153–175, Hanover, August 1979. Published by North Holland, Amsterdam. 1982.

10. A. Pettorossi. A powerfull strategy for deriving programs by transformation. In *ACM Lisp and Functional Programming Conference*, pages 405–426, 1984.

11. S. Koenig R. Paige. Finite Differencing of Computable Expressions. *ACM Transformation on Functional Programming Languages and Systems*, 4:pp. 405–454, 1982.

12. H. Tamaki and T.Sato. Transformational logic program synthesis. In *Proceedings of the International Conference on Fifth Generation Computer Systems*. ICOT, 1984.

13. P. Wadler. Deforestation: Transforming Programs to Eliminate Trees. In *Proceedings of European Symposium on Programming*, pages 344–358. Nancy, France, 1988.

14. S.S. Wainer. Computability - logical and recursive complexity, July 1989.

Datalog and TwoGroups and C++

Greg Butler

Centre Interuniversitaire en Calcul Mathématique Algébrique
Department of Computer Science
Concordia University
Montreal, Quebec, H3G 1M8 Canada
Email: gregb@cs.concordia.ca

Abstract. *Datalog* is the language of deductive databases, a first step towards intelligent relational databases. *TwoGroups* is a database of p-groups, primarily the groups of order dividing 2^8, which has seen actually use by research mathematicians. *TwoGroups* was originally developed on the NU-Prolog platform. We report on our efforts to implement a *Datalog* interpreter in C++, and to implement *TwoGroups* in C++ on top of the *Datalog* platform. Although *TwoGroups* is very relational in style, its use of range queries means that *Datalog* is not the perfect platform on which to implement *TwoGroups*. However, *TwoGroups* can directly utilise the multi-attribute retrieval algorithms within the *Datalog* implementation which do support range queries.

1 Introduction

TwoGroups is a database of p-groups, primarily the groups of order dividing 2^8, which has seen actually use by research mathematicians. *TwoGroups* was originally developed on the NU-Prolog platform. The data is essentially relational so this approach was successful. However, there were a number of limitations: primarily the need to discretize range queries, a lack of access to the underlying mechanisms for retrieval, and a lack of extensibility of the system. In addition, our other work on group-theoretic knowledge-bases had to deal with complex mathematical objects, whch would have to integrate with the *TwoGroups* database. *Datalog* is the language of deductive databases, a first step towards intelligent relational databases. We report on our efforts to implement a *Datalog* interpreter in C++, and to implement *TwoGroups* in C++ on top of the *Datalog* platform. Although *TwoGroups* is very relational in style, its use of range queries means that *Datalog* is not the perfect platform on which to implement *TwoGroups*. However, *TwoGroups* can directly utilise the multi-attribute retrieval algorithms within the *Datalog* implementation which do support range queries.

This work is part of a long tradition seeking to make mathematical knowledge more accessible, to mathematicians and non-mathematicians alike. Historically, the print media has been the way to disseminate mathematical data. The data includes lists of prime numbers, the digits of π, tables of integrals or combinatorial identities, or tables of function values such as statistical distributions.

The printed literature on group-theoretical data includes the Hall and Senior catalogue of the 2-groups of order dividing 64, the list of primitive permutation groups of small degree, the *Atlas* of the sporadic simple groups, and the perfect groups of order up to one million. The use of computers to disseminate group-theoretical data is much more recent: such use began about 1975. The first examples were on-line collections of files; these are used as input data to algorithmic computations; they are kept for ease of updating as errors are detected and corrected; or they are browsed through or printed. The collections are searched or queried in much the same fashion as the printed literature. The libraries of group data for the computer algebra systems *Cayley* and GAP are used in the same manner as these on-line collections of files. *TwoGroups* is the first mathematical database. *SmallSimpleGroups* [4] is the first knowledge base for mathematics, though still experimental. It contains objects, parameterised data, and heuristics.

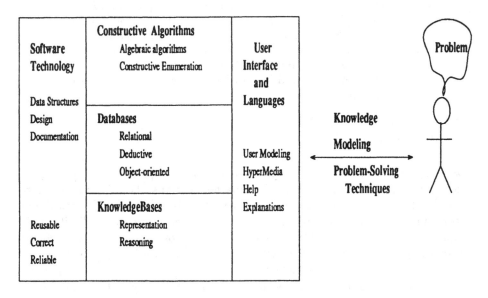

Fig. 1. Integrated Mathematical Knowledge Based System

Our aim at the moment is to construct a C++ platform integrating the software components necessary to continue the research into mathematical knowledge based systems. This paper describes the initial steps in constructing that platform.

In Figure 1, we depict the main components of future computer algebra systems, which will be problem-solving environments for engineers, scientists, and students. Communication between the system and the user will involve accurate description of a model of the problem, and any other problem-specific information, such as problem-solving techniques, known to the user. The tools brought to bear on the problem will include constructive algebraic algorithms, database query processing, and reasoning. Modern software engineering practices will be needed to successfully complete such integrated intelligent systems for mathematical problem-solving.

This paper briefly covers the background on query processing and retrieval algorithms. Multi-attribute indexing algorithms and data structures are of particular importance. Then we review *Datalog* and the design of our interpreter written in C++; describe the *TwoGroups* system as it was implemented in NU-Prolog; and the version of *TwoGroups* built upon the *Datalog* interpreter. A complete implementation of query optimisation, and the study of the benefits of various disk retrieval algorithms is not yet complete. The complete study will be reported in a later communication.

2   Query Processing

Different databases and application domains — such as business accounts, engineering designs, geometric and graphic data, text documents, scientific data, and hypermedia documents — have different requirements as to storage and retrieval capabilities [10, 29, 30]. Some types of queries include

Exact Match Queries specify a literal value (also called a ground value) for an attribute, and a *match* of that value is expected. A predicate (or relation) may have several attributes. A *fully ground query* specifies a literal value for each attribute, and requires confirmation whether this fact is in the database. An exact match may also refer to the case where a literal value is given for the primary key of a relation, and the retrieval of the complete record with the given key is required.

Partial Match Queries specify a literal value for some attributes, and a *partial match* is required — that is, a match is required for each of the attributes that have a literal value specified, but the other attributes can have any value: that is, the attributes match a "wild card".

Range Queries specify a range of values for an attribute. The ranges may be
   - an open interval range, such as $low < attribute < high$;
   - a closed interval range, such as $low \leq attribute \leq high$;
   - a half-open interval range, such as $low <= attribute < high$, or $low < attribute \leq high$;
   - a semi-infinite interval range, such as $low < attribute$, or $attribute \leq high$;

Best Match Queries specify a literal value for some attributes but do not require that an exact match for each of the specified literals be found. In the event that such an exact match does not exist in the database, then the fact in the database which comes "nearest" to matching the query should be retrieved. One metric for "nearness" is the number of attributes whose value matches the value specified in the query.

Other Queries covers a broad range of queries such as:
   - *string matching query* in a textual database;

- *Boolean property query* where the attributes take only Boolean values, which indicate the presence or absence of some property.

This list should be extended to include navigational queries common in object-oriented databases, and libraries providing persistence.

2.1 Retrieval Algorithms

Ordinary retrieval algorithms [11] match a single attribute at a time, using indexing on the primary key, or secondary indexing on other attributes. For an attribute without an index, a search must be performed. Index data structures for single attribute retrieval have been well-studied, and are usually based on hashing or B-trees.

Exact matches have been much more studied than range queries, but for single attribute retrieval using B*-trees (but not hashing) it is easy to adapt the exact match algorithm for range queries.

Multi-attribute (also called multi-dimensional) retrieval algorithms match several attributes at once. Superimposed coding [26, 27] is based on hashing. The other methods are generalisations of B-trees in that each key in a B-tree divides the 1-dimensional key space into three intervals: those less than the key, those greater than the key, and those equal to the key. A data structure handling n attributes simultaneously is matching against an n-tuple of keys, which divides an n-dimensional space into three parts along each dimension. Examples of such retrieval algorithms are *grid files* [20, 21], *bang files* [12, 13], *kd-trees* [1, 2], and *hB-trees* [17]. The latest word on the subject seems to be Reference [19].

3 Datalog

Datalog extends relational databases by allowing rules to derive additional rows or tuples of relations, or to derive entirely new relations. The book of Maier and Warren [18] discusses the three languages Proplog, *Datalog*, and Prolog through database and knowledge-base examples. They present interpreters, written in Pascal, for each of the languages. The architecture of the *Datalog* interpreter is presented in Figure 2. A formal treatment of *Datalog* semantics, and the optimisation of *Datalog* queries can be found in [7]. The major aim of optimization of *Datalog* queries is to eliminate recursion in the rules, so that retrieval from the database of facts is not repeated unnecessarily.

Proplog handles propositional calculus, which has constants but not variables. Datalog handles predicate calculus, which allows variables for attribute values. These are implicitly quantified: universally quantified in the head of a rule, and existentially quantified in the body of a rule. An object model [25] of the language concepts is presented in Figure 3. The DDL (Data Definition Language) allows the definition of rules and facts as clauses. The DML (Data Manipulation Language) only allows queries, which are posed as (headless) clauses: that is, as a list of literals.

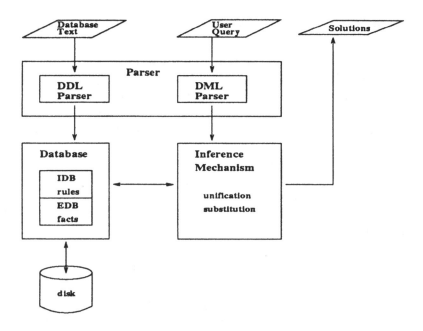

Fig. 2. Overview of Datalog Architecture

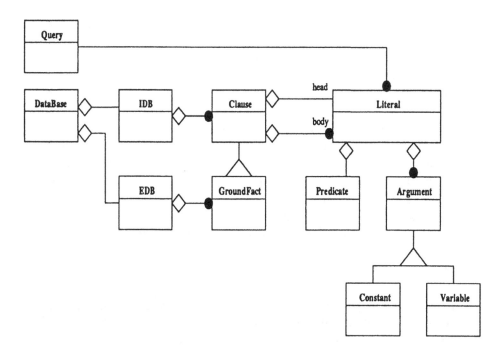

Fig. 3. Datalog Object Model

3.1 Data Dictionary for Datalog

Atomic Constant is a primitive value and is indicated by an identifier which starts with a lower case letter.

Binding associates a *variable V* with a value, which may be either a *constant* or another *variable W*. If *V* is bound to *W* and *W* is bound to a value, then both variables share the same value. A variable may be *unbound*: that is, not associated with any value. A binding is part of a *substitution*.

Body is a list of *literals*, and forms part of a *clause*.

Clause consists of a *head* and a *body*. A clause could be read as a rule, "if the body is true then the head is true". It is one part of the definition of the predicate of the head. The definition of the predicate is the "or" of each of the clauses.

Constant same as *Atomic Constant*

Extensional Data Base (EDB) is the collection of the *facts* explicitly stated as part of the database.

Fact a *clause* which has an empty *body*. A fact may contain variables as arguments, but more often a fact is fully ground: that is, all arguments are constants.

Goal same as *Query*.

Head is a literal, and forms part of a *clause*.

Intensional Data Base (IDB) is a collection of clauses which define those facts which may be derived from the *EDB*. Often a *program* is viewed as precisely that part of the IDB needed to answer a specific query.

Literal is a reference to a predicate, which specifies the arguments of the predicate as either constants or variables.

Predicate is a relation.

Predicate Name the symbolic name of a predicate.

Program is a set of clauses (or equivalently predicates) which define the relationship between the predicates in a *query* and those in the *database*.

Query is a list of literals. It may also be viewed as a clause with an empty head.

Relation same as *Predicate*

Solution is a set of all facts that can be derived from the database, and that satisfy the query. The solution set may also be viewed as a set of substitutions for the variables in the query.

Substitution is a collection of bindings.

Symbol same as *Atomic Constant*

Unification is a process which matches literals. Unification determines the most general substitution, called the most general unifier, which, when applied to the literals being matched, gives an identical literal.

Variable stands as a place holder for values. It may be bound to different values, or be unbound. A variable is indicated by an identifier which starts with an upper case letter.

3.2 Datalog Interpreter

The grammar for the *Datalog* language is given below. The input of the database and the queries is translated by a recursive descent parser. The (naive) inference engine for the interpreter uses backward-chaining with unification to process the *IDB* rules, and partial match retrieval of the *EDB* facts. The retrieval of the facts is implemented using the hB-tree [17] data structure when the data is stored on disk, and is implemented using a simple list structure when facts are stored in-memory.

database	::= *database clause*
	| empty
clause	::= *head :- body* .
clause	::= *head* .
head	::= *literal*
	| empty
body	::= *literal_list*
literal	::= *predicate_name* (*argument_list*)
argument	::= *constant* | *variable*
query	::= :- *literal_list* .

4 TwoGroups

The *TwoGroups* database [5, 6] provides access to a large body of information on the 58 761 groups of order dividing 256. It also provides a convenient interface for mathematical users via a "natural" set-theoretic query language [3]. Information on other lists of p-groups can be included in the database without major difficulty or alteration.

The database is built using a NU-Prolog implementation platform [24]. It provides an inference engine, a translator for definite-clause grammars which constructs a parser written in Prolog, and a partial match retrieval scheme for external databases [27]. The two-level retrieval scheme, called SIMC, is based on superimposed codewords. The query language is based on set constructors and is a natural and convenient generalisation of the proposed language for *Cayley, version 4* [3]. A set-theoretic query is translated and optimised to an equivalent Prolog goal. Figure 4 presents an overview of query processing in *TwoGroups*, again using the notation of [25]. The database has an extensive help facility which provides contextual information on the available data, the Prolog predicates, and the syntax and functionality of the set-theoretic query language.

Each group of order O is identified by a unique identifier $O\#N$, where N is its index position in the sequence of groups of this order. For each of the groups, the database stores the following information: in four relations:

- the group identifier $O\#N$;
- a power-commutator presentation encoded as a binary number;

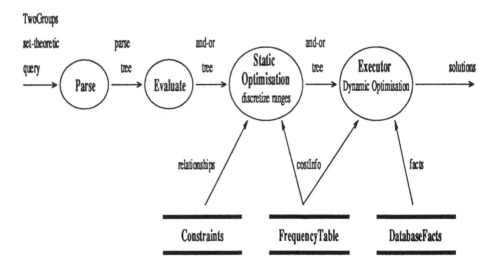

Fig. 4. Data Flow Diagram for TwoGroups (NU-Prolog version)

- the order of the automorphism group;
- the number of conjugacy classes of each length;
- the number of elements of each order;
- the number of defining generators and the number of power-commutator presentation generators;
- the exponent-p class and the nilpotency class;
- the centre, the central quotient, the commutator subgroup, and the parent group;

in four relations called *group_info, seq, elements_of_order,* and *num_of_classes*. Definitions of unfamiliar terms may be found in O'Brien [22].

The query language is based on the familiar notation of set theory. It provides statements to query the database and to print solution sets or to print individual groups. The set theoretic queries can also be used to refine existing solution sets – a process we call *incremental goal refinement*. A query may either directly access stored information or require the derivation of some new information: the exponent and number of conjugacy classes of a group and the order of its inner automorphism group. The database may be queried using the statement syntax

$$set\_id := \{ \ group\_id \ \textbf{in} \ \texttt{twgps} \ | \ boolean\_expr \ \};$$

Here

set_id	is an identifier for the set of groups that satisfy the query,
group_id	symbolises a general group in the database,
twgps	is the name of the NU-Prolog database, and
boolean_expr	specifies the condition that the group must satisfy.

The identifiers begin with a letter, and consist of letters and digits. The group identifier must begin with an upper case letter. A constant in the boolean expression may be a literal integer or a group literal. The simplest group literal has the form *Order#Number*, where both *Order* and *Number* are literal integers. However, it is possible for either or both of *Order* and *Number* to be an upper case letter. For example, the family of cyclic groups is denoted by $C\#1$ and the groups of order 32 are denoted by $32\#Y$.

The syntax for boolean expressions is

boolean_expr	::= *boolean_expr boolean_op boolean_expr*
	| *integer_expr int_relop integer_expr*
	| *group_expr group_relop group_expr*
	| not *boolean_expr*
boolean_op	::= and | or
int_relop	::= eq | ne | gt | ge | lt | le | divides
group_relop	::= eq | ne
integer_expr	::= *integer_expr int_op integer_expr*
	| *int_function_call*
	| *integer_literal*
int_op	::= + | - | * | /
group_expr	::= *group_function_call*
	| *group_literal*
group_function_call	::= parent(*group_expr*)
	| centre(*group_expr*)
	| central quotient(*group_expr*)
	| commutator subgroup(*group_expr*)
int_function_call	::= #classes(*group_expr*)
	| #classes length(*group_expr, integer_literal*)
	| #elements order(*group_expr, integer_literal*)
	| #defining generators(*group_expr*)
	| order(*group_expr*)
	| exponent pclass(*group_expr*)
	| nilpotency class(*group_expr*)
	| exponent(*group_expr*)
	| automorphism group order(*group_expr*)
	| auto(*group_expr*)
	| inner automorphism group order(*group_expr*)
	| inner(*group_expr*)

The meaning of each function should be clear from its name: the symbol # means "number of", and **inner** and **auto** are simply abbreviations for the functions listed immediately above them.

The query can also refine a previous solution set by

$$set\_id := \{ \; group\_id \; in \; [set\_id_1, set\_id_2, \cdots] \mid boolean\_expr \; \};$$

where the identifier **twgps** has been replaced by a list of identifiers, set_id_1, $set\_id_2, \cdots$, of previous solution sets within square brackets, "[" and "]".

The solution set may be printed using the **print** *set_id*; statement. We can display all properties of a given group $O\#N$, where O and N are integer literals, using the **print all** $O\#N$; statement. It is also possible to print all properties of all groups in a set using the **print all** *set_id*; statement.

5 TwoGroups Using Datalog

We will describe the C++ implementation of *TwoGroups* upon the *Datalog* interpreter using an example. After examination of the available parser generator tools, we chose *Flex++* [16, 23] and *Bison++* [8, 9] to translate the *TwoGroups* grammar into a C++ lexer and parser class. The translation to *Datalog* follows the original NU-Prolog version, however, it uses the *Iterator* and *Visitor* patterns [14] to walk over the parse tree and the and-or tree. Consider the query

s := { G in twgps | order(G) le 4 and #elements order(G,2) eq 3 };

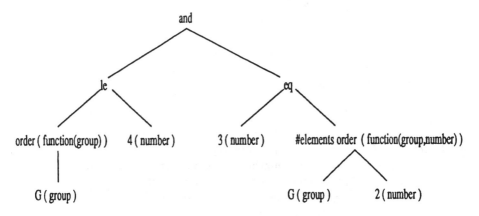

Fig. 5. Corresponding parse tree

The parser will check the syntax of the query and produce a parse tree. The parse tree will also denote the type of every element of the query, in order to check the validity of the query. Figure 5 shows the parse tree of the example query. This parse tree is then processed by the *evaluator*. The role of the evaluator is to build another tree, called an *and-or tree*. The contents of this tree are closer to the Datalog goal. Figure 6 shows the and-or tree of our example.

The and-or tree is then processed by the optimizer, which will reorganize and optimize the current tree: that is, put it in a *normal disjunctive form* and optimize it as in the NU-Prolog version.

The last stage of the query process generates a *Datalog program* from the and-or tree. The Datalog program of our example is shown in Figure 7.

For static execution of the *Datalog* query, we need an interface to *Datalog* which accepts a query for a database, and returns the set of solutions — actually

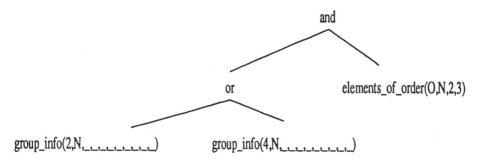

Fig. 6. Corresponding and-or tree

```
soln(2,N):-group_info(2,N,_,_,_,_,_,_,_,_,_),
           elements_of_order(2,N,2,3).
soln(4,N):-group_info(4,N,_,_,_,_,_,_,_,_,_),
           elements_of_order(4,N,2,3).
goal(O,N):-soln(O,N).
```

Fig. 7. Corresponding Datalog program

an iterator which traverses the set of substitutions for the *goal* predicate of the query. At the moment, all queries return a set of group identifiers, $O\#N$. Incremental goal refinement uses the whole set of solutions.

The **print** and **print all** commands in *TwoGroups* require an interface which given a group identifier returns a description of the group suitable for printing.

For dynamic execution, the *TwoGroups* interpreter mixes optimisation phases with static execution of subqueries. The solutions to subqueries are more diverse than just sets of group identifiers. The executor needs to apply the substitutions (from the solution of subqueries) to unsatisfied subgoals as part of the optimisation phases.

6 Conclusion

Datalog is not the best target language for *TwoGroups*, but direct reuse of the software components of a *Datalog* implementation resolves the problems. The main problems are range queries, and dynamic execution and optimisation of *TwoGroups* queries. One can fudge range queries in *Datalog* by defining the integers via the *succ* function of the Peano axioms for the integers, and explicitly defining the *gt* relation. But, this is highly inefficient and unsatisfactory. *TwoGroups* depends heavily on range queries, so it is best to go directly to the heart of the *Datalog* interpreter, bypassing the *Datalog* query language, and utilising the facilities of hB-trees to answer range queries.

Acknowledgements: My students have contributed greatly to implementing this work, and to stimulating the discussions: Mohan Rao Tadisetty implemented the first versions of the interpreter for Proplog and Datalog; Dorel Baluta worked on a Z specification of the relational data model, and (potentially) a Z model for deductive databases; Khanh Tuan Vu partially implemented the hB-tree retrieval mechanism; Deming Li implemented a recursive descent parser for the Datalog language; Gilles Charles wrote the translator from the *TwoGroups* set-theoretic query language to *Datalog*. Naturally, this work owes a great debt to the NU-Prolog version of my student Sridhar Iyer.

References

1. J.L. Bentley, *Multidimensional binary search trees used for associative searching*, CACM **18**, 9 (September 1975) 509–517.

2. J.L. Bentley, *Multidimensional binary search trees in database applications*, IEEE Trans. Software Eng. **SE-5**, 4 (July 1979) 333–340.

3. G. Butler & John Cannon, *Cayley, version 4: the user language*, **ISSAC'88**, (Proceedings of 1988 International Symposium on Symbolic and Algebraic Computation, Rome, July 4-8), Lecture Notes in Comput. Sci. **358**, 456–466, Springer-Verlag, Berlin, 1989.

4. G. Butler and S.S. Iyer, *An experimental knowledge base of simple groups*, submitted.

5. G. Butler, S.S. Iyer & E.A. O'Brien, *A database of groups of prime-power order*, The Australian National University, Mathematics Research Report 021-92, 1992.

6. G. Butler, S.S. Iyer & E.A. O'Brien, *TwoGroups : a database for group theory*, The Australian National University, Mathematics Research Report 005-93, 1993. (also Notices of the AMS **40**, 7 (September 1993) 839–841.)

7. Stefano Ceri, Georg Gottlobb, Letizia Tanca, *What you always wanted to know about Datalog (and never dared ask)*, IEEE Trans. KDE 1, 1 (March 1989) 146–166.

8. Alain Coëtmeur (1993), *Bison ++, Flex ++*, coetmeur@icdc.fr.

9. Charles Donnelly and Richard Stallman (1992), *Bison : The YACC-compatible Parser Generator*, Free Software Foundation.

10. Ramez Elmasri, Shamkant B. Navathe, **Fundamentals of Database Systems**, Benjamins/Cummings Inc., Redwood City, California, 1989.

11. Michael J. Folk and Bill Zoellick, **File Structures**, 2nd edition, Addison-Wesley, 1992, (xviii)+590 pages.

12. Michael W. Freeston, *The BANG file: a new kind of grid file*, Proc ACM SIGMOD Conference, San Francisco, 1987, pp.260–269.

13. M.W. Freeston, *Advances in the design of the BANG file*, in **Foundations of Data Organisation and Algorithms**, (Proc. 3rd International Conference on Foundations of Data Organisation and Algorithms, Paris, France, June 21–23, 1989), W. Litwin and H.-J. Schek (eds), Lecture Notes in Comput. Sci. **369**, Springer-Verlag, Berlin, 1989, pp.322–338.

14. Erich Gamma, Richard Helm, Ralph Johnson, John Vlissides, *Design patterns: Abstraction and reuse of object-oriented design*, ECOOP'93, O.M. Nierstrasz (ed.), Lecture Notes in Computer Science **707**, Springer-Verlag, 1993, 406–431.

15. Scott Robert Ladd, **C++ Components and Algorithms**, M&T Publishing, San Mateo, CA, 1992, (xv)+779 pages.

16. M. E. Lesk and E. Schmidt, *Lex — A Lexical Analyzer Generator*, Bell Laboratories, Murray Hill, New Jersey 07974.

17. D.B. Lomet and B. Salzberg, *The hB-tree: a multiattribute indexing method with good guaranteed performance*, ACM TODS **15**, 1(1990) 625–658.

18. David Maier and David S. Warren, **Computing with Logic**, The Benjamin/Cummings Publishing Company, Inc, 1988.

19. Y. Nahamura, S. Abe, Y. Oshawa, M. Sakauchi, *A balanced hierarchical data structure for multi-dimensional data with highly efficient dynamic characteristics*, IEEE Trans. KDE **5**, 4 (August 1993) 682–694. (Special Issue on Multimedia Information Systems)

20. J. Nievergelt, H. Hintenberger, K.C. Sevcik, *The grid file: implementation and case studies of applications*, Internal Report No. 46, Institut fuer Informatik, ETH Zuerich, December 1981.

21. J. Nievergelt, H. Hintenberger, K.C. Sevcik, *The grid file: an adaptable, symmetric multikey file structure*, ACM TODS **9**, 1 (1984) 38–71.

22. E.A. O'Brien *The n-group generation algorithm, J. Symbolic Comput., 0, 677 000.*

23. Vern Paxson, *Flex — Fast Lexical Analyzer Generator*, Lawrence Berkeley Laboratories, University of California.

24. K. Ramamohanarao, J. Shepherd, I. Balbin, G. Port, L. Naish, J. Thom, J. Zobel and P. Dart, *An overview of the NU-Prolog deductive database system*, in **Prolog and Databases**, P.M.D. Gray and R.J. Lucas (Eds.), pp. 212–250. Ellis Horwood, 1988.

25. J. Rumbaugh, M. Blaha, W. Premerlani, F.Eddy, W. Lorenson, **Object-Oriented Modelling and Design**, Prentice Hall, 1991.

26. R. Sacks-Davis, A. Kent, and K. Ramamohanarao, *Multikey access methods based on superimposed coding techniques.* ACM Trans. Database Systems **12**, 4 (1987) pp. 655–696.

27. John Andrew Shepherd, *Indexing Schemes and File Organisations for the NU-Prolog Deductive Database System.* Ph.D. Thesis, University of Melbourne, 1989.

28. Bjarne Stroustrup, **The C++ Programming Language**, 2nd edition Addison-Wesley, 1991.

29. Jeffrey D. Ullman, **Principles of Database and Knowledge-base systems**, Volume I, Computer Science Press, Inc., Rockville, Madison, 1988.

30. Jeffrey D. Ullman, **Principles of Database and Knowledge-base systems**, Volume II, Computer Science Press, Inc., Rockville, Madison, 1989.

Linear Logic and Real Closed Fields : a Way to Handle Situations Dynamically

Pierre Jumpertz[1,2]

[1] Equipe de Logique Mathématique (ura 753) Université de Paris VII, 2 place Jussieu, case 7012, 75251 Paris cedex 05, France
[2] Ecole Nationale des Techniques Avancées, 32 Bd Victor, 75015 Paris, France

Abstract. The formalisation of proofs and algorithms in the theory of real closed fields yields a control over the structures handled by computations. At first we study the behaviour of the connectors and give a system which links proofs in linear logic to manipulations of polynomial structures. We extend the proof system with proper axioms which are results obtained from the completeness of the theory, and we show that the system is sound and that it controls when the computation may be done. Then we study the case of sign assignments upon roots of polynomials, and give a way to control cylindrical algebraic decompositions and thus to mix two cell decompositions.

1 Introduction

The theory of real closed fields (Rcf),i.e. the theory of ordered fields with intermediate value ($[P(x) > 0 \wedge P(y) < 0 \wedge x < y] \rightarrow [\exists a\, P(a) = 0 \wedge x < a < y]$), is expressed in the ring language $(0, +, -, .)$ with formulas as Boolean combinations of polynomial equalities or inequalities. This is the "natural" frame for many geometrical problems . Tarski was interested in completeness and decidability of this theory (to achieve logical results [Tar]). Collins found an effective (and computer written) quantifier elimination method [Col]. Algebraic computations construct a cylindrical decomposition, where sample points allow to decide whether a formula is true or not. The double exponential complexity bound of this algorithm was lowered to a single exponential bound for a single block of quantifiers [Ggv], and a double exponential bound in the number alternation of blocks [Ren] [HRS], which is now the best known bound.

All these decision methods make computations based on polynomials extracted from the formula which has to be decided. At first, an algebraic "complex" of roots is constructed, and then formulas are decided with the help of the "complex", and with the actual connectors and quantifiers.

An algorithm may interweave these two phases : the algorithm of [BKR] gives a more algebraic way to decide sign assignments, with a phase of "structure" calculus (with tensor products) and with a phase of decision and simplification (solving matrix systems and computing Sturm's sequences). The proof of the algorithm constructs a Boolean algebra, and the algorithm constructs the atoms of this algebra by successive combinations. We generalize [BKR]'s method into a deductive proof system, neither looking for a new root determination algorithm, nor for a new complexity lower bound. Real closed field theory is analyzed distinguishing the "structure" constructing phase and the decision (upon these "structures") phase.

The sample points of Collins' decomposition, and the Boolean algebra of sign assignments are both "structures". A structure is : a Boolean algebra of semi-algebraic sets (a "complex"), linked to a "context" (a set of semi-algebraic formulas). Such "structures" may be combined by combining their respective "complex" and "contexts". These combinations are determined by the connectors. Linear logic is a frame to handle more carefully the context, and the "objects" of the Boolean algebra : it is natural to search how linear logic expresses some properties of these sample points or Boolean "structures".

For many applications, like robot planning, a deduction system would be better than a decision method : a decision method gives a yes/no answer, but a deduction method gives the way to achieve a result. Linear logic is a frame for a certain kind of planning [Tol].

It would be useful to have a system that deals with symbolic mathematic computation, to have a single planning system for instance and to handle proofs in reasoning. We propose a extended linear logic proof system where the proof of a formula gives an algorithm and a semi-algebraic "structure" to decide this formula. So each semi-algebraic structure is constructed by an algorithm given by the proof of the associated formulas. The order of computations may be altered, depending on the actual connectors of the formula. Moreover, the formula to be decided needs not to be known in advance, but may be constructed dynamically.

In section 2, the linear logic applied to real closed fields is explained and linear logic rules are extended with some proper axioms linked to the models of the theory of real closed fields. In section 3, the soundness is shown. In section 4, the deduction system is applied to the [BKR]'s algorithms, that are expressed in a natural way, and so, every formula of real closed field theory may be decided in the deduction system. In section 5, the [BKR]'s matrix system is extended as result of the linear proof. This gives an elegant fashion to handle the logical proofs. Then, in section 6, a new dynamic approach of cylindrical algebraic decompositions is given as combination of cell "structures".

2 Semi-algebraic Sets and Geometry

The theory of real closed fields is expressed with linear logic [Gir] connectors, so that, in the frame of proof theory, decision rules [see annex A and B] coincide with constructing algorithms in Rcf. A extended linear logic proof system (Rcf+LL) is constructed, where combinations of sets, at the geometrical semi-algebraic level, as well as at the level of algorithms, are denoted by connectors.

2.1 Language

Symbols. The language is : $0, 1$ as constant symbols, x, y, z...as symbols of variables (first order), $+, ., -$ as binary function symbols. There are two binary relation symbols : $>, =$. Formulas are constructed with quantifiers $\exists, \forall$ and connectors of linear logic : $\tilde{}, \otimes, \wp, \oplus, \&$, and with the constant logical symbol $\vee$.

Terms and Formulas. Terms are constructed with these symbols (as in Rcf theory), atomic predicates (with two terms and a binary relation), then logical formulas are constructed with quantifiers, and linear connectors.

Particular Cases. A sub-algebra may be defined, in a sub-language (for instance, the language of real closed fields without $>$) or with a sub-structure (for instance, sign assignments upon the roots of polynomials). The general frame developed here remains valid.

2.2 Semi-algebraic Sets, Space Decomposition

Semi-algebraic Sets. Formulas define semi-algebraic sets.

Definition 1. A semi-algebraic set is defined in $\mathbb{R}^n$ by a formula of the real closed fields

examples : $x^2 + y^2 < 1$, the disk of centre $(0;0)$ and of radius 1, or $\exists x\ x.y - 1 = 0$, the projection of an hyperbola...
  A semi-algebraic set will be expressed in Rcf+LL as a multiplicative formula. Multiplicative connectors are linked with the expression of a semi-algebraic set[3], $\begin{cases} x^2 < 5 \\ y^2 < 3 \end{cases}$ so it gives : $(x^2 < 5) \otimes (y^2 < 3)$

"Structures" and Decompositions. Since linear logic is applied to real closed fields, each sequent $\Gamma \vdash A, \Delta$ and thus polynomial set, defines a semi-algebraic structure, with a context (linked to Γ, Δ), and a complex of semi-algebraic sets (associated with some formulas, here A, that are studied in a special way, with a decomposition, and a Boolean algebra linked with them).

Boolean Algebra and "Complex". Semi-algebraic sets have a Boolean algebra structure (with union, intersection...) which is linked to the Boolean connectors of the formulas (or, and, negation...). Thus, the connectors define the usual set operations : $\begin{smallmatrix} \wedge & \cap \\ \vee & \cup \end{smallmatrix}$. The negation is linked with the complementary (in a context, or a closure) : $\neg$ comp. The existential quantifier is a projection, which reduces the space dimension by one : $\exists x$ pr.$_x$. The universal quantifier is obtained as complementary of the projection set of the negation of the formula. Such a set of semi-algebraic sets is called a "complex".

---

[3] There is a difference between an object (a semi-algebraic set) and consequences upon such a set. So the multiplicative connectors link formulas together as an "object", for instance a cell or a semi-algebraic set. The additive connectors state that several objects may be handled simultaneously.

Context. The context defines a space of work too, where semi-algebraic sets are. A semi-algebraic set lays in a space which dimension is given by the number of free variables of the formula and of the context. The "and" may be interpreted differently as an intersection from sets already in $\mathbb{R}^2$: $\begin{cases} x+y=0 \\ x-y=0 \end{cases}$, or as the Cartesian product of two spaces : $\begin{cases} x=0 \\ y=0 \end{cases}$. Thus, the interpretation of connectors is linked with a working space (or a context).

Decomposition and "Plus". A decomposition is a partition of a "space" into semi-algebraic sets (C_i) which have some special properties (sign invariance,...) that we call a "complex". A decomposition is related to a "context", Γ. Thus, $\Gamma \vdash \oplus_{i\in I} C_i$. The "plus"($\oplus$) denotes that several (semi algebraic)"sets" are independent in a given situation. With this notation, as a disjunctive formula, simpler objects (for instance simpler polynomials) may be handled simultaneously to achieve a global treatment : $\Gamma \vdash (x^2 - 1) = 0 \oplus (x - 4) = 0$, or a decomposition may be expressed as a formula.

Atoms and Basis. The set obtained from a decomposition may be useful to decide other formulas. For a set of formulas A, a special decomposition, called basis for A, may give a way to decide all the Boolean combinations of formulas of A, with a check of all the formulas upon the basis. Some new terms may be computed and new atoms obtained, along the process (an atom is the expression of a set of the basis in a multiplicative way, i.e it is an atom of the Boolean algebra of formulas).

Composition. Multiplicative connectors $(\tilde{\ }, \otimes, \wp)$ indicate combination operations (as composition...) upon structures which are associated to sequents.

Composition. The tensor $(\otimes)$ means the composition (of the contexts and of the semi-algebraic complex) of two structures :

$$\frac{\Gamma_1 \vdash \oplus_{i\in I} f_i \quad \Gamma_2 \vdash \oplus_{j\in J} g_j}{\Gamma_1, \Gamma_2 \vdash (\oplus_{i\in I} f_i) \otimes (\oplus_{j\in J} g_j)}$$

For instance, two structures may be composed to obtain all the semi-algebraic sets constructed by two sets of formulas :

$$\frac{\Gamma_1 \vdash \begin{pmatrix} (x<1) \oplus (x=1) \\ \oplus(1<x<2) \oplus (x=2) \oplus (2<x) \end{pmatrix} \quad \Gamma_2 \vdash \begin{pmatrix} (y<3) \\ \oplus(y=3) \oplus (3<y) \end{pmatrix}}{\Gamma_1, \Gamma_2 \vdash \begin{pmatrix} (x<1) \oplus (x=1) \\ \oplus(1<x<2) \oplus (x=2) \oplus (2<x) \end{pmatrix} \otimes \begin{pmatrix} (y<3) \\ \oplus(y=3) \oplus (3<y) \end{pmatrix}}$$

Distributivity. The link between the composition of two structures and locally constructed consequences of this composition, is given by distributivity [see annex B] of "tensor" upon "plus". It gives a natural way to compute :

$$\frac{\Gamma_1, \Gamma_2 \vdash (\oplus_{i\in I} f_i) \otimes (\oplus_{j\in J} g_j)}{\Gamma_1, \Gamma_2 \vdash \oplus_{i,j\in I\times J} (f_i \otimes g_j)}$$

2.3 Demonstrations and Algorithms

Formulas in this system (Rcf+LL) are provable in a special fashion. There is a two level system of proof : one that deals with the structural aspect which is handled by linear logic rules, and the other one that deals with the computations and decisions in real closed fields which is controlled by some proper axioms. Thus, a proof of a formula constructs a "basis" to decide this formula and calls some proper axioms to check the consistency of the formula.

Linear Logic Rules. Rules[4] of linear logic (see annex A) are kept to deal with the structural aspect.

Proper Axioms. Since the theory of real closed fields, is complete, the proof system is extended with new axioms[5] that are true in the real closed field theory, and that are constructed from formula computations and decisions in the real closed fields. The proof system needs some proper axioms as the real closed field theory does.

Satisfiability. If an atomic sentence A is inconsistent in the real closed field theory (when the connectors are interpreted as classical connectors) then a "semantical decision" may be noted in the proof system : $A \tilde{\vdash} \lor$. Inconsistent formulas (as $\exists x \; P(x) = 0$) may be found with Sturm's sequences. The axiom $\exists x \; P(x) = 0 \tilde{\vdash} \lor$ is added if there is no root or $\exists x \; (P(x) = 0) \otimes (a < x < b) \tilde{\vdash} \lor$ if there is no root in the interval]a,b[. Since a sign assignment (for one or several polynomials) upon the roots of an other polynomial may also be found inconsistent, the following axiom is added $\exists x \; (P(x) = 0) \otimes (Q(x) > 0) \tilde{\vdash} \lor$.

Composition. The connectors are linked to some semantical meanings. Combinations of structures, denoted by connectors, induce algebraic operations. An atom from a new algebra is constructed with a new algebraic expression from atoms of two combined algebras. New factors are constructed like a kind of subformula. Proper axioms give a way to introduce some sub-terms, for example : gcd is a common factor of two polynomials but not a sub-formula.

Implication and Properties. Some properties (of a semi-algebraic set) that are true in the real closed field theory but introduce a kind of weakening, may be found, which we call a "semantical deduction". For instance, with the properties of fields, the sequent, $P(x) = 0 \tilde{\vdash} Q(x).P(x) = 0$, may be stated, or with the ordering $P(x) > 0 \tilde{\vdash} Q(x)^2.P(x) > 0$ and $P(x) > 0 \tilde{\vdash} Q(x)^2 + P(x) > 0$, or a simplified expression as $P(x)^2 = 0 \tilde{\vdash} P(x) = 0$, or a weakening as $(P(x) > 0) \otimes (Q(x) > 0) \tilde{\vdash} P(x) > 0$

This kind of properties is noted in the proof system in the following way : $A \tilde{\vdash} a$ where a is shown upon A.

---

[4] the ? and the ! are excluded of the system as the proper axioms deal with this kind of things.

[5] As axioms, they may be stated as often as necessary

Proof System and Algorithm.

Local Properties. Each object have properties, that can be verified in a smaller context. These properties may be denoted for an object, for instance S (a semi-algebraic set) which has property $\Phi : S \to \Phi$

Every formula Φ of a set Γ will be true or false (in a global way) on a semi-algebraic set S of a decomposition constructed to fit Γ. So $\frac{S \vdash \Phi}{S \vdash (\Phi \oplus \Phi)}$ or $\frac{S \vdash \Phi}{S \vdash (\Phi \oplus \Phi)}$

This notation is useful to reason upon a decomposition. If Sigma and Tau are formulas associated to semi-algebraic sets where the formula Delta is true, we could write :

$$\frac{\Sigma \vdash \Delta \quad T \vdash \Delta}{\Sigma \oplus T \vdash \Delta}$$

Composition by Cut. The connector $\oplus$ has an extension property :

$$\frac{\Gamma \vdash (x^2 - 1) = 0 \oplus (x - 4) = 0}{\Gamma \vdash (x^2 - 1) = 0 \oplus (x - 4) = 0 \oplus (x - 5) = 0}$$

Thus, locally handled questions may be included with a "cut" to a global setting,

$$\frac{\Gamma \vdash A \oplus B, \Delta \quad \dfrac{A \vdash \vee \quad B \vdash B}{\dfrac{A \vdash \vee \oplus B \quad B \vdash \vee \oplus B}{A \oplus B \vdash B}}}{\Gamma \vdash B, \Delta}$$

This cut done with an inconsistent formula is called "simplification" : it may be done to discard inconsistent sets. There is no lose of information.

But if "semantical deduction" are done in a more extensive way, there are lost informations,

$$\frac{\Gamma \vdash A \oplus B, \Delta \quad A \vdash a}{\Gamma \vdash a \oplus B, \Delta}$$

as in proof with weakening.

Adequacy. Algebra have to be appropriated to the connectors : if a formula is associated with a void "object" (in a structure) then all tensorizations from this and other formulas have to be associated to void objects too. This is called adequacy rule : $A \otimes \vee \vdash \vee$ This means that the logic proof system is appropriated to the semantic of real closed fields, and gives the soundness of the system with simplification.

3 Commutation and Soundness

The proof system handles context in linear logic, and may handle properties given in the real closed fields, as proper axioms. These proper axioms commute with the other rules, and this extended linear logic proof system remains sound.

3.1 Commutation Lemmas

Permutation lemmas show that a "semantic deduction" may be done also after linear rules, without altering the system : the same sequent is obtained as conclusion of the proof, and the proof is derived from the same proofs δ_1 and δ_2. All these results are also true for simplification ($A\widetilde{\vdash}F$ has to be replaced by $A\widetilde{\vdash}\vee$)

Permutation Lemmas. We state here only one permutation lemma for "tensor", but they are similar for "par", "plus" , "with", "non"

Lemma 2. *permutation of semantic deduction "cut" and tensorization $(\otimes)$.*

$$
\cfrac{
\cfrac{\delta_1}{\Gamma_1 \vdash \oplus_{i\in I}A_i} \quad
\cfrac{
\cfrac{A_j \vdash A_j}{A_j \vdash T\oplus F} \quad \cfrac{A_k \vdash F}{A_k \vdash T\oplus F}
}{\oplus_{i\in I}A_i \vdash T\oplus F}
}{
\cfrac{\Gamma_1 \vdash \left(\oplus_{i\in I-\{k\}}A_i\right)\oplus F \quad \cfrac{\delta_2}{\Gamma_2 \vdash \oplus_{j\in J}B_j}}{\Gamma_1,\Gamma_2 \vdash \left(\oplus_{i,j\in(I-\{k\})\times J}(A_i\otimes B_j)\right)\oplus\left(\oplus_{j\in J}(F\otimes B_j)\right)}
}
$$

is tranformed into

$$
\cfrac{
\cfrac{\delta_1}{\Gamma_1 \vdash \oplus_{i\in I}A_i} \ \cfrac{\delta_2}{\Gamma_2 \vdash \oplus_{j\in J}B_j}
}{\Gamma_1,\Gamma_2 \vdash \oplus_{i,j\in I\times J}(A_i\otimes B_j)} \quad
\cfrac{
\cfrac{\cfrac{A_i\vdash A_i \ B_j\vdash B_j}{A_i\otimes B_j \vdash A_i\otimes B_j} \ \cfrac{A_k\vdash F \ B_l\vdash B_l}{A_k\otimes B_l \vdash F\otimes B_l}}{\cfrac{(A_i\otimes B_j)\vdash S\oplus R \ (A_k\otimes B_l)\vdash S\oplus R}{\oplus_{i,j\in I\times J}(A_i\otimes B_j)\vdash S\oplus R}}
}{}
$$
$$
\overline{\Gamma_1,\Gamma_2 \vdash \left(\oplus_{i,j\in(I-\{k\})\times J}(A_i\otimes B_j)\right)\oplus\left(\oplus_{j\in J}(F\otimes B_j)\right)}
$$

where T stands for $\left(\oplus_{i\in I-\{k\}}A_i\right)$ and $S\oplus R$ stands for

$$\left(\oplus_{i,j\in(I-\{k\})\times J}(A_i\otimes B_j)\right)\oplus\left(\oplus_{j\in J}(F\otimes B_j)\right)$$

Simplified Form Lemma.

Lemma 3. *simplified forms may be composed to retrieve more complex equivalent forms.*

If $A\widetilde{\vdash}(A\otimes B)\&(A\otimes C)$ then

$$
\cfrac{\cfrac{A\widetilde{\vdash}A\otimes B \quad C\vdash C}{A\otimes C \vdash A\otimes B\otimes C} \quad A\widetilde{\vdash}A\otimes C}{A \vdash A\otimes B\otimes C}
$$

Remark. If a simplified form is introduced by a "cut" then the initial form may be retrieve by an other "cut", as a semantic deduction works on both ways. Both expressions are equivalent and, by permutation theorem, the "cuts" may be done at the same level and then eliminated.

3.2 Downward and Upward Moves in Proof

If the adequacy rule holds, combined algebra systems behave in the same way whether simplifications are done before or after the combinations.

Theorem 4. *if a "cut", obtained from a "semantic deduction", is possible, then it may be done after all the following rules.*

Corollary 5. *Upwards deduction moves. A simplification or a semantic deduction may be moved upwards as far as the deduction in the real closed fields remains true.*

If $F \otimes G \overset{\sim}{\vdash} H$ and if $F \overset{\sim}{\vdash} K$ and if $K \otimes G \overset{\sim}{\vdash} H$ then :

$$\frac{\dfrac{\Gamma_1 \vdash F \quad \Gamma_2 \vdash G}{\Gamma_1, \Gamma_2 \vdash F \otimes G} \quad F \otimes G \overset{\sim}{\vdash} H}{\Gamma_1, \Gamma_2 \vdash H}$$

is transformed in (moving "upward" the cut) :

$$\frac{\dfrac{\dfrac{\Gamma_1 \vdash F \quad F \overset{\sim}{\vdash} K}{\Gamma_1 \vdash K} \quad \Gamma_2 \vdash G}{\Gamma_1, \Gamma_2 \vdash K \otimes G} \quad K \otimes G \overset{\sim}{\vdash} H}{\Gamma_1, \Gamma_2 \vdash H}$$

Remark. In this system the "cut" elimination is not achieved. "Cuts" are necessary for "semantic deductions", as results (solving an equation system, or deciding a sign assignment) are consequences of several conditions : the upward move is then stopped.

Remark. Semantical deductions are a kind of weakening, so the former structure may be not retrieved. There are lost informations. But, if semantical deductions are restrained to simplifications (it gives what we call a "global" structure), then structures, which have all the properties of the context (for instance decision features), are kept and constructed.

Lemma 6. *global structures and simplifications. If two structures $\Gamma_1 \vdash \oplus_{i \in I} A_i$ and $\Gamma_2 \vdash \oplus_{j \in J} B_j$ are "global" (i.e constructed only with simplification "cuts" and linear logic rules) then $\Gamma_1, \Gamma_2 \vdash \oplus_{i,j \in I \times J} (A_i \otimes B_j)$ is global, and the simplifications done to this system construct a global system too.*

3.3 Soundness

These lemmas show that the proof system is sound (if the adequacy rule holds).

Corollary 7. *Form of a proof. A proof may be written with all the calculus done in linear logic, without "cut" and then all "semantic cuts" are done.*

Corollary 8. *The system is sound*

Proof. The proof is written with all the linear logic rules at the top and then "semantic cuts". No contradiction may be derived in the linear logic part. And no contradiction may be derived in the "cut" part as the semantic deduction may be translated with the usual connectors in the theory of real closed fields, which is sound[6].

4 System of Polynomial Roots and Sign Assignments

The linear logic handling of structures gives a frame to computations and proofs for decompositions. The proof system is a natural way to express algorithms to compute a relatively prime polynomial decomposition, or to decide all the consistent sign assignments. In this fashion all the consistent semi-algebraic sets may be found in the extended Rcf+LL proof system, and any decomposition may be constructed with the proof system : so all formulas of real closed fields may be decided in this extended proof system.

4.1 Algebra of Polynomial Roots and Sign Assignments

In this part, we study sign assignments upon roots for a given polynomial set (and to simplify, the polynomials are with the same variable). We study two kinds of algebra, the root algebra (the formulas are only with equalities), and the sign assignment algebra.

Algebra of Polynomial Roots. A set of square free polynomials defines a set of roots (in the real closure). The Boolean algebra upon these roots is defined in the following way : let I be a set of square free polynomials and $\theta(p)$ be the set of roots of a polynomial p

$$\bigcap_{p \in I} \theta(p) = \theta(\gcd(I))$$

$$\bigcup_{p \in I} \theta(p) = \theta(\mathrm{lcm}(I))$$

$$\neg\theta(p) = \theta(\mathrm{lcm}(\Sigma)/p)$$

where Σ is the set of all polynomials taken into account.

The set of the atoms of this algebra is a pairwise relatively prime polynomial decomposition of Sigma, and in the linear logic analysis there is an other "atom" which gives an extension possibility to the algebra.

---

[6] The linear proof system is a more acute proof system than the one with classical connectors. In the Gentzen's sequent formalism it is clear that all linear proofs hold as classical ones.

Algebra of Sign Assignments. A sign assignment is the n-upple formed with 1 or −1 or 0, where 1 in position i denotes that, for α a real, $(q_i(\alpha) > 0)$, and -1 that $(q_i(\alpha) < 0)$, and 0 that $(q_i(\alpha) = 0)$. The study deals with the sign assignments upon p (a square free polynomial in $\mathbb{R}[X]$)of polynomial sets $\{q/q_i \in \mathbb{Q}[X], i \in \Sigma\}$ where Σ is a finite part of $\mathbb{N}$ (integer), and the q_i are square-free polynomials. A set of roots of p is associated to each sign assignment (it may be void).

A Boolean algebra is constructed on roots of p as in [BKR]). Let :

- $c_{p\otimes q+} = \{a/p(a) = 0 \text{ and } q(a) > 0\}$
- $c_{p\otimes q-} = \{a/p(a) = 0 \text{ and } q(a) < 0\}$
- $c_{p\otimes q} = \{a/p(a) = 0 \text{ and } q(a) = 0\}$

All the atoms of (Σ) are given by the choice of three sets I , I' and I'' such that : $I \cup I' \cup I'' = \Sigma$ and $I \cap I' = I' \cap I'' = I \cap I'' = \vee$. Atoms are :

$$\bigcap_{q\in I} c_{p\otimes q+} \cap \bigcap_{q\in I'} c_{p\otimes q-} \cap \bigcap_{q\in I''} c_{p\otimes q}$$

which are generators of a Boolean algebra B.

4.2 Proper Axioms and Connectors

In this frame, proper axioms give an interpretation of the multiplicative connectors, and parallel combinations may be expressed. This is a generalization of [BKR]'s result into a formal proof system.

Decision. Sturm's sequences give the number of roots of a square-free polynomial, and so give a way to decide formula as : $\exists x\ f(x) = 0$. The sequent $S(P, P', a, b) > 0 \vdash \exists x\ (P(x) = 0) \otimes (a < x < b)$ holds. If the roots are ordered, we get "a primary deduction", for instance : $(x^2 - 1) = 0 \vdash (x-1) = 0 \oplus (x+1) = 0$. If there is no root, we get : $(x^2 + 1) = 0 \vdash \vee$ which may be simplified. The following axioms are added to deal with inequalities and to decide of sign assignments upon the roots of a polynomial Q, for instance :

$$\begin{pmatrix} (P(x) = 0) \\ \oplus (P(x) > 0) \\ \oplus (P(x) < 0) \end{pmatrix} \otimes (Q(x) = 0) \widetilde{\vdash} \qquad \begin{matrix} (P(x) = 0) \otimes (Q(x) = 0) \\ \oplus \vee \\ \oplus (P(x) < 0) \otimes (Q(x) = 0) \end{matrix}$$

Composition.

tensor $\otimes$. As greatest common divisor (gcd) for two polynomials may be computed in the real closed field theory, the following axiom is added : $(P(x) = 0) \otimes (Q(x) = 0) \widetilde{\vdash} \gcd(P, Q)(x) = 0$

example : $(x^2 - 1)(x - 4) = 0 \otimes (x^2 - 1)(x - 5) = 0 \widetilde{\vdash} (x^2 - 1) = 0$. Thus, expressions may be simplified, and if the polynomials are relatively pairwise prime it gives 1, and as $1 = 0$ is void $(1 = 0) \widetilde{\vdash} \vee$, a simplification may be done.

the "par" $\wp$. It is interpreted as the least common multiple (lcm) of two polynomials, which is computable in the theory of real closed fields : $(P(x) = 0)\wp(Q(x) = 0)\tilde{\vdash}\mathrm{lcm}(P,Q)(x) = 0$

example : $(x^2 - 1)(x - 4) = 0\wp(x^2 - 1)(x - 5) = 0\tilde{\vdash}(x^2 - 1)(x - 4)(x - 5) = 0$

negation . Negation is interpreted as Euclid's division on polynomials (when it can be computed) : $(P(x) = 0) \otimes \Upsilon(Q(x) = 0)\tilde{\vdash}\frac{P(x)}{\gcd(P,Q)(x)} = 0$. Otherwise ($\Upsilon(P(x) = 0) \otimes \Upsilon(Q(x) = 0)$) has a formal meaning to compute system extensions. So negation has two meanings : subtraction of roots and system extension possibility.

example : $(x^2 - 1)(x - 4) = 0 \otimes^\sim[(x^2 - 1)(x - 5) = 0]\tilde{\vdash}\frac{(x^2-1)(x-4)}{(x^2-1)} = 0$. Negation is compatible with "par" and verifies that $\Upsilon(g\wp h)$ is equivalent to $\tilde{g} \otimes \tilde{h}$.

Proposition 9. *Multiplicative equality formulas are equivalent to an atomic formula (a polynomial equation, called positive atom, or the negation of a polynomial equation, called negative atom)*

Adequacy. Adequacy rule : $A \otimes \vee \tilde{\vdash} \vee$ is verified as the void set is associated to $1 = 0$, and as for any polynomial f $\gcd(1, f) = 1$ (thus, $(1 = 0) \otimes (f(x) = 0)\tilde{\vdash}\vee$ which give simpler expressions[7]).

4.3 Decomposition

The algorithm of decomposition of a set of square free polynomials into a set of relatively pairwise prime polynomials is obtained following the proof of the sequent : $(f \oplus \tilde{f}) \otimes (g \oplus \tilde{g}) \ldots \otimes (h \oplus \tilde{h}) \vdash (f \otimes g \ldots \otimes h) \oplus (\tilde{f} \otimes g \ldots \otimes h) \oplus (f \otimes \tilde{g} \ldots \otimes h) \oplus (\tilde{f} \otimes \tilde{g} \ldots \otimes h) \oplus \ldots (\tilde{f} \otimes \tilde{g} \ldots \otimes \tilde{h})$ where $f, g \ldots h$ are atomic predicates (like $P(x) = 0$), and the right hand side of the sequent is the set of all the relatively pairwise prime polynomials expressed as multiplicative formulas and a formula which gives a way to further extensions. Each sequent is associated to a structure : a decomposition complex of semi-algebraic sets and the context of this decomposition. For sign assignments, the form is the same but the first step differs (there are three cases) and inequalities appear. Linear logic gives a way to handle several "objects"(of a decomposition) and thus, to combine two decompositions.

Logical Construction. The set of atoms, for a given context, is constructed in the following way :

1. first step : for each polynomial, the sequent $f \oplus \tilde{f} \vdash f \oplus \tilde{f}$ defines all the atoms in the context of a polynomial with the possibility to extend the structure.

---

[7] In this way, the proof system is alike classical logic, but linear logic gives a way to handle several polynomials together. Thus proofs of simplification and of correctness for algorithms are achieved.

2. tensorization : Two structures may be tensorized and by distributivity (from $\otimes$ upon $\oplus$) the atoms of the new structure are computed (in the right part of the sequent) :

$$\frac{\Gamma_1 \vdash (\oplus f_i) \oplus 0_1 \quad \Gamma_2 \vdash (\oplus g_j) \oplus 0_2}{\Gamma_1, \Gamma_2 \vdash ((\oplus f_i) \otimes (\oplus g_j)) \oplus ((\oplus f_i) \otimes 0_2) \oplus (0_1 \otimes (\oplus g_j)) \oplus (0_1 \otimes 0_2)}$$

where $0_1, 0_2$ mean the purely negative formulas got for each system, f_i and g_j are the atoms of the two former structures.

Simplified expressions may be obtained when "cuts" deduced from proper axioms are possible.

example : the decomposition of $h(x) = (x+1)(x-4)(x-7)$ and $k(x) = (x^2-1)(x-6)$ is given by :

$$(h \oplus \tilde{h}) \otimes (k \oplus \tilde{k}) \mid ((x+1) = 0) \oplus ((x-1)(x-6) = 0) \uplus ((x-4)(x-7) = 0) \oplus (\tilde{h} \otimes \tilde{h})$$

If it is combined with the result of the tensorization of $f(x) = (x^2-1)(x-4)$ and $g(x) = (x^2-1)(x-5)$,

$$(f \oplus \tilde{f}) \otimes (g \oplus \tilde{g}) \vdash ((x^2-1) = 0) \oplus ((x-5) = 0) \oplus ((x-4) = 0) \oplus (\tilde{f} \otimes \tilde{g})$$

It yields :

$$\begin{array}{c} (f \oplus \tilde{f}) \otimes (g \oplus \tilde{g}) \\ \otimes (h \oplus \tilde{h}) \otimes (k \oplus \tilde{k}) \end{array} \vdash \begin{bmatrix} ((x^2-1) = 0) \\ \oplus ((x-5) = 0) \\ \oplus ((x-4) = 0) \\ \oplus (\tilde{f} \otimes \tilde{g}) \end{bmatrix} \otimes \begin{bmatrix} ((x+1) = 0) \\ \oplus ((x-1)(x-6) = 0) \\ \oplus ((x-4)(x-7) = 0) \\ \oplus (\tilde{h} \otimes \tilde{k}) \end{bmatrix}$$

which is reduced to :

$$\begin{array}{llllll}
(f \oplus \tilde{f}) & ((x+1) = 0) & \oplus ((x-1) = 0) & \oplus \vee & & \oplus \vee \\
\otimes (g \oplus \tilde{g}) & \oplus \vee & \oplus \vee & \oplus \vee & & \oplus ((x-5) = 0) \\
\otimes (h \oplus \tilde{h}) \ \vdash & \oplus \vee & \oplus \vee & \oplus ((x-4) = 0) & & \oplus \vee \\
\otimes (k \oplus \tilde{k}) & \oplus \vee & \oplus ((x-6) = 0) & \oplus ((x-7) = 0) & \oplus (\tilde{f} \otimes \tilde{g} \otimes \tilde{h} \otimes \tilde{k})
\end{array}$$

Decomposition Fit to a Formula. A complete decomposition may be useless, it would be more relevant to construct an atomic decomposition fit to a formula. For instance, with the formula $(f(x) = 0 \otimes g(x) = 0) \oplus h(x) = 0 \oplus k(x) = 0$, decomposition with the formulas, $f(x) = 0 \otimes g(x) = 0$, $h(x) = 0$, $k(x) = 0$ would be better than to handle independently all polynomials f, g, h, k . So, with the polynomials of the former decomposition, the following decomposition is obtained :

$$\begin{bmatrix} (h \oplus \tilde{h}) \\ (k \oplus \tilde{k}) \end{bmatrix} \otimes [(f \otimes g) \oplus (f \otimes g)] \vdash \begin{bmatrix} ((x+1) = 0) \\ \oplus ((x-1)(x-6) = 0) \\ \oplus ((x-4)(x-7) = 0) \\ \oplus (\tilde{h} \otimes \tilde{k}) \end{bmatrix} \otimes [((x^2-1) = 0) \oplus (\tilde{f} \tilde{\rho} \tilde{g})]$$

and we get :

$$\left[\begin{matrix}(h \oplus \widetilde{h}) \\ (k \oplus \widetilde{k})\end{matrix}\right] \otimes \left[(f \otimes g) \oplus (\widetilde{f} \otimes g)\right] \vdash \begin{matrix} (x+1) = 0 \\ \oplus(x-1) = 0) \\ \oplus\vee \\ \oplus\vee \end{matrix} \quad \begin{matrix} \oplus\vee \\ \oplus((x-6) = 0) \\ \oplus((x-4)(x-7) = 0) \\ \oplus(\widetilde{f} \otimes \widetilde{g} \otimes (\widetilde{f\wp}g)) \end{matrix}$$

instead of the former expression. So the relevance and meaning of linear negations are emphasized.

5 Matrix Form of Logical Operation

A matrix interpretation, constructed to fit to the linear logic decomposition, may be expressed. The linear system has the same simplification properties as the proof system (i.e if the cardinal number of a set associated to a formula is null, this formula may be simplified in the matrix).

5.1 Matrix Expression of the Pairwise Prime Polynomial Decomposition

The formula $p \oplus \widetilde{p}$, where p is the formula $P(x) = 0$, is such that the cardinal number of p, $|c_p|$, may be computed but the cardinal number of $\widetilde{p}$, $|c_{\widetilde{p}}|$, can not be computed. Thus, for a polynomial, the following matrix equality holds (where T is an expression such that $T(1,1)$ [8] can not be computed, and $T(p,q) = S(p, p'.q)$ the usual Sturm's sequence) :

$$\begin{pmatrix} 1 & 0 \\ 1 & 1 \end{pmatrix} \begin{pmatrix} |c_p| \\ |c_{\widetilde{p}}| \end{pmatrix} = \begin{pmatrix} S(p, p') \\ T(1,1) \end{pmatrix}$$

Then, with matrix tensor product, the following expression is obtained :

$$\begin{pmatrix} 1 & 0 & 0 & 0 \\ 1 & 1 & 0 & 0 \\ 1 & 0 & 1 & 0 \\ 1 & 1 & 1 & 1 \end{pmatrix} \begin{pmatrix} |c_{p \otimes q}| \\ |c_{p \otimes \widetilde{q}}| \\ |c_{\widetilde{p} \otimes q}| \\ |c_{\widetilde{p} \otimes \widetilde{q}}| \end{pmatrix} = \begin{pmatrix} S(p \otimes q, (p \otimes q)') \\ S(p, p') \\ S(q, q') \\ T(1,1) \end{pmatrix}$$

which computes the same atoms as the "logical tensorization".

Solving partially the linear system gives, for each formula, whether there are some roots or not. It is a kind of decision. This matrix expression may be simplified too, as in [BKR]. Formulas associated with the void set, may be discarded. If an expression is of cardinal number 0, the corresponding column, and a linked line may be deleted. The last line is a formal one that can not be simplified because it denotes an infinity of potential roots.

---

[8] It is a notation of Sturm's sequences, with two places : $T(p,q) = S(p, p'.q)$. The first place is expressed in linear logic (here, 1 means $1 \otimes A \equiv A$,) the second with the usual polynomials (here, 1 is such that $1.f = f$).

5.2 Proper Axioms and Connectors as a Matrix Expression for Sign Assignments

The proper axioms for sign assignments may be associated to computations with the following matrix.

Matrix form for Axiom. This linear system may be generalized to compute any atom with the following first step matrix :

$$
\begin{pmatrix} 1 & 0 & 0 \\ 1 & 1 & 1 \\ 0 & 1 & -1 \end{pmatrix}
\begin{pmatrix} |c_p| \\ |c_{p+}| \\ |c_{p-}| \end{pmatrix}
=
\begin{pmatrix} S(p,p') \\ T(1,1) \\ T(1,p) \end{pmatrix}
$$

Composition. [BKR] have shown that the tensorization of two matrix systems of sign assignments upon the same polynomial (and it holds also upon the greatest common divisor (gcd) [9] of the two polynomials upon which the sign assignment matrix were constructed) constructs a new system, that gives the cardinal number of the tensorization (in the logic meaning) of atoms. In the general case, with square free polynomials, we get after tensorization :

$$
\begin{pmatrix}
1 & 0 & 0 & & & & & \\
1 & 1 & 1 & 0 & & & 0 & \\
0 & 1 & -1 & & & & & \\
1 & 0 & 0 & 1 & 0 & 0 & 1 & 0 & 0 \\
1 & 1 & 1 & 1 & 1 & 1 & 1 & 1 & 1 \\
0 & 1 & -1 & 0 & 1 & -1 & 0 & 1 & -1 \\
 & & & 1 & 0 & 0 & -1 & 0 & 0 \\
0 & & & 1 & 1 & 1 & -1 & -1 & -1 \\
 & & & 0 & 1 & -1 & 0 & -1 & 1
\end{pmatrix}
\begin{pmatrix}
|c_{p\otimes q}| \\
|c_{p\otimes q+}| \\
|c_{p\otimes q-}| \\
|c_{p+\otimes q}| \\
|c_{p+\otimes q+}| \\
|c_{p+\otimes q-}| \\
|c_{p-\otimes q}| \\
|c_{p-\otimes q+}| \\
|c_{p-\otimes q-}|
\end{pmatrix}
=
\begin{pmatrix}
T(p\otimes q,1) \\
T(p\otimes 1,1) \\
T(p\otimes 1,1.q) \\
T(q,1) \\
T(1,1) \\
T(1,q) \\
T(q,p) \\
T(1,p) \\
T(1,pq)
\end{pmatrix}
$$

It gives [BKR] 's matrix system as a sub-matrix which computes the cardinal number of roots of p for a given sign of q (a sub matrix gives signs of p upon roots of q as well) and so on. The decision sequent associated to

$$
\left[\begin{array}{l}
((q_1(x)>0)\otimes(q_2(x)>0)) \quad \oplus ((q_1(x)>0)\otimes(q_2(x)<0)) \\
\oplus ((q_1(x)<0)\otimes(q_2(x)>0)) \oplus ((q_1(x)<0)\otimes(q_2(x)<0))
\end{array} \right] \otimes (p(x)=0)
$$

has a matrix equivalent

$$
\begin{pmatrix}
1 & 1 & 1 & 1 \\
1 & -1 & 1 & -1 \\
1 & 1 & -1 & -1 \\
1 & -1 & -1 & 1
\end{pmatrix}
\begin{pmatrix}
|c_{p\otimes q_1^+} \cap c_{p\otimes q_2^+}| \\
|c_{p\otimes q_1^-} \cap c_{p\otimes q_2^+}| \\
|c_{p\otimes q_1^+} \cap c_{p\otimes q_2^-}| \\
|c_{p\otimes q_1^-} \cap c_{p\otimes q_2^-}|
\end{pmatrix}
=
\begin{pmatrix}
T(p,1) \\
T(p,q_1) \\
T(p,q_2) \\
T(p,q_1 \cdot q_2)
\end{pmatrix}
$$

---

[9] Since all roots of the gcd of A and B are roots of A, the matrix form for A is adapted to gcd(A,B)

Matrix form for Decisions. There is no way to compute the cardinal number of a single atom of the sign assignment algebra, as in the root algebra, but solving a matrix system gives the cardinal number of several atoms simultaneously. The following linear system computes the cardinal number of $(p(x) = 0) \otimes (q(x) > 0)$ and $(p(x) = 0) \otimes (q(x) < 0)$:

$$\begin{pmatrix} 1 & 1 \\ 1 & -1 \end{pmatrix} \begin{pmatrix} |c_{p \otimes q^+}| \\ |c_{p \otimes q^-}| \end{pmatrix} = \begin{pmatrix} S(p, p') \\ S(p, p'.q) \end{pmatrix}$$

if p and q are relatively prime.

The tensorization of such matrix, constructs a linear system which solutions gives a way to find consistent or non-consistent sets. Then, simplifications may be done to discard inconsistent sets.

The matrix system is a partial matrix system of the composition matrix . But in this matrix, some expressions can not be computed as there is no root upon which the sign assignment is defined yet, but they may be useful in a subsequent tensorization.

Adequacy. All the simplification properties hold, as in [BKR]. The matrix simplification of void formulas is linked to a logical simplification and it is fitted to the system. Adequacy rule holds.

So we get a global matrix representation of both the algebra of roots and the algebra of sign assignments.

6 Dynamic Cylindrical Algebraic Decomposition

A cylindrical algebraic decomposition handles many "objects", polynomials and cells which may be combined into a structure. We state properties of this decomposition with the linear connectors, and we construct a dynamic logic system to combine cell decompositions.

6.1 Cells and their Properties

Cells. In the language of Rcf+LL we study intervals and connected sets of $\mathbb{R}^n$. We introduce a slightly different notion of cell than the Collins'one.

Definition 10. A cell is a connected semi-algebraic set.

If S is a cell in $\mathbb{R}^n$ and F is the formula with $x_1 \ldots x_n$ as free variables associated to the cell then $\overline{x} \in S \leftrightarrow F(\overline{x})$

Remark. A cell is a connected set, so the associated formula has to include some coding as in Thom's lemma or other sub-formulas that assert the sign invariance of coefficients.

Remark. As for a semi-algebraic set, the dimension of a cell is less or equal than the number of free variables in the context.

Here all formulas are assumed without quantifier, or in prenex form and with the variables in a fixed order.

Sign Invariance upon a Cell or a Complex of Cells

Definition 11. a cell S is f-invariant if the polynomial f has a constant sign upon S, $\forall x, y \in S \ f(x).f(y) = 0$ or $\forall x, y \in S \ f(x).f(y) > 0$.

A formula of Rcf+LL states the sign invariance. If Σ defines the cell S and S is sign invariant for f then :

$$\frac{\Sigma \vdash f(x) = 0}{\Sigma \vdash (f(x) = 0) \oplus (f(x) < 0) \oplus (f(x) > 0)}$$

or

$$\frac{\Sigma \vdash f(x) < 0}{\Sigma \vdash (f(x) = 0) \oplus (f(x) < 0) \oplus (f(x) > 0)}$$

or

$$\frac{\Sigma \vdash f(x) > 0}{\Sigma \vdash (f(x) = 0) \oplus (f(x) < 0) \oplus (f(x) > 0)}$$

The formula $(f(x) = 0) \oplus (f(x) < 0) \oplus (f(x) > 0)$ defines sign invariance for a polynomial f, and tensorizations of such formulas define it for several polynomials.

$$[(f(x) = 0) \oplus (f(x) < 0) \oplus (f(x) > 0)] \otimes [(f(x) = 0) \oplus (f(x) < 0) \oplus (f(x) > 0)]$$

A cell S that verifies such a formula is invariant for the set $\{f, g\}$.

Definition 12. let A be a set of polynomials ; a set S is A-invariant if each polynomial of A has a constant sign upon S.

A set S, sign-invariant for a polynomial set A, verifies a given sign assignment for polynomials of A.

Function Fit to a Domain. Linear logic enables to work with functions linked to a definition domain (a context). So each polynomial is associated to a domain (a validity cell) and thus to a context, where it is useful and computable : $C \to B$. We get formula as : $A \to (C \to B)$ which is equivalent to $A \otimes C \to B$, in linear logic, so the context usefulness may be composed. As $(A \to B) \otimes (C \to D) \vdash (A \otimes C) \to (B \otimes D)$ the usual tensorization combines the domains, for instance to compute common roots for two functions.

6.2 Proper Axioms and Connectors.

The former proper axioms with Sturm's sequences remain but the system is extended with delineability to handle each root independently.

Delineability and Conditions of Signs.

Delineability and Conditions of Signs for a Function.

Definition 13. The roots of f are delineable upon a cell S, if there is a set of functions ζ_i such that :

1. the ζ_i are continuous functions from S to $\mathbb{C}$ (the complex numbers)
2. for all i $\exists e_i$ such that $\zeta_i(a_1, \ldots, a_{n-1})$ is the root of $f(a_1, \ldots, a_{n-1}, x)$ for $(a_1, \ldots, a_{n-1}) \in S$ of multiplicity e_i
3. if b is a root of $f(a_1, \ldots, a_{n-1}, x)$ for $(a_1, \ldots, a_{n-1}) \in S$ then $\exists \zeta_i$ s. t. $\zeta_i(a_1, \ldots, a_{n-1}) = b$
4. $\exists k$ such that $\zeta_1, \ldots, \zeta_k$ are real functions and $\zeta_{k+1}, \ldots, \zeta_m$ are non-real ones.

A similar definition is stated with linear connectors in the following way[10] :

$$\vdash \forall y \forall x \, S(y) \rightarrow \left[\begin{array}{c} f(x,y) = 0 \rightarrow \left[\bigoplus_{i=1}^{k}(\xi_i(y) = x)\right] \otimes S(y) \\ \& \, (\xi_i(y) = x) \rightarrow f(x,y) = 0 \\ \& \, \&_{i=1}^{k-1} \, ([\xi_i(y) < \xi_{i+1}(y)]) \end{array} \right]$$

The functions ξ_i for $1 \leq i \leq k$ are totally ordered[11]. They induce a sign assignment in the cylinder $S \times \mathbb{R}$

$$\vdash \left[\begin{array}{c} \left(f(x,y) = 0 \rightarrow \left[\bigoplus_{i=1}^{k}(\xi_i(y) = x)\right] \otimes S(y) \right) \\ \& \, (\xi_i(y) = x) \rightarrow f(x,y) = 0 \\ \& \, \&_{i=1}^{k-1} \, ([\xi_i(y) < \xi_{i+1}(y)]) \end{array} \right. \\ \left. \rightarrow \left[\begin{array}{c} (f(x,y) = 0 \oplus f(x,y) > 0 \oplus f(x,y) < 0) \\ \rightarrow [\bigoplus_{i=1}^{k}(\xi_i(y) = x) \oplus \bigoplus_{i=0}^{k}(\xi_i(y) < x < \xi_{i+1}(y))] \otimes S(y) \\ \& \, (\xi_i(y) = x) \rightarrow f(x,y) = 0 \\ \& \, \&_{i=0}^{k} \, ([\xi_i(y) < x < \xi_{i+1}(y)] \rightarrow [f(x,y) > 0 \oplus f(x,y) < 0]) \end{array} \right] \right]$$

Some sub-formulas as

$$\vdash S(y) \rightarrow ([\xi_i(y) < x < \xi_{i+1}(y)] \rightarrow [f(x,y) > 0 \oplus f(x,y) < 0])$$

or

$$\vdash S(y) \rightarrow ((\xi_i(y) = x) \rightarrow f(x,y) = 0)$$

are useful to find a cell[12] which verifies a sign invariance formula, with the transformation :

$$S(y) \otimes [\xi_i(y) < x < \xi_{i+1}(y)] \vdash [f(x,y) > 0 \oplus f(x,y) < 0]$$

---

[10] This is an axiom scheme for any polynomial and it gives a result about some other semi-algebraic functions (the root functions ξ_i). For $i = 1$ to k $\zeta_i = \xi_i$.

[11] for conveniance $\xi_0(y) = -\infty$ and $\xi_{k+1}(y) = +\infty$

[12] The totality is obtained with the reverse formula :

$$S(y) \vdash (f(x,y) = 0 \oplus f(x,y) > 0 \oplus f(x,y) < 0)$$
$$\rightarrow \left[\bigoplus_{i=1}^{k}(\xi_i(y) = x) \oplus \bigoplus_{i=0}^{k}(\xi_i(y) < x < \xi_{i+1}(y))\right] \otimes S(y)$$

it is equivalent to construct new cells with one more dimension :

$$S(y) \otimes (f(x,y) = 0 \oplus f(x,y) > 0 \oplus f(x,y) < 0)$$
$$\vdash \left[\bigoplus_{i=1}^{k}(\xi_i(y) = x) \oplus \bigoplus_{i=0}^{k}(\xi_i(y) < x < \xi_{i+1}(y))\right] \otimes S(y)$$

Conditions of Delineability for a Function. Sign invariance conditions for the "main" coefficients of polynomial upon a cell C assert delineability upon C of this polynomial [Col].

Lemma 14. *On a cell C, if a polynomial A is of constant degree and the number of roots is constant then roots of A are delineable on C.*

The actual degree of a polynomial and the number of roots (or of multiple roots) is equivalent to the sign invariance of the actual head coefficients of A and of $\gcd(A, A')$. So if $P(x, y) = \sum_{i=0}^{n} a_i(y).x^i$, whenever a cell C verifies one of the following formulas, $a_n(y) \neq 0, a_n(y) = 0 \otimes a_{n-1}(y) \neq 0, \ldots, a_n(y) = 0 \otimes a_{n-1}(y) = 0 \ldots \otimes a_0(y) \neq 0$ then the head coefficient is constant. Moreover, upon such a cell C, if the $\gcd(P, P') = \sum_{i=0}^{m} g_i(y).x^i$, and the cell verifies one of the following formula, $g_m(y) \neq 0, g_m(y) = 0 \otimes g_{m-1}(y) \neq 0, \ldots, g_m(y) = 0 \otimes g_{m-1}(y) = 0 \ldots \otimes g_0(y) \neq 0$ then P is delineable on this cell C. To state these properties the following proper axioms are added to the system.

$$\begin{array}{l} (a_n(y) = 0 \otimes a_{n-1}(y) = 0 \ldots \otimes a_k(y) \neq 0) \\ \otimes (g_m(y) = 0 \otimes g_{m-1}(y) = 0 \ldots \otimes g_l(y) \neq 0) \end{array} \vdash del(P)$$

Delineability for Several Functions. The delineability axiom is :

$$\vdash \exists k \, S(y) \rightarrow \left[\begin{array}{c} (f(x,y) = 0 \oplus g(x,y) = 0) \rightarrow \left[\bigoplus_{i=1}^{k} (\xi_i(y) = x) \right] \otimes S(y) \\ \& \, (\xi_i(y) = x) \rightarrow (f(x,y) = 0 \oplus g(x,y) = 0) \\ \& \, \&_{i=1}^{k-1} \left([\xi_i(y) < \xi_{i+1}(y)] \right) \end{array} \right]$$

Delineability is an additive formula. If two functions f and g have no common root upon $S(y)$ then the roots of both may be ordered[13], and then delineability of each of them imply delineability of both together.

The delineability of several polynomials gives a way to construct cell with the sign invariance property :

$$S(y) \vdash (\xi_i(y) = x) \rightarrow \left(\begin{array}{c} [f(x,y) = 0 \oplus f(x,y) < 0 \oplus f(x,y) > 0] \\ \otimes [g(x,y) = 0 \oplus g(x,y) < 0 \oplus g(x,y) > 0] \end{array} \right)$$

Condition of Delineability for Several Polynomials Together.

Lemma 15. *If roots of P and Q are independently delineable upon C and if the number of their common roots is constant (i.e the degree of the $\gcd(P, Q)$ is constant, i.e sign invariance of the actual head coefficient of gcd) then roots of P and Q are delineable together on C.*

---

[13] Two functions may be delineable upon a cell, and not be delineable together. If they have a common root, it should be upon the whole cell $S(y)$. A pairwise relatively prime decomposition may be done to find all the polynomials that have to be of constant degree and of constant root number. If it is so upon a cell, functions of A and B are delineable together. As the roots of $\gcd(A, B)$ are roots of A, they remain of constant multiplicity on $S(y)$ as roots of A do. If all the common roots remain of same degree upon $S(y)$ the delineability of the whole is possible.

As polynomials are linked to conditions (as their degrees or number of distinct real roots), these conditions are combined and the conditions of the constant common root number is added. If $C_1 \vdash del(P(x,y))$ and if $C_2 \vdash del(Q(x,y))$ and if the $\gcd(P,Q) = \sum_{i=0}^{r} k_i(y).x^i$ then for each actual degree of the gcd $(k_r(y) = 0 \otimes k_{r-1}(y) = 0 \ldots \otimes k_l(y) \neq 0)$, the following proper axiom asserts delineability for $\{P, Q\}$:

$$C_1 \otimes C_2 \otimes (k_r(y) = 0 \otimes k_{r-1}(y) = 0 \ldots \otimes k_l(y) \neq 0) \vdash del(\{P, Q\})$$

Decompositions of Cells and Sign Decisions.

Delineability for functions in one variable. With polynomials in one variable, Sturm's sequences compute the number of their roots, and their roots are ordered,

$$(f(x) = 0)\oplus(f(x) > 0)\oplus(f(x) < 0) \vdash (\oplus_{i=0}^{k}(\xi_i < x < \xi_{i+1}))\oplus(\oplus_{i=1}^{k}(x - \xi_i = 0))$$

Sign of a function upon a cell. If a point of a cell is known, then the sign of all the sign invariant function upon this cell may be found :

$$C(a), (C(x) \rightarrow \begin{bmatrix} f(x) = 0 \\ \oplus f(x) < 0 \\ \oplus f(x) > 0 \end{bmatrix}) \vdash \begin{array}{l} f(a) = 0 \rightarrow [C(x) \rightarrow f(x) = 0] \\ f(a) < 0 \rightarrow [C(x) \rightarrow f(x) < 0] \\ f(a) > 0 \rightarrow [C(x) \rightarrow f(x) > 0] \end{array}$$

6.3 Cell Decomposition

Cylindrical algebraic decomposition is a sign invariant decomposition for all the cells.

Logical Composition. A cell X which verifies some sign condition A for a polynomial set is found. A is a sign assignment formula such that, if a cell verifies this assignment then a function F is delineable upon this cell. The delineability gives new cells (of one more dimension) which verify a sign invariance formula :

$$X(y) \otimes (\xi_i(y) < x < \xi_{i+1}(y)) \vdash (f(x,y) > 0 \oplus f(x,y) < 0)$$

Thus the following deduction is done,

$$\frac{\dfrac{X \vdash A \quad A \vdash del(F)}{X \vdash del(F)} \quad del(F) \vdash \left(\begin{array}{c} \xi_i(y) < x < \xi_{i+1}(y)) \\ \rightarrow (f(x,y) > 0 \oplus f(x,y) < 0) \end{array} \right)}{\dfrac{X \vdash (\xi_i(y) < x < \xi_{i+1}(y)) \rightarrow (f(x,y) > 0 \oplus f(x,y) < 0)}{X \otimes (\xi_i(y) < x < \xi_{i+1}(y)) \vdash (f(x,y) > 0 \oplus f(x,y) < 0)}}$$

With a sample point, the actual sign of the function is found, and some cells are kept with the useful sign conditions.

The first step is done with the axioms of delineability for functions with one variable.

Series of Projections. Conditions of delineability are linked with the actual form of a polynomial upon a cell.

$$\begin{array}{c}(a_n(y) = 0 \otimes a_{n-1}(y) = 0 \ldots \otimes a_k(y) \neq 0) \\ \otimes (g_m(y) = 0 \otimes g_{m-1}(y) = 0 \ldots \otimes g_l(y) \neq 0)\end{array} \vdash del(P)$$

and each condition is necessary for the next set of sign assignments.

$$A \vdash del(\{a_n(y), \ldots a_k(y), g_m(y), \ldots g_l(y)\})$$

Thus,

$$\left(\begin{array}{c}(a_n(y) = 0 \otimes a_{n-1}(y) = 0 \ldots \otimes a_k(y) \neq 0) \\ \otimes (g_m(y) = 0 \otimes g_{m-1}(y) = 0 \ldots \otimes g_l(y) \neq 0)\end{array}\right) \otimes A$$
$$\vdash del(P) \otimes del(\{a_n(y), \ldots a_k(y), g_m(y), \ldots g_l(y)\})$$

is get by the tensorization of delineability conditions. For each variable, there is a tree of conditions, as the actual form of polynomials may changed.

Remark. As a cell is associated to polynomials, the associated polynomials are delineable on the cell (what happens on the cell to the other polynomials is unknown).

Dynamic Construction by Tensorization. With this method, a decomposition may be achieved for a polynomial family A, and then composed with the decomposition obtained for an other family B. The ordering of the variables in the quantifiers has to be the same for the two decompositions.

Each decomposition is linked with a serie of sequents that denote the proper axioms to assert delineability[14] : $A_2 \vdash del(A_1)$, $A_1 \vdash del(A_0)$ $B_2 \vdash del(B_1)$, $B_1 \vdash del(B_0)$.

The new "functions" needed by the decomposition are added recursively.

At the top level, the delineability condition $C(1) = \gamma_1$ is added to assert the delineability of the two families together : $A_1 \otimes B_1 \otimes \gamma_1 \vdash del(A_0, B_0)$ where γ_1 express the actual degree of the gcd for all couple from $A_0 \times B_0$. We get three families, one coming from A, the other from B, the third $C(1)$ being a new one.

For the following step, this new family has to be taken into account. Suppose that $C_2 \vdash del(C(1))$ and that $A_2 \otimes B_2 \otimes \gamma_2^1 \vdash del(A_1, B_1)$, $A_2 \otimes C_2 \otimes \gamma_2^2 \vdash del(A_1, C(1))$, $B_2 \otimes C_2 \otimes \gamma_2^3 \vdash del(B_1, C(1))$, then $A_2 \otimes B_2 \otimes C_2 \otimes \gamma_2^1 \otimes \gamma_2^2 \otimes \gamma_2^3 \vdash del(A_1, B_1, C(1))$. All the new functions are put together in a family $C(2)$ for the next step. $C(2) = C_2, \gamma_2^1, \gamma_2^2, \gamma_2^3$

All the simplifications done in the decomposition of A (or B) hold in this new decomposition.

---

[14] to simplify sets of polynomials which have to be delineable are noted with the same symbol as the particular sign assignment upon these polynomials.

7 Conclusion

A finer analysis of semi-algebraic sets is achieved and thus :

- simplifications may be done
- good sub-structures may be found
- a "natural" interpretation of semi-algebraic sets as "natural" objects is achieved. Semi-algebraic sets may be handled in a global way. For instance, an obstacle in a robotic scene may be handled globally, and not as a collection of polynomials which describe it.
- by handling formulas as object and with the possibility of "leading" algorithms by proofs, dynamic reasoning may be achieved, such as the combination of two situations, or computation with new objects.

A Sequent Calculus

Gentzen sequent theory formalizes proof in the following way :

- a sequent, $\Gamma \vdash \Delta$, is formed of two sets of formulas Γ, Δ , such that Γ yields Δ.
- a rule is a relation between sequents such that the lower sequent may be obtained from the upper sequent(s). example : "and" $\frac{\Gamma \vdash A \;\; \Gamma \vdash B}{\Gamma \vdash A \wedge B}$ or "the modus ponens" $\frac{\Gamma \vdash A \;\; \Gamma \vdash A \rightarrow B}{\Gamma \vdash B}$

In this formalism, proofs are "trees" with hypothesis at the top and conclusions at the bottom.

B Rules of Linear Logic

Linear logic [Gir] tries to analyze the deduction rules, and introduces new connectors to distinguish the connection rules.

axiom

$$A \vdash A$$

cut

$$\frac{\Gamma \vdash \Delta, A \quad \Gamma', A \vdash \Delta'}{\Gamma, \Gamma' \vdash \Delta, \Delta'}$$

tensor

$$\frac{\Gamma \vdash \Delta, A \quad \Gamma' \vdash \Delta', B}{\Gamma, \Gamma' \vdash \Delta, \Delta', A \otimes B}$$

$\otimes$

$$\frac{\Gamma, A, B \vdash \Delta}{\Gamma, A \otimes B \vdash \Delta}$$

par

$$\frac{\Gamma \vdash \Delta, A, B}{\Gamma \vdash \Delta, A \wp B}$$

$\wp$

$$\frac{\Gamma, A \vdash \Delta \quad \Gamma', B \vdash \Delta'}{\Gamma, \Gamma', A \wp B \vdash \Delta, \Delta'}$$

non

$$\frac{\Gamma \vdash \Delta, A}{\Gamma, \tilde{\ }A \vdash \Delta}$$

$\sim$

$$\frac{\Gamma, A \vdash \Delta}{\Gamma \vdash \Delta, \tilde{\ }A}$$

with

$$\frac{\Gamma \vdash \Delta, A \quad \Gamma \vdash \Delta, B}{\Gamma \vdash \Delta, A \& B}$$

&

$$\frac{\Gamma, A \vdash \Delta}{\Gamma, A \& B \vdash \Delta}$$

plus

$$\frac{\Gamma \vdash \Delta, A}{\Gamma \vdash \Delta, A \oplus B}$$

$\oplus$

$$\frac{\Gamma, A \vdash \Delta \quad \Gamma, B \vdash \Delta}{\Gamma, A \oplus B \vdash \Delta}$$

The $\to$ is defined : $(\tilde{\ }G \wp K) \equiv (G \to K)$

The distributivity of "tensor" upon "plus" is an equivalence in linear logic, $(G \otimes K) \oplus (H \otimes K) \equiv [(G \oplus H) \otimes K]$. There is an other distributivity property, $(G \wp K) \& (H \wp K) \equiv [(G \& H) \wp K]$. The following implications hold : $G \otimes (H \& K) \to (G \otimes H) \& (G \otimes K)$ and $(G \wp H) \oplus (G \wp K) \to G \wp (H \oplus K)$

References

[BKR] M. Ben-Or, D. Kozen and J. Reif. : The Complexity of Elementary Algebra and Geometry, J. of Comput. and Syst. Sci. **32** (1986) 251–264

[Col] G.E. Collins : Quantifier elimination for real closed fields by cylindrical algebraic decomposition, proc.2nd GI Conference on Automata Theory and Formal languages", Lect. Notes in Comput. Sci. **35** (1975) 134–183

[Gir] J.Y. Girard : Linear Logic, theoretical Computer Sci. (1987) 1–102

[Ggv] D. Grigor'ev : Complexity of deciding Tarski algebra, J. Symbolic Computation **5** (1988) 65–108

[HRS] J. Heintz, M.F. Roy, P. Solernó: Sur la complexiti du principe de Tarski-Seidenberg, Bull. Soc. Math. France. **118** (1990) 101–126

[Ren] J. Renegar : On the computational complexity and geometry of the first-order theory of the reals, J. Symbolic Computation **13** (1992) 255–352

[Tar] A. Tarski : The completeness of elementary algebra and geometry, schedulded to appear in Paris 1940 (1967)

[Tol] C. Tollu : Planification en Logique Linéaire et Langage de Requêtes Relationnelles avec Compteurs, thèse, Paris, (1993) 13–30

A Proof Environment for Arithmetic with the Omega Rule

Siani Baker[1] and Alan Smaill[2]

[1] Cambridge University, Cambridge CB2 3QG, UK
e-mail: siani@hplb.hpl.hp.com
[2] Edinburgh University, Edinburgh EH1 1HN, UK
e-mail: smaill@aisb.ed.ac.uk

Abstract. An important technique for investigating derivability in formal systems of arithmetic has been to embed such systems into semi-formal systems with the ω-rule. This paper exploits this notion within the domain of automated theorem-proving and discusses the implementation of such a proof environment, namely the CORE system which implements a version of the primitive recursive ω-rule. This involves providing an appropriate representation for infinite proofs, and a means of verifying properties of such objects. By means of the CORE system, from a finite number of instances a conjecture for a proof of the universally quantified formula is automatically derived by an inductive inference algorithm, and checked for correctness. In addition, candidates for cut formulae may be generated by an explanation-based learning algorithm. This is an alternative approach to reasoning about inductively defined domains from traditional structural induction, which may sometimes be more intuitive.

1 Introduction

Normally, proofs considered in theorem-proving are finite; however, there is a reasonable notion of infinite proof involving the ω-rule, which infers a proposition from an infinite number of individual cases of that proposition. The ω-rule involves the use of infinite proofs, and therefore poses a problem as far as implementation is concerned.

With the goal of automatic derivation of proofs within some formalisation of arithmetic in mind, an (implementable) representation for an arithmetical system including the ω-rule is proposed. The implemented system is useful as a proof environment (and incidentally also as a guide to generalisation in the more usual formalisation of arithmetic [3]).

The following sections present a formalisation of arithmetic with the ω-rule ($PA_{c\omega}$), discuss how this was correctly implemented to produce the framework of the CORE proof environment, and establish soundness of the implementational system with respect to $PA_{c\omega}$. We also give an indication of how the system can be used.

2 The Constructive Omega Rule

In this section a constructive version of the infinitary ω-rule is introduced as an interesting alternative to induction proofs in arithmetic. A standard form of the ω-rule is

$$\frac{A(0), A(\underline{1}) \ldots A(\underline{n}) \ldots}{A(x)}$$

where $\underline{n}$ is a formal numeral, which for natural number n consists in the n-fold iteration of the successor function applied to zero, and A is formulated within the language of arithmetic. This rule is not derivable in Peano Arithmetic (PA)[3], since for example, for the Gödel formula $G(x)$, for each natural number n, $PA \vdash G(\underline{n})$ but it is not true that $PA \vdash G(x)$. This rule together with Peano's axioms gives a complete theory - - the usual incompleteness results do not apply since this is not a formal system in the usual sense.

However, this is not a good candidate for implementation since there are an infinite number of premises. It would be desirable to restrict the ω-rule so that the infinite proofs considered possess some important properties of finite proofs. One suitable option is to use a **constructive ω-rule**. The ω-rule is said to be constructive if there is a recursive function f such that for every n, $f(n)$ is a Gödel number of $P(n)$, where $P(n)$ is defined for every natural number n and is a proof of $A(\underline{n})$ [20]. This is equivalent to the requirement that there is a uniform, computable procedure describing $P(n)$, or alternatively that the proofs are recursive (in the sense that both the proof-tree and the function describing the use of the different rules must be recursive) [22]. There is a primitive recursive counterpart[4] which is also a candidate for implementation. Note that in particular these rules differ from the form of the ω-rule (involving the notion of provability) considered by Rosser [15] and subsequently Feferman [7].

Various theoretical results are known for these systems. Shoenfield has shown that '$PA + \omega$-rule' (PA_ω)[5] is equivalent to 'PA + recursively restricted ω-rule' [19]. The sequent calculus enriched with the recursively restricted ω-rule in place of the rule of induction (let us call it $PA_{r\omega}$[6]) has cut elimination, and is complete [19].

The primitive recursive variant has also been shown to be complete by Nelson [12]. If one has the rule of repetition $\frac{\Gamma \vdash \Delta}{\Gamma \vdash \Delta}$ in PA_ω, any recursive derivation can be "stretched out" to a primitive recursive derivation using the same rules of inference, plus this rule [10, P169]. Since our implementation is developed using effective operations over representations of object-level syntax (where effectiveness is an analogous concept to primitive recursion), and PA with the unrestricted ω-rule forms a conservative extension of this system, the (classical)

---

[3] See for example [18] for a formalisation.

[4] In other words, such that there is a primitive recursive function f for which, for every n, $f(n)$ is a Gödel number of the proof of $A(\underline{n})$, the nth numerator of the ω-rule.

[5] See Section 3 below for description.

[6] For a more formal description see [1].

system PA with a primitive recursive restriction on the proof-trees was chosen as a basis for implementation.

In the context of theorem proving, the presence of cut elimination for these systems means that generalisation steps are not required. In the implementation, although we do not claim completeness, some proofs that normally require generalisation can be generated more easily in $PA_{c\omega}$ than PA.

3 $PA_{c\omega}$: Arithmetic with the Constructive Omega Rule

The system PA_ω is essentially PA enriched with the ω-rule in place of the rule of induction. The derivations are then infinite trees of formulae; a formula is demonstrated in PA_ω by "exhibiting" a proof-tree labelled at the root with the given formula. Syntactical details about this system PA_ω are given in [10, P162] (see [14, P266–267] for a natural deduction representation). PA_ω has been described by Schütte as a semi-formal system to stress the difference between this and usual formal systems which use finitary rules [17, P174].

For implementational purposes, infinite proofs must be thought of in the constructive sense of being generated, rather than absolute. It is necessary to place a restriction on the proof-trees of PA_ω such that only those which have been constructively generated are allowed, in order to capture the notion of infinite labelled trees in a finite way. The normal approach when dealing with a system with infinitary proofs such as PA_ω is to work with numeric codes for the derivations rather than using the derivations themselves. See [18, P886] for further details, including the case of the ω-rule. By adding the provability relation and numeric encoding, a reflection system which necessarily extends the original one may be formed [9, P163]. However, the necessity of using this Gödel numbering approach may be avoided by following Tucker in defining primitive recursion ("effectiveness") over various data-types that are better adapted to computational purposes [21].

If an arithmetical encoding method were to be used, the primitive recursive constraint could be attached directly to the ω-rule. However, without using such an approach the restriction must be placed on the shape of the proof tree in which the ω-rule appears: only derivations which are "effective" will be accepted.

Hence we define

$$\vdash_{PA_{c\omega}} \varPhi \qquad \textit{iff} \qquad \exists f.\ f \text{ is an 'effective' proof-tree of } PA_{c\omega} \text{ with } \varPhi \text{ as initial sequent.}$$

$PA_{c\omega}$ may be defined as a (semi-)formal system by further specifying axioms and rules of inference (in this case, corresponding to those of PA: "this wholesale carry over of derived rules from predicate logic is one of the special virtues of cut free infinite proofs" [9, P166]).

The alternative, standard approach is to use numeric encoding (using notation $\ulcorner . \urcorner$) and strengthen PA by adding an arithmetic schema of the form:

$$(\exists \varPi\ (\text{proof-tree}(\varPi) \ \wedge\ \text{conc}(\varPi) = \ulcorner \varPhi \urcorner)) \rightarrow \varPhi.$$

We must now provide some means for reasoning about primitive recursive infinite proof trees. The objects of interest are recursive (possibly infinite) proof-trees (in the sense of López-Escobar [10]), labelled with formulae (namely, the sequents to be proved at each point) and rules. The notion of effectiveness of a tree, which corresponds to primitive recursion, is defined in [1]. In addition, a (proof) tree must be well-founded, in the sense that it does not have an infinitely deep branch. The rules that relate the formulae between node and subnode are the standard rules for the logical connectives, the extra ω-rule with subgoals $\Phi(\underline{0}), \Phi(\underline{1}), \ldots$, and substitution. A formula in PA is demonstrated in the extended theory by exhibiting a proof-tree labelled at the root with the given formula. Properties of such primitively recursively defined trees can be proved using induction principles associated with the datatypes, as we see in section 5. These are the sorts of proofs that have been automated by [5], and we are able to automate the simpler proofs that arise here. This involves, for example, giving a proof that a given rewrite applied a given number of times to a formula schema yields a particular formula schema.

3.1 Definition of Effective Prooftrees for $PA_{c\omega}$

These notions are now formalised. We define prooftrees as functions

$$f : \text{Position of Node} \mapsto (\text{Sequent at Node, Rule used at Node}),$$

where the range specifies labels, or symbols, in the tree associated with each node, and the position is represented by lists of natural numbers.

Definition 1 Effective prooftree for $PA_{c\omega}$. f is an effective prooftree for $PA_{c\omega}$ if and only if f is an effective function $f : nat\ list \rightarrow seq \times rule$ such that f is well-founded and correct.

Definition 2 Order in tree. Define the relation $\leq$ on $nat\ list \times nat\ list$ by:

$$Pos1 \leq Pos2 \leftrightarrow \exists l\ Pos2 = Pos1 <> l,\ l \in nat\ list$$

Definition 3 Empty node. $\check{\epsilon}$, the empty label, is shorthand for (dummy seq, dummy rule), and indicates that there is nothing at a particular node.

Definition 4 Derivation in $PA_{c\omega}$. $f : nat\ list \rightarrow seq * rule$ describes a derivation in $PA_{c\omega}$ for the sequent Φ, where Φ is the sequent at the top of the tree (viz. at the node [ ]) if:

1. $\{p : nat\ list | f(p) \neq \check{\epsilon}\}$ is a well-founded tree according to $\leq$.
2. If $f(p) \neq \check{\epsilon}$, then $q_2^1(f(p))$ is a sentence associated with the node p (namely the sequent to be proved), and $q_2^2(f(p))$ is the name of a rule of $PA_{c\omega}$ used to produce its immediate successors, where q_i are projection functions such that $q_2^1(A, B) = A$ and $q_2^2(A, B) = B$.
3. If p is a bottommost node in the tree, ie. $f(q) = \check{\epsilon}$ for all q such that $p \leq q$, and $f(p) \neq \check{\epsilon}$, then either $q_2^1(f(p))$ is an axiom of $PA_{c\omega}$ and $q_2^2(f(p))$ is $axiom$, or else $f(p)$ is set to incomplete to indicate that the tree is incomplete.

4. If $Pos <> [K]$ ($K \in IN$) is not a bottommost node, then $q_2^1(f(Pos <> [K]))$ is the Kth subgoal of $q_2^2(f(Pos <> [K]))$ applied to $q_2^1(f(Pos))$.

Definition 5 Incomplete tree. The derivation is incomplete if not all the leaves are closed ie. if incomplete is associated with any node in the tree.

Definition 6 Prooftree. The derivation will be a prooftree if it is a complete derivation, in other words if all its leaves are axioms (and the others marked as dummy nodes, if appropriate, since there is infinite branching at each node).

Definition 7 Subgoals. The *subgoals* of p may be defined as
$$subgoals(p) = \{q_2^1(f(p <> [n]))|n \in IN, f(p <> [n]) \neq \check{\varepsilon}\}.$$

Consequences of this Definition. Properties of the tree will be:

1. Defining $br(Rule)$, where $Rule$ is a rule of $PA_{c\omega}$, as the number of subgoals of $Rule$ (ie. $br(\forall r_\omega) = \omega$, $br(\rightarrow r = 1$, $br(\rightarrow l) = 2$ etc.), if $L \neq \omega$, where $br(q_2^2(f(Pos))) = L, L \in IN$, then $f(Pos <> [M]) = \check{\varepsilon}$ $\forall M \geq L, M \in IN$.[7] Since $br(axiom) = 0$, if $p \leq q$ and $p \neq q$ for some position representation p,q, where p is a bottommost node in the tree, then $f(q) = \check{\varepsilon}$.
2. If $f(Pos <> [I]) = \check{\varepsilon}$, then $f(Pos <> [J]) = \check{\varepsilon}$ $\forall J \geq I \in IN$.

The approach described above is suitable for automation, since it generates the subgoals of the ω-rule, rather than having to check their presence.

3.2 Correctness of Prooftrees

There are two main notions of correctness of well-founded trees, namely 'local' correctness, which checks that an appropriate rule is applied at each node of the tree, and 'global' correctness, which is concerned with whether the tree is well-founded.

To check for local correctness, structural induction is used over the prooftrees. In this case, the corresponding meta-induction (for *nat list*) is used for the tree-defining function $f : nat\ list \rightarrow string \times string$. Thus:

$$\frac{f([\ ])\quad f(Pos) \Rightarrow f(Pos <> [k])}{\forall x\ f(x)}$$

where $Pos, x \in nat\ list$ and $k \in nat$. That is to say that given an initial sequent (and initial rule used), plus a way of obtaining from a sequent at some position the sequent at a node directly below that position, then the sequent at each node of the tree is defined. This process of obtaining sequents at subnodes, given a sequent at a node, is carried out by applying the rule associated with the node to the sequent at the node, and is described above. The result is uniquely determined, given a rule and sequent, and hence the prooftrees are locally correct.

---

[7] $\geq$, since the natural numbers, and the subnodes, are taken to start at 0.

Transfinite induction over the partial ordering $\leq$ of the tree representation is allowable if numeric encoding is used at each node [9, P163].

At this stage it could be objected that there might be circularity, for although the ω-rule is used instead of induction, meta-induction is being introduced here, which might result in there being no advance. However, the generalisation problem does not in practice occur in the theory of trees, and so there is a gain after all.

4 ω-Proofs

This section deals with the issues involved in making the ω-rule into a rule for machine proof. One use of the constructive ω-rule is to enable automated proof of formulae, such as $(x + x) + x = x + (x + x)$, which cannot be proved in the normal axiomatisation of arithmetic without recourse to the cut rule, which is the logical justification of any generalisation step. In these cases the correct proof could be extremely difficult to find automatically. However, it is possible to prove this equation using the ω-rule since the proofs of the instances $(0 + 0) + 0 = 0 + (0 + 0), (1 + 1) + 1 = 1 + (1 + 1), \dots$ are easily found, and the general pattern determined by inductive inference. The algorithm used to automatically recognise the general pattern generalises an initial set of rewrite rules describing an individual proof, and then updates this generalisation according to other individual proof examples until the general proof representation (ω-proof) satisfies all of the (large number of) cases considered; any appropriate inductive inference algorithm, such as Plotkin's least general generalisation [13], or that of Rouveirol, who has tackled the problem of controlling the hypothesis generation process to get only the most relevant candidates [16], could be used to guess the ω-proof from the individual proof instances. In general, the complexity of the algorithm needed to guess an ω-proof from non-uniformly generated examples is exponential, whereas the stages of checking the ω-proof and suggesting a cut formula are less complex, and this is reflected in the time taken to produce the result. As an alternative, the user may bypass this whole stage by specifying the ω-proof directly. Meta-induction is used to ensure that the proposed general rule applications do indeed give a proper proof when applied to the general case of the sequent to be proved. Note that such inductive inference algorithms for generating generalisation produce a proof for an arbitrary instance: the penultimate section suggests how this can relate to finding the proper induction formulae for inductive theorem provers.

Such ω-proofs are not simply disguised PA-proofs, since the system $PA_{c\omega}$ is a logically stronger system than that of PA [19]. Moreover, these ω-proofs may be considered to be more intuitive than standard inductive proofs of the same theorems, in the sense of corresponding more closely to the way in which people convince themselves of the correctness of the proof. Philosophical induction (from "trivial" test cases) may sometimes be the means of construction by humans of a generic proof for case n. In addition, the generic proof itself might have some psychological validity. For instance, in order to show that $\forall l\ rotate(length(l), l) = l$

where

$$rotate(0, l) \Rightarrow l \text{ and } rotate(s(n), h :: t) \Rightarrow rotate(n, append(T, h :: nil))$$

a human might provide the following explanation

> "Imagine applying the definition of rotate n times. The elements pop off the front of the list in order and stack up at the end in the same order. Eventually you get back to the original expression."

rather than " Suppose it were true for n. Now consider the case $n + 1 \ldots$ ".

For the implementation it is necessary to provide (for the nth case) a description for the ω-proof in a constructive way which captures the notion that each $P(n)$ is being proved in a uniform way (from parameter n). This is done by manipulating $A(\underline{n})$, where $A(x)$ is the sequent to be proved, and using recursively defined function definitions of PA as rewrite rules, with the aim of reducing both sides of the equation to the same formula. The primitive recursive function sought is described by the sequence of rule applications, parameterised over n. In practice, the first few proofs will be special cases, and it is rather the correspondence between the proofs of $P(99)$, say, and $P(100)$, which should be captured. The processes of generation of a (recursive) ω-proof from individual proof instances, and the (metalevel) checking that this is indeed the correct proof have been automated (see [2]). Further details of the algorithms and representations used, together with the correspondence between the adopted implementational approach and the formal theory of the system are described in [1].

Thus, the ω-proof representation represents $P(n)$, the proof of the nth numerator of the constructive ω-rule, in terms of rewrite rules applied $f(n)$ or a constant number of times to formulae (dependent upon the parameter n). As an example, the implementational representation of the ω-proof for $(x+x)+x = x+(x+x)$ takes the form given in Figure 1 (although it may be represented in a variety of ways) presuming that, within the particular formalisation of arithmetic chosen, one is given the axioms of addition of Figure 1.

By $s^n(0)$ is meant the numeral $\underline{n}$, ie. the term formed by applying the successor function n times to 0. The next stages use the axioms as rewrite rules from left to right, and substitution in the ω-proof, under the appropriate instantiation of variables, with the aim of reducing both sides of the equation to the same formula. The ω-proof represents, and highlights, blocks of rewrite rules which are being applied. Meta-induction may be used (on the first argument) to prove the more general rewrite rules from one block to the next: for example, $\forall n \; s^n(x) + y = s^n(x + y)$ corresponds to n applications of axiom (2) above. We now describe in more detail how this is done.

5 Showing Correctness of ω-Proofs.

By an effective proof-tree, as discussed in Section 3, we understand a function which returns for each potential position in the tree either a pair representing

Axioms

$$0 + y = y \tag{1}$$
$$s(x) + y = s(x + y) \tag{2}$$

Proof

$$(\underline{n} + \underline{n}) + \underline{n} = \underline{n} + (\underline{n} + \underline{n})$$

$\underline{n} \equiv s^n(0)$	$(s^n(0) + s^n(0)) + s^n(0) = s^n(0) + (s^n(0) + s^n(0))$
(2) n TIMES ON LEFT	$s^n(0 + s^n(0)) + s^n(0) = s^n(0) + (s^n(0) + s^n(0))$
(1) ON LEFT	$s^n(s^n(0)) + s^n(0) = s^n(0) + (s^n(0) + s^n(0))$
(2) n TIMES ON RIGHT	$s^n(s^n(0)) + s^n(0) = s^n(0 + (s^n(0) + s^n(0)))$
(1) ON RIGHT	$s^n(s^n(0)) + s^n(0) = s^n(s^n(0) + s^n(0))$
(2) n TIMES ON LEFT	$s^n(s^n(0) + s^n(0)) = s^n(s^n(0) + s^n(0))$
	EQUALITY

Fig. 1. An ω-Proof of $(x + x) + x = x + (x + x)$

the sequent and rule associated with that position, or a token indicating that the position is outside the tree. Positions are given by a list of positive integers referring to the path through the tree from the root. We thus have a partial function

$$tree : list(int) \rightarrow sequent \times rule$$

where we must define $rule, sequent$.

5.1 Schematic Syntax

The representation of sequents as a function of position is achieved as follows. For simplicity, we will concentrate on the single goal formula of a sequent, though the discussion extends to the full sequent.

We consider formulae simply as strings[8]. A function

$$form : int \rightarrow string$$

can be considered as representing formulae schematically, provided that the function always takes values among the formulae of the language.

For example, using ML syntax we can define the following functions (where ^ is infix string concatenation; nth_suc implements $s^n(t)$ above):

```
val zero = "0"
fun plus x y = "plus(" ^ x ^ "," ^ y ^ ")"
fun nth_suc 0 x = x
  | nth_suc n x = "s(" ^ (nth_suc (n-1) x) ^ ")"
```

---

[8] Alternatively, and more elegantly, we could have made use of abstract syntax here.

```
fun eq x y = x ^ " = " ^ y
fun goal n =
    let val m = nth_suc n zero
    in
        eq (plus (plus m m) m) (plus m (plus m m))
    end;
```

The function **goal** here returns a string representing the *n*th subgoal to the ω-rule application in our example above:

```
- goal 0;
"plus(plus(0,0),0) = plus(0,plus(0,0))" : string
- goal 1;
"plus(plus(s(0),s(0)),s(0)) = plus(s(0),plus(s(0),s(0)))" : string
```

In this way a proof-tree can be built up by assigning such formulae to positions. There are constraints on which positions have attached formulae, to ensure that the tree is indeed a tree, and is well-founded. The *rule* specification is likewise given by a string, corresponding to the name of the rule applied at this position, and any parameters used. The function gives an ω-proof if, at every node of the tree, the formulae at the sub-nodes and the formula at the node are correctly related according to the rule of inference associated with the node.

It would be extremely cumbersome to build up such tree functions explicitly, and we do not do this in practice. However, the system is intended to ensure that such a tree is constructible whenever an ω-rule application is shown correct.

5.2 Rewrite Rules as Derived Rules of Inference

It is customary to use recursion equations as rewrite rules in order to evaluate expressions. Since we have a substitution rule, we can show that such applications to quantifier-free formulae are sound. For example, the axiom (2) gives us the rewrite

$$s(X) + Y \Rightarrow s(X + Y). \tag{3}$$

The rewrite rule of inference then, given a goal and a specified sub-term that matches the left hand side of the rewrite, yields as the single subgoal the rewritten goal. Note that any use of this derived rule could be expanded into a small proof tree of fixed shape in our original theory.

In reasoning about ω-proofs, we will chiefly use this derived rule.

5.3 Multiple Rule Applications

In the simpler examples of the use of the ω-rule, there is only one application of that rule, and by using the rewrite inference rule we can regard each subsequent branch of the proof as linear, with no further branching of the tree.

In this case, what we want to know is that the sequence of rewrites along the branch is correct, and leads to an axiom. To capture the generality of the shape of the tree, we need to be able to reason about such things as the application of a given rewrite n times in a particular position of the nth formula. (More generally, the position and the number of applications can both be functions of n.)

Supposing that we have defined what it is for one-step rewriting to apply using a given rule at a given position, we can define n-fold rewriting by:

```
fun nfoldRewrite 0 rule pos wff = wff
  | nfoldRewrite n rule pos wff =
        rewrite rule pos (nfoldRewrite (n-1) rule pos)
```

Notice that our definitions of schematic syntax and multiple rewriting are primitive recursive. In order to prove properties of them, we can use standard techniques for proof of primitive recursively defined functions. For example, as part of the correctness proof for our example, we want to show that n applications of (0) on the left hand side of

$$(s^n(0) + s^n(0)) + s^n(0) = s^n(0) + (s^n(0) + s^n(0))$$

yields

$$s^n(0 + s^n(0)) + s^n(0) = s^n(0) + (s^n(0) + s^n(0)).$$

Writing the second expression as a function of n as goal2(n), calling the rewrite rule plus2, and noting that the left hand side is picked out by position [2], what has to be shown is that

```
nfoldRewrite n plus2 [2] (goal n) = goal2 n.
```

The proof of this proceeds by induction; show that

```
nfoldRewrite 0 plus2 [2] (goal 0) = goal2 0
```

and that

```
nfoldRewrite n plus2 [2] (goal n) = goal2 n
  ⊢ nfoldRewrite (n+1) plus2 [2] (goal (n+1)) = goal2 (n+1)
```

This is the proof that is carried out automatically. When each of the rewrite steps has been verified in this way, and the leaf checked as an axiom, we conclude that there is a correct ω-proof.

Next we move on to consider further the relationship between the ω-proof representation and the system $PA_{c\omega}$.

5.4 Relationship between ω-Proof Representation and $PA_{c\omega}$

In the representation of $PA_{c\omega}$, a derived form of primitive recursive function-defining equations as inference rules is required to allow rewriting of formulae (which takes place in ω-proof examples, and hence an analogy is needed). These are of the type $\mathcal{I}$ below — it is just convenient to define these as single steps, as shorthand in order to avoid going through all the equivalent tedious steps of PA.

$$\frac{\Gamma \vdash \forall xy R(x,y) = S(x,y) \quad \Delta \vdash T(R(p,q))}{\Gamma, \Delta \vdash T(S(p,q))} \mathcal{I}$$

An example of such a derived step would be the justification of $s(0+0) = s(0)$ from $s(0) + 0 = s(0)$, given rewrite rule (3) above. $\mathcal{I}$ is a derived rule of PA, as it is equivalent to the following:

$$\frac{\dfrac{\dfrac{\rule{2cm}{0.4pt}}{R(p,q) = S(p,q) \vdash R(p,q) = S(p,q)} \; axiom}{\forall xy \, R(x,y) = S(x,y) \vdash R(p,q) = S(p,q)} \forall l \quad \Gamma \vdash \forall xy \, R(x,y) = S(x,y)}{\dfrac{\Gamma \vdash R(p,q) = S(p,q)}{\Gamma, \Delta \vdash T(S(p,q))} \quad \Delta \vdash T(R(p,q))} \; cut \; subst}$$

One may deduce a more general rule $\mathcal{I}'$, in a similar manner, where $\mathcal{I}'$ is the following:

$$\frac{\Gamma \vdash \forall \boldsymbol{x} R(\boldsymbol{x}) = S(\boldsymbol{x}) \quad \Delta \vdash T(R(\boldsymbol{p}))}{\Gamma, \Delta \vdash T(S(\boldsymbol{p}))} \mathcal{I}'$$

where if the arity of $\boldsymbol{x}$ is k, then $\boldsymbol{p} = \{p_1, \ldots p_k\}$ for terms p_i $(1 \leq i \leq k)$ in the language of $PA_{c\omega}$.

The derived rule $\mathcal{I}$ above corresponds to "apply $\mathcal{R}$ once", where $\mathcal{R}$ is the rewrite rule $R(x,y) \Rightarrow S(x,y)$. "Apply $\mathcal{R}$ a constant number of times, k" just repeats $\mathcal{I}$ k times. $\mathcal{I}$ corresponds to the basic formulation of $PA_{c\omega}$, so if the given procedure about tree construction from rules is followed, a recursive prooftree will be obtained. Note that in the tree there is infinite branching, so dummy subtrees will have to be inserted if necessary.

In the ω-proof, one starts with something to be proved, and ends with equality, corresponding to an axiom. In practice, only rewrite rules are usually used, but it is possible to use logical rules which might cause the ω-proof to split. However, such case splits of conditionals do not pose any problem, because they correspond exactly to the logical rules of $PA_{c\omega}$.

Note that if such derived rules were to be used in the prooftree representation, the whole equivalent subtree should be substituted instead, since the new node position otherwise would not be a direct subnode (in terms of its position representation) of the original node at which such a rule was applied.

The other case to consider is when a rewrite rule of the form "apply $\mathcal{R}$ some function of n times" is applied in the ω-proof. This corresponds to the (derived) rule $\mathcal{J}$ below, which may be proven in $PA_{c\omega}$, with the use of induction.

$$\frac{\Gamma \vdash \forall x \forall y R(x,y) = S(x,y) \quad \Delta \vdash T(R^{f(n)}(p,q))}{\Gamma, \Delta \vdash T(S^{f(n)}(p,q))} \mathcal{J}$$

where $S^{f(n)}(p,q)$ is the result of applying the rule $\mathcal{R}$ $f(n)$ times to $R^{f(n)}(p,q)$ (assuming that such a rule application may be carried out). One such example is that $s^n(x) + y \Rightarrow s^n(x + y)$ corresponds to n applications of the rewrite rule $s(x) + y \Rightarrow s(x + y)$. It is also possible to derive a more general version $\mathcal{J}'$, which is analogous to $\mathcal{I}'$.

A summary of the above consideration of the relationship of the ω-proof representation with $PA_{c\omega}$ is that the proof of $A(\underline{n})$ in the ω-rule $\frac{A(0),...A(\underline{n})...}{\forall x A(x)}$ is represented in the ω-proof using notation which is not that of $PA_{c\omega}$, but which is more suitable for implementation. However, the notation (represented by derived rules $\mathcal{I}$ and $\mathcal{J}$), may be converted to that of $PA_{c\omega}$, as may be seen by the fact that the rules $\mathcal{I}$ and $\mathcal{J}$ correspond to proofs in $PA_{c\omega}$. Hence, the ω-proof can be accounted for in terms of $PA_{c\omega}$.

5.5 Correctness of the ω-Proof Representation

Global correctness necessitates showing that for each ω-proof construct there is a primitive recursive function which indicates what is at any particular node in the tree representation described above. This is possible by inspecting the tree diagrams for the derived inference rules $\mathcal{I}$ and $\mathcal{J}$ above. The primitive recursive function would be analogous to the definition of f above. When "apply rule $\mathcal{R}$ k times" appears in the ω-proof, the function would generate the tree as previously described, but using the rules from $\mathcal{I}$ (repeated as appropriate), or $\mathcal{J}$, and generating the rest of each infinite layer as dummy variables. Of course, the layer is not literally filled in, as this will be a non-terminating process. It is merely necessary to note that, for example, $f(Pos <> []) = \check{e} \ \forall l \in nat \ list$, and that any particular individual position could be checked as yielding $\check{e}$. This is enough to give correctness of the tree. A tree with infinite branching points may be generated using the constructive ω-rule (ie. from the ω-proof). This should be using a depth-first generation, because the tree is required to be well-founded; in this way the tree would in essence be completed — after a certain point with generating the subgoals of the ω-rule, the case for the kth subtree could be given if termination was required.

Therefore, in order to convert from ω-proofs to a recursive prooftree form, it is necessary to substitute k for $\underline{n}$, where $k \in I\!N$, to get the kth subtree. Rewrite rules should be converted to the appropriate rules of $PA_{c\omega}$ (from the trees given above for $\mathcal{I}$ and $\mathcal{J}$) and applied as appropriate; this also covers the case of the application of rules $f(n)$ times.

In this section a justification for the implemented representation of ω-proofs has been given. The following section describes the overall structure of the implemented system.

6 A Proof Environment for the Constructive Omega Rule

This section describes the Constructive Omega Rule Environment (CORE), which is a proof development environment in which a (constructive) version of the ω-rule may be used as a rule of inference, and a system in which ω-proofs may be displayed and investigated. The implementation allows both the automatic or incremental construction of ω-proofs, and the validations of descriptions of ω-proofs. It is carried out within the framework of an interactive theorem-prover with Prolog as the tactic language, namely Oyster, which is a reimplementation of NuPRL [4]. This embodies a higher-order, typed constructive logic in sequent-calculus form. Within the Oyster framework, the object-level logic is replaced with Peano and Heyting arithmetic and the rules that can be applied are those of the sequent calculus axiomatisation of first order logic given in [6, P133], together with mathematical induction. The search for a proof must be guided either by a human user or by a proof tactic. Each proof is built up in the form of a tree, and every stage of the tree may be displayed on the screen with information as to the hypotheses, goals, position in the tree and whether the subtree is proved below it.

Theorem	Cut Formula
$\forall x \ (x+x)+x = x+(x+x)$	$\forall x \forall y \forall z \ (x+y)+z = x+(y+z)$
$\forall x \ x+s(x) = s(x+x)$	$\forall x \forall y \ x+s(y) = s(x+y)$
$\forall x \ x+s(x) = s(x)+x$	$\forall x \forall y \ x+s(y) = s(x)+y$
$\forall x \ x.(x+x) = x.x+x.x$	$\forall x \forall y \forall z \ x.(y+z) = x.y+x.z$
$\forall x \ (x+x).x = x.x+x.x$	$\forall x \forall y \forall z \ (x+y).z = y.z+x.z$
$\forall x \ (2+x)+x = 2+(x+x)$	$\forall x \forall y \ (2+x)+y = 2+(x+y)$
$\forall x \ \forall y \ (x+y)+x = x+(y+x)$	$\forall x \forall y \forall z \ (x+y)+z = x+(y+z)$
$\forall x \ x \neq 0 \rightarrow p(x)+s(s(x)) = s(x)+x$	$\forall x \forall y \ x \neq 0 \rightarrow p(x)+s(s(y)) = s(x)+y$
$\forall x \ even(x+x)$	$\forall x \ even(2.x)$
$\forall l \ len(rev(l)) = len(l)$	$\forall l \ len(rev(l) <> a) = len(rev(a) <> l)$
$\forall l \ rotate(len(l),l) = l$	$\forall l \ rotate(len(l),l <> a) = a <> l$
$\forall l \ rev(rev(l) <> y :: nil) = y :: rev(rev(l))$	$\forall l \ rev(a <> y :: nil) = y :: rev(a)$
$\forall l \ rev2(l,nil) = rev(l)$	$\forall l \ rev2(l,a) = rev(l) <> a$
$\forall l \ (l <> l) <> l = l <> (l <> l)$	$\forall l \forall p \forall q \ (l <> p) <> q = l <> (p <> q)$

Table 1. Cut Formulae Suggested by Guiding Method for Various Examples

Within CORE, any finitely large number of individual instances of proofs of a proposition may be generated automatically by the use of various tactics. The general representation of the proofs is provided by an inductive inference algorithm, which starts with an initial generalisation and then works by updating this ω-proof using the other individual proofs, until the ω-proof seems to have reached a stable form. This ω-proof is then automatically checked to see if it is indeed the correct one, as described above. There are two options which are allowable from a goal $\Gamma \vdash \forall x P(x)$. One is to ask to use the constructive ω-rule, whereby the system will check to see whether it can find a correct ω-proof, and then return to the former system and close the branch, or else report failure. The user may then continue to investigate other positions in the proof-tree. The other option, which shall be discussed more fully in the following section, is to ask for an appropriate cut to be carried out in PA (the cut being worked out by the system from the ω-proof), with a further option to complete the tree as far as possible (using standard theorem-proving techniques). The ω-proof may be provided automatically, but there is an option in each case to switch temporarily to another system which will allow for the description, manipulation and display of the ω-proof. The user may specify the proof incrementally, in terms of applications in positions in the tree, plus induction over a distinguished parameter, or all at once — and this is checked. The system builds up a recursive function description of the ω-proof, and is able to display individual proofs in addition to the general case.

[2] provides details of the proof development systems upon which the implementation is based; representation of the ω-rule and its subgoals; generation of individual proofs; the application of rewrite rules; provision of a ω-proof; correctness checking of the ω-proof (using meta-induction); generalisation, and finally, the interactive system, and how to use it.

7 An Application

As mentioned above, the CORE system provides implementation of a new generalisation method (described in [3]). A cut formula is automatically suggested from ω-proofs using an implementation based on the method of explanation-based generalisation, which is a technique for formulating general concepts on the basis of specific training examples, first described in [11]. In general terms the process works by generalising a particular solution to the most general possible solution which uses the rules of the original solution. It does this by applying these rules, making no assumptions about the form of the generalised solution, and using unification to fill in this form. The method is applied in this instance to a new domain, namely that of ω-proofs. It is of course possible to carry out explanation-based generalisation upon proofs in PA in order to produce a more general expression. However, if inductive proof is not possible (because induction is blocked) and therefore use of the cut rule is required, such a method will not work, and another proof (such as the ω-proof) must be used in order to obtain a generalisation. Further details regarding such a generalisation method, including details of how ω-proofs may be "linearised" to suggest inductive proofs, are given in [3].

The resulting system has been tested on a variety of arithmetical examples: cut formulae are automatically suggested for examples including all the arithmetical examples of Table 1. Although the examples listed in the table are of a similar simple form, this method may also be applied to complicated examples containing nested quantifiers, etc., for the ω-rule applies to arbitrary sequents. The seventh example provides an instance of nested use of the ω-rule, which carries through directly. If an ω-proof is provided, even without a generalisation being suggested, something has still been achieved, in the sense that a pattern might still emerge for the user. For example, the cut formula of $even(2.x)$ could possibly be extracted by a user from the form of the ω-proof for $even(x + x)$, which is an improvement over other generalisation methods. Thus the generalisation method may still be useful within a co-operative environment if it breaks down. This contrasts with alternative methods of generalisation, which do not provide much information if they fail. Moreover, because the suggested method explicitly exploits general patterns, it has a higher-level structure and thus greater potential for extension than other more special-purpose approaches.

Hence a new method for generalisation has been proposed which is robust enough to capture in many cases what the alternative methods can do (in some cases with less work), plus it works on examples on which they fail (cf. Table 1, many of the examples in which pose a problem for other theorem provers). This same generalisation method can be used in a more general context for lemma generation, regardless of the way an ω-proof was obtained, so long as one can represent in the particular system of interest the notions of nth-successor, parametrised applications of rewrite rules and also the correctness check. Hence, although the learning device for cut formulae is not reliant upon the given proof system involving use of the ω-rule, in practice suitable proof environments such as HOL [8] would require extension, and moreover the user would have to input the proof directly unless some other method of generation could be provided.

8 Conclusions

Implementation of a system of arithmetic with the ω-rule has been carried out within the framework of an interactive theorem-prover with Prolog as the tactic language. This can provide a useful aid to automated deduction. The approach suggested in this paper is of general relevance, both regarding lemma generation in systems for reasoning about inductive domains, and also the representation of infinite and schematic proofs.

Acknowledgements. The research reported in this paper was supported by EPSRC.

References

1. Baker, S.: *Aspects of the Constructive Omega Rule within Automated Deduction.* PhD thesis, University of Edinburgh (1992)
2. Baker, S.: CORE manual. Technical Paper 10, Dept. of Artificial Intelligence, Edinburgh (1992)
3. Baker, S.: A new application for explanation-based generalisation within automated deduction. In A. Bundy, editor, *12th International Conference on Automated Deduction,* Lecture Notes in Artificial Intelligence, Springer-Verlag (1994) 177–191. Also available from Cambridge as Computer Laboratory Technical Report 327.
4. Bundy, A., van Harmelen, F., Horn, C., and Smaill, A.: The Oyster-Clam system. In M.E. Stickel, editor, *10th International Conference on Automated Deduction,* Lecture Notes in Artificial Intelligence, Springer-Verlag **449** (1990) 647–648
5. Bundy, A., Stevens, A., van Harmelen, F., Ireland, A., Smaill, A.: Rippling: A heuristic for guiding inductive proofs. *Artificial Intelligence,* **62** (1993) 185–253

6. Dummett, M.: *Elements of Intuitionism.* Oxford Logic Guides. Oxford Univ. Press, Oxford (1977)

7. Feferman, S.: Transfinite recursive progressions of axiomatic theories. *Journal of Symbolic Logic* **27** (1962) 259–316

8. Gordon, M.: HOL: A proof generating system for higher-order logic. In G. Birtwistle and P.A. Subrahmanyam, editors, *VLSI Specification, Verification and Synthesis*, Kluwer (1988)

9. Kreisel, G.: Mathematical logic. In T.L. Saaty, editor, *Lectures on Modern Mathematics*, John Wiley and Sons III (1965) 95–195

10. Löpez-Escobar, E.G.K.: On an extremely restricted ω-rule. *Fundamenta Mathematicae* **90** (1976) 159–72

11. Mitchell, T.M.: Toward combining empirical and analytical methods for inferring heuristics. Technical Report LCSR-TR-27, Laboratory for Computer Science Research, Rutgers University (1982)

12. Nelson, G.C.: A further restricted ω-rule. *Colloquium Mathematicum* **23** (1971)

13. Plotkin, G.: A note on inductive generalization. In D. Michie and D. Meltzer, editors, *Machine Intelligence*, Edinburgh University Press **5** (1969) 153–164

14. Prawitz, D.: Ideas and results in proof theory. In J.E. Fenstad, editor, *Studies in Logic and the Foundations of Mathematics: Proceedings of the Second Scandinavian Logic Symposium*, North Holland **63** (1971) 235–307

15. Rosser, B.: Gödel-theorems for non-constructive logics. JSL **2** (1937) 129–137

16. Rouveirol, C.: Saturation: Postponing choices when inverting resolution. In *Proceedings of ECAI-90*, Stockholm (1990) 557–562

17. Schütte, K.: *Proof Theory.* Springer-Verlag (1977)

18. Schwichtenberg, H.: Proof theory: Some applications of cut-elimination. In Barwise, editor, *Handbook of Mathematical Logic*, North-Holland (1977) 867–896

19. Shoenfield, J.R.: On a restricted ω-rule. Bull. Acad. Sc. Polon. Sci., Ser. des sc. math., astr. et phys. **7** (1959) 405–7

20. Takeuti, G.: *Proof theory.* North-Holland, 2 edition (1987)

21. Tucker, J.V., Wainer, S.S., Zucker, J.I.: Provable computable functions on abstract-data-types. In M.S. Paterson, editor, *Automata, Languages and Programming*, Lecture Notes in Computer Science, Springer-Verlag **443** (1990) 660–673

22. Yoccoz, S.: Constructive aspects of the omega-rule: Application to proof systems in computer science and algorithmic logic. Lecture Notes in Computer Science, Springer-Verlag **379** (1989) 553–565

Using Commutativity Properties for Controlling Coercions

Stephan A. Missura[*1] and Andreas Weber[**2]

[1] Institute for Theoretical Computer Science
ETH Zürich, CH-8092 Zürich, Switzerland
E-mail: missura@inf.ethz.ch
[2] Wilhelm-Schickard-Institut für Informatik
Universität Tübingen
72076 Tübingen, Germany
E-mail: weber@informatik.uni-tuebingen.de

Abstract. This paper investigates some soundness conditions which have to be fulfilled in systems with coercions and generic operators. A result of Reynolds on unrestricted generic operators is extended to generic operators which obey certain constraints. We get natural conditions for such operators, which are expressed within the theoretic framework of category theory.

However, in the context of computer algebra, there arise examples of coercions and generic operators which do not fulfil these conditions. We describe a framework — relaxing the above conditions — that allows distinguishing between cases of ambiguities which can be resolved in a quite natural sense and those which cannot. An algorithm is presented that detects such unresolvable ambiguities in expressions.

1 Introduction

Reynolds [10] uses category theory to investigate the problems of the interaction of coercions (implicit conversions) and generic operators (also called overloaded operators). He concludes with the global requirement that all possible coercions and generic operators have to commute to get a sound system. We refine and extend the work of Reynolds in two directions:

First, we allow *constraints* on the generic operators using the type system described in [12, 13, 15]. These papers show that this formalism is well suited to express an important subset of the type system supported by the computer algebra environment AXIOM [5].

Second, we show how the conditions can be relaxed without losing safety: Reynolds' requirements as well as our refinements are global conditions, needed already at *design* time of the coercions and generic operators. Hence, they are sometimes too restrictive for computer algebra systems, especially for interactive ones. We relax the conditions and get a more refined control of the interaction between coercions and generic operators through the use of information about commutativity. Hence, our refinement of

---

[*] Supported by the Swiss National Science Foundation.
[**] Supported by the *Deutsche Forschungsgemeinschaft*, grant Lo 231/5-1.

Reynolds' global condition is applied only in a much more local context and some common constructions often arising in computer algebra become possible which would be forbidden in Reynolds' framework. We present an algorithm which uses the commutativity information and which detects those ambiguities in an expression that cannot be resolved by any natural criterion.

2 Preliminaries

2.1 Category Theory

For the notions of category theory we refer the reader to the literature. [7] is a comprehensive source. [8, 10] define the needed notions, too.

2.2 Types and Type Classes

A *type* is — as in [3, 12, 13, 15] — just an element of the set of all order-sorted terms over a regular signature $(S, \leq, \Sigma)$ freely generated by some family of infinite sets $V = \{V_\sigma \mid \sigma \in S\}$.

The formalism is well suited to express the subset of the type system of AXIOM [5], in which only non-parameterized categories[3] are considered and the properties correspond to the non-parameterized categories.

The sorts of the order-sorted signature are the constraints on the types and correspond to AXIOM categories [5] and Haskell type classes [4]. They were also called *properties* [3, 13]. We use several of these names depending on the context, avoiding the term "category" for type classes since categories in AXIOM differ substantially from categories in category theory.

As in [13] we assume that we have a *semantics* for the ground types which satisfies the following conditions:

- The ground types correspond to mathematical objects in the meaning of universal algebra or model theory.
- Functions between ground types are set theoretical functions. If we say that two functions $f, g : t_1 \longrightarrow t_2$ are equal ($f = g$) then we mean equality between them as set theoretic objects.

The obvious interpretation of the types as set theoretic objects is adequate because we only need a set theoretic semantics for *ground types* and functions between them.

2.3 Generic Operators with Constraints

Type classes and parameterized type classes (i.e. AXIOM categories) can be used to declare generic operators using the type classes and its parameters as *constraints* on the generic operators. Hence, a generic operator *op* has in general the following *profile*:[4]

---

[3] "Category" in the sense of the AXIOM language, not in the sense of category theory!

[4] For simplicity, we exclude in the following discussion arbitrary polymorphic types that are different from type variables. We do not consider higher-order functions, since they do not play a central role in computer algebra although they are useful (see [14] Sec. 3.2.3).

$$op : \xi_1 \times \cdots \times \xi_n \longrightarrow \xi_{n+1},$$

where each ξ_i is either a type variable constrained by a type class, a parameter of a parameterized type class or a ground type. A precise definition will be given in section 3.

In the following we refer only to the profiles of the constraint generic operators. It is irrelevant whether they are obtained syntactically within a type class or by another mechanism.

2.4 Coercions

Coercions are *implicit conversion* operations [2, 10] that can be used by the interpreter or compiler without explicit user request. A coercion $\phi : A \longrightarrow B$ between the types A and B implies that every element $a \in A$ can be viewed also as an element of B because the coercion can be (implicitly) applied to a.

We assume we have a mechanism to declare arbitrary function as coercions. If there is a coercion $\phi : A \longrightarrow B$ then we write $A \unlhd B$. The relation "$\unlhd$" can be seen as a *subtype* relation on the types A and B which should be at least a pre-order: It is reflexive because the identity function on a type is treated as a coercion and transitive because the composition of two coercions gives a new coercion.

Using the set theoretic semantics on the ground types an important property of the system is *coherence* [13].

Complete Sets of Minimal Upper Bounds. If for two types t_1 and t_2 there is a type t such that $t_1 \unlhd t$ and $t_2 \unlhd t$ then t is called a *common upper bound* of t_1 and t_2.

A *minimal upper bound* $mub(t_1, t_2)$ of two types t_1 and t_2 is a type t satisfying the following conditions.

1. The type t is a common upper bound of t_1 and t_2.
2. If t' is a type which is a common upper bound of t_1 and t_2 such that $t' \unlhd t$, then $t \unlhd t'$.

A *complete set of minimal upper bounds* for two types t_1 and t_2 is a set $\mathrm{CSMUB}(t_1, t_2)$ such that

1. all $t \in \mathrm{CSMUB}(t_1, t_2)$ are a minimal common upper bound of t_1 and t_2, and
2. for every type t' which is a common upper bound of t_1 and t_2 there is a $t \in \mathrm{CSMUB}(t_1, t_2)$ such that $t \unlhd t'$.

In [15] it is shown that *finite* complete sets of minimal upper bounds exist for many classes of coercions which occur in the context of computer algebra.

3   A Category Theoretic Framework

Let

$$op : \overbrace{v_\sigma \times \cdots \times v_\sigma}^{n} \longrightarrow v_\sigma$$

be an n-ary operator defined on a type class σ and let $A \trianglelefteq B$ be types belonging to σ and let

$$\phi : A \longrightarrow B$$

be the coercion function. Moreover, let op_A and op_B be the instances of op in A and B, respectively. For $a_1, \ldots, a_n \in A$ the expression

$$op(a_1, \ldots, a_n)$$

might denote different objects in B, namely

$$op_B(\phi(a_1), \ldots, \phi(a_n))$$

or

$$\phi(op_A(a_1, \ldots, a_n)).$$

The requirement of a unique meaning of

$$op(a_1, \ldots, a_n)$$

just means that ϕ has to be a *homomorphism* for σ with respect to op.

The typing of op in the example above is only one of several possibilities. In general if σ is a type class having $p_{\tau_1}, \ldots, p_{\tau_k}$ as parameters — i. e. p_{τ_i} is a type variable of sort τ_i — then a n-ary first-order operation op defined in σ can have the following profile:

$$op : \xi_1 \times \cdots \times \xi_n \longrightarrow \xi_{n+1},$$

where ξ_i, $1 \leq i \leq n+1$, is either v_σ, or p_{τ_l}, $l \leq k$, or a ground type t_m.

Let C_σ be the category of ground types of sort σ as objects and the coercions as arrows.[5] For a ground type t let C_t be the subcategory which has t as single object and has thus the identity on t as single arrow.[6] Now let

$$C_i = \begin{cases} C_\sigma, & \text{if } \xi_i = v_\sigma, \\ C_{\tau_l}, & \text{if } \xi_i = p_{\tau_l}, \\ C_{t_m}, & \text{if } \xi_i = t_m \text{ for a ground type } t_m. \end{cases}$$

Let Γ_{op} be a functor from $C_1 \times \cdots \times C_n$ into C_{n+1}. If $(\zeta_1, \ldots, \zeta_n)$ is an object of $C_1 \times \cdots \times C_n$, i. e.

$$\zeta_i = \begin{cases} A_\sigma, & \text{if } \xi_i = v_\sigma \text{ and } A_\sigma \text{ is a ground type belonging to } \sigma, \\ A_{\tau_l}, & \text{if } \xi_i = p_{\tau_l} \text{ and } A_{\tau_l} \text{ is a ground type belonging to } \tau_l, \\ t_m, & \text{if } \xi_i = t_m, \end{cases}$$

---

[5] It is easy to check that the axioms of a category are fulfilled.
[6] If the type system is not *coherent* in the sense of [13] then this subcategory might have more than one arrow.

then $\Gamma_{op}(\zeta_1, \ldots, \zeta_n)$ is an object of $\mathcal{C}_{n+1}$, i.e. a ground type belonging to σ resp. $\tau_{l'}$, or is a ground type $t_{m'}$ depending on the value of ξ_{n+1}.

Informally Γ_{op} can be used to specify the type of the range of an instantiation of op if instantiations of σ and the parameters of σ are given. We need a functor Γ_{op} because of the following reason: Given two instantiations of the type class which can be described by $(\zeta_1, \ldots, \zeta_n)$ and $(\zeta_1', \ldots, \zeta_n')$ such that

$$\zeta_i \trianglelefteq \zeta_i' \quad \forall i \leq n$$

it is necessary that

$$\Gamma_{op}(\zeta_1, \ldots, \zeta_n) \trianglelefteq \Gamma_{op}(\zeta_1', \ldots, \zeta_n').$$

Otherwise, if a_i is an object of type ζ_i, $1 \leq i \leq n$, the expression

$$op(a_1, \ldots, a_n)$$

has the types $\Gamma_{op}(\zeta_1, \ldots, \zeta_n)$ and $\Gamma_{op}(\zeta_1', \ldots, \zeta_n')$ for which a coercion has to be defined in order to give the expression a unique meaning.

If σ is a non-parameterized type class *any* mapping assigning an appropriate type to a tuple $(\zeta_1, \ldots, \zeta_n)$ can be extended to a functor. So the requirement that Γ_{op} is a functor is only a restriction for parameterized type classes.

Since there are unique coercions between types in a coherent type system, we omit the names of the coercions and write

$$\Gamma_{op}(\zeta_1 \trianglelefteq \zeta_1', \ldots, \zeta_n \trianglelefteq \zeta_n')$$

for the image of the single arrow between the objects

$$(\zeta_1, \ldots, \zeta_n) \text{ and } (\zeta_1', \ldots, \zeta_n')$$

in the category

$$\mathcal{C}_1 \times \cdots \times \mathcal{C}_n$$

under the functor Γ_{op}. Thus $\Gamma_{op}(\zeta_1 \trianglelefteq \zeta_1', \ldots, \zeta_n \trianglelefteq \zeta_n')$ is an arrow in $\mathcal{C}_{n+1}$.

Let SET be the category of all sets as objects and functions as arrows.[7] By the assumption of set theoretic ground types and coercion functions we can assign to any object of $\mathcal{C}_\sigma$ an object of SET and to any arrow in $\mathcal{C}_\sigma$ an arrow of SET in a functorial way. We write $\mathbf{T}_{\mathcal{C}_\sigma}$ for the functor defined by this mapping.

We use the notation $\zeta_i \trianglelefteq \zeta_i'$ to denote the single arrow between ζ_i and ζ_i' in $\mathcal{C}_i$. Thus

$$\mathbf{T}_{\mathcal{C}_1} \times \cdots \times \mathbf{T}_{\mathcal{C}_n}(\zeta_1 \trianglelefteq \zeta_1', \ldots, \zeta_n \trianglelefteq \zeta_n')$$

is an arrow in the category

$$\underbrace{\text{SET} \times \cdots \times \text{SET}}_{n}.$$

Since n-tuples of sets are sets there is a functor from SET^n into SET which is denoted by $\mathbf{F}_n$.

---

[7] Note that the category SET is quite different from AXIOM's `SetCategory`.

If $(\zeta_1, \ldots, \zeta_n)$ is an object in $C_1 \times \cdots \times C_n$ we are now ready to formalize a requirement on the instantiation of op given by $(\zeta_1, \ldots, \zeta_n)$. We do not impose this condition directly on $op_{(\zeta_1, \ldots, \zeta_n)}$. It is convenient to regard the set-theoretic interpretation

$$T_{C_1} \times \cdots \times T_{C_n}(\zeta_1, \ldots, \zeta_n)$$

of $(\zeta_1, \ldots, \zeta_n)$ instead this n-tuple of types itself. Then the set-theoretic interpretation of $op_{(\zeta_1, \ldots, \zeta_n)}$ induces a function between

$$F_n(T_{C_1} \times \cdots \times T_{C_n}(\zeta_1, \ldots, \zeta_n))$$

and

$$T_{C_{n+1}}(\Gamma_{op}(\zeta_1, \ldots, \zeta_n)),$$

which is denoted by $O_{op}(\zeta_1, \ldots, \zeta_n)$.

Given $(\zeta_1, \ldots, \zeta_n)$ and $(\zeta_1', \ldots, \zeta_n')$ such that

$$\zeta_i \trianglelefteq \zeta_i' \quad \forall i \leq n$$

we require the following diagram to be commutative:

$$F_n(T_{C_1} \times \cdots \times T_{C_n}(\zeta_1, \ldots, \zeta_n)) \xrightarrow{O_{op}(\zeta_1, \ldots, \zeta_n)} T_{C_{n+1}}(\Gamma_{op}(\zeta_1, \ldots, \zeta_n))$$

$$\left\downarrow L \qquad\qquad\qquad\qquad \right\downarrow R$$

$$F_n(T_{C_1} \times \cdots \times T_{C_n}(\zeta_1', \ldots, \zeta_n')) \xrightarrow{O_{op}(\zeta_1', \ldots, \zeta_n')} T_{C_{n+1}}(\Gamma_{op}(\zeta_1', \ldots, \zeta_n'))$$

In the diagram above we have set

$$L = F_n(T_{C_1} \times \cdots \times T_{C_n}(\zeta_1 \trianglelefteq \zeta_1', \ldots, \zeta_n \trianglelefteq \zeta_n'))$$

and

$$R = T_{C_{n+1}}(\Gamma_{op}(\zeta_1 \trianglelefteq \zeta_1', \ldots, \zeta_n \trianglelefteq \zeta_n')).$$

This requirement on O_{op} can be read that O_{op} is a *natural transformation* between the functor

$$F_n \circ (T_{C_1} \times \cdots \times T_{C_n})$$

and the functor

$$T_{C_{n+1}} \circ \Gamma_{op}.$$

Thus for any n-ary first-order operator op the requirements are:

1. Given instantiations of a type class and its parameters the assignments of a range type for an operation has to be "functorial".
2. The instantiation of the operator has to correspond to a natural transformation between functors given the set-theoretic interpretations of the ground types and the coercions between them.

This guarantees that type classes and coercions interact nicely and that expressions involving *op* have a unique meaning.

A brief inspection of the examples of parameterized type classes occurring in AXIOM has suggested that there is no example violating the first requirement which always holds in non-parameterized type classes. Nevertheless, a formal requirement for a computer algebra language seems to be useful to ensure that no such violation will occur in future extensions.

The second requirement is formulated as one on the possible instantiations of operators. However, it can also be read that given the instantiations only certain coercions between base types are allowed, namely only coercions for which the interpretation is a natural transformation. In the next section we show that using this view most coercions have to be injective.

Remark. Our conditions imposed above on the combination of type classes and coercions are an adaptation of the work of Reynolds [10] on *category-sorted algebras*. The difference is that Reynolds allows each operator to be generic, i.e. that it may be instantiated with any type in any position. We allow type-class polymorphism at some position and do not allow polymorphism at all in other positions which seems to be the natural way to describe many important examples.

An extension of the work of Reynolds to higher-order functions can be found in [6].

4 Some Applications of the Framework

4.1 Injective Coercions

An important type class is the one on which a test for equality of objects can be performed in the system (called Eq in Haskell and SetCategory in AXIOM). In this type class the operator

$$= \; : \; t_{\mathsf{Eq}} \times t_{\mathsf{Eq}} \longrightarrow \texttt{Boolean}$$

is used to denote the system test for equality. In order to distinguish between the "system equality" and "true equality" we use

$$\texttt{isequal} \; : \; t_{\mathsf{Eq}} \times t_{\mathsf{Eq}} \longrightarrow \texttt{Boolean}$$

for the system equality in the following.

Given types A and B, both being instances of the "equality type class", and a coercion ϕ between A and B, then the following diagram has to commute (where $\mathrm{id}_{\texttt{Boolean}}$ is the identity coercion on booleans):

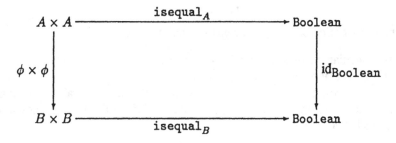

Hence, the boolean values of

$$\texttt{isequal}(a_1, a_2)$$

and

$$\texttt{isequal}(\phi(a_1), \phi(a_2))$$

have to be the same. Especially, if the latter evaluates to `true` then the former should also evaluate to `true`. In analogy to the definition of injectivity this means that ϕ has to be an injective function with respect to system equality (usually, the definition of injective involves true equality). Thus, coercions between types belonging to the equality type class need to be injective.

Of course, a non-injective coercion function would not violate our requirements, if A and B do not use the same operator symbol as a test for equality. Therefore, defining two different type classes EqA and EqB with operators `isequalA` resp. `isequalB` as tests for equality and having A of type class EqA and B of type class EqB would allow to define a non-injective function to be a coercion between A and B. However, such a construction is non-intuitive and we lose the advantages of type classes.

Since it is an undecidable problem to check whether a given recursive function is injective, it is not possible in general to enforce by a compiler that coercions are injective if functions defined by arbitrary code can be declared as coercions. Nevertheless, it seems to be useful to state this requirement as a guideline for a programmer.

Some frequent examples of coercions in a computer algebra system which are not injective are the following:

- The mapping from the (arbitrary precision) integers $\mathbb{Z}$ into a residue class ring $\mathbb{Z}_m$ for $m \in \mathbb{N}$.
- The mapping of the arbitrary precision integers into floating point numbers. Even in the case of range arithmetic instead of floating point numbers (see e. g. [1]), this mapping is not injective, because at some point the decimal width of the range has to be given.

4.2   Greatest Common Divisors

In a principal ideal domain P the gcd of two elements $a, b \in P$ is the element generating the ideal (a, b). Since a field contains only the two trivial ideals (0) and (1) – being therefore also a principal ideal domain – the gcd can be defined in an any field as

$$\gcd(a, b) = \begin{cases} 0, & \text{if } a = b = 0 \\ 1, & \text{else} \end{cases}$$

Hence, given the canonical coercion ϕ from $\mathbb{Z}$ into its quotient field $\mathbb{Q}$, the following diagram does not commute:

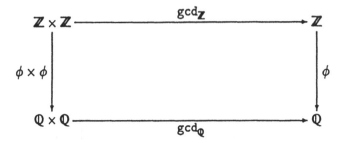

If we want to stay within Reynolds' framework we have to

- use different identifiers for the gcd operations on $\mathbb{Z}$ and $\mathbb{Q}$, respectively, or
- avoid the coercion from $\mathbb{Z}$ into $\mathbb{Q}$.

Both possibilities are unsatisfactory.

4.3 Examples of Languages and Systems Contradicting Reynolds' Condition

This subsection discusses some examples which arise in concrete languages and systems that provide coercions and generic operators.

An example of non-intuitive behaviour can be found in the language C++ [11]: the division operator denotes for floating-point values division of floating-point numbers but for integers the quotient operation! Furthermore, there is a coercion from integers to floating point numbers. With this coercion and a context where the identifier d is declared to denote a double, the following assignment is legal:

```
d = 1 / 2;
```

To give the identifier d a unique meaning – it is the floating point representation for 0 and not the equally senseful value 0.5 – some ad hoc coercion rules are given to prevent ambiguities.

The following example of opaque behaviour occurs in AXIOM V. 1.1: The input

```
t : Fraction(Integer) := gcd(2,4)
```

results in

```
1
                                    Type: Fraction Integer
```

This means that the arguments of the gcd operation are coerced to rational numbers *before* the application is evaluated. Hence, the gcd for rational numbers is used, which returns always 1 (which is in fact false for the argument $(0,0)$). But it would be equally intuitive to apply the integer gcd operation first and then coercing the result to rationals resulting in $2/1$. In any case, the expression is ambiguous, and therefore it should be handled with care, say by emitting a warning or prohibiting it.

As opposed to the above example the following one, in which the operator gcd is replaced by the operator +, does not lead to any problem:

```
t : Fraction(Integer) := 1 + 2
```

resulting in

3

```
                              Type: Fraction Integer
```

because addition of integers and rationals, respectively, commutes with the canonical embedding of $\mathbb{Z}$ into $\mathbb{Q}$.

The examples show that some mechanism is needed that is less restrictive than Reynold's condition but not so ad hoc and non-intuitive as in C++ or AXIOM. In the next section such a mechanism is presented.

5 Relaxing the Strict Conditions

Let op be a n-ary operation. We assume that op is given a profile of the form

$$op : \xi_1 \times \cdots \times \xi_n \longrightarrow \xi_{n+1},$$

where ξ_i, $1 \leq i \leq n + 1$, is either a type variable v_{τ_l}, $l \leq k$, or a ground type $\bar{t}_i$. Given objects $o_1, \ldots, o_n$ having types $t_1, \ldots, t_n$, respectively, the expression

$$op(o_1, \ldots, o_n)$$

is well-typed having type ξ_{n+1} if and only if the following conditions are satisfied:

1. If $\xi = \bar{t}_i$ for some ground type $\bar{t}_i$ then $t_i \trianglelefteq \bar{t}_i$.
2. If $\xi_i = \xi_j = v_{\tau_l}$ for some $i \neq j$ then there is a type $t : \tau_l$ such that $t_i \trianglelefteq t$ and $t_j \trianglelefteq t$.
3. If $\xi_i = v_{\tau_k}$ then there is a type $t : \tau_k$ such that $t_i \trianglelefteq t$.

If we require that all objects have ground types — if they have a type at all — then algorithms solving the problems imposed by the above conditions can be used to solve the type inference problem using a bottom-up process.[8] We refer to [15] for algorithms which solve the given subproblems for important cases of coercions occurring in the context of computer algebra.

In the following we assume that the type of an expression can be determined by such a bottom-up type inference algorithm and that for any two ground types t_1 and t_2 there is a *finite* set of minimal upper bounds.

Then the algorithm in Fig. 5 nicely distinguishes between cases in which an ambiguity can be resolved and cases in which it cannot. It uses given "commutativity tuples", which are 4-tuples of operators and their types:

$$(op_1 : \alpha \to \beta, \ \phi_1 : \beta \to \delta, \ \phi_2 : \alpha \to \gamma, \ op_2 : \gamma \to \delta)$$

expressing the commutativity of the diagram

---

[8] This was already seen in [3, Sec. 4] and in [9].

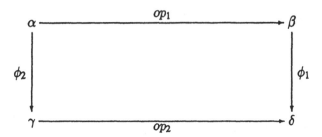

We assume that these commutativity tuples are given, for instance by explicit user declaration or by theorem proving.

> **Input:** an expression e that typechecks correctly.
> **Output:** true if e does not contain any unresolvable ambiguous operators, false otherwise.
>   (1) Perform a bottom-up type inference for e assigning the complete set of minimal upper bounds A_n to each node n of the tree g_e representing the expression e.
>   (2) For each node n in the tree representing the expression e let B_n be the union of all sets A_m where m is a node on the path between the root of g_e and n (including n).
>   (3) For all nodes n in g_e perform the following check:
>   (*) If there are commutativity tuples for the generic operator assigned to node n and for all types in B_n there are commutativity tuples then the check holds for node n otherwise it fails.
>   (4) If the check (*) holds for all nodes n in g_e then return true else return false.

Fig. 1. Algorithm "check for resolvable ambiguity"

Provided the canonical coercion between $\mathbb{Z}$ and $\mathbb{Q}$, generic operators, and commutativity tuples are available, the algorithm yields true if it is applied to the following expressions:

```
gcd(5,10)
gcd(5,2/3)
```

This makes sense because in the former expression only the integer gcd can be meant without applying a coercion. In the latter example, the gcd operator can only denote the gcd on the rationals.

However, in the following example the algorithm would return false, because we do not know if we should coerce *before* or *after* the applications:

```
gcd(5,10)/5
```

Hence, the algorithm forbids expressions which do not have a unique meaning.

5.1 Forcing an Expression to a Certain Type

Let force_to_type be a generic operator with profile

$$\text{force\_to\_type} : v \times v \longrightarrow v$$

where v is a type variable without a constraint, i.e. having the topmost sort of the order-sorted signature as a constraint.[9] We interpret force_to_type as the projection on the first coordinate, i.e.

$$\text{force\_to\_type}(e_1, e_2) = e_1$$

for all expressions e_1 and e_2 such that

$$\text{force\_to\_type}(e_1, e_2)$$

typechecks.

If for any type t we have an expression k_t of type t in the system then we can interpret an AXIOM assignment statement

$$r \; : \; ResultType := expression$$

in the following way:

"assign

$$\text{force\_to\_type}(expression, k_{ResultType})$$

to the variable r."

Thus the algorithm described above can also be applied in these cases and detects unresolvable ambiguities. For instance, in the following case an error would occur:

```
t : Fraction(Integer) := gcd(14, 21)
```

6 Conclusions and Acknowledgements

We extended the work of [10] to generic operators with constraints and showed that the resulting framework can be too restrictive, especially in the context of computer algebra. Hence, we presented a relaxed condition including an algorithm which treats most ambiguities occurring in expressions in a natural way.

We would like to thank M. Monagan for drawing our attention to certain aspects of the problem which we have investigated. Furthermore, we thank G. Grivas, R. Maeder and N. Mannhart for reviewing a draft version of this paper.

Some of the results presented above are part of the second author's PhD-thesis [14] written under the supervision of Prof. R. Loos at the University of Tübingen. He is indebted to Prof. Loos for initiating and supervising his research.

The macros of M. Barr were used to produce the diagrams.

---

[9] Without losing generality we can assume that such a topmost sort does exist.

References

1. O. Abarth and M. J. Schaefer. Precise computation using range arithmetic, via C++. *ACM Trans. Math. Software*, 18:481–491, 1992.

2. A. V. Aho, R. Sethi, and J. D. Ullman. *Compilers — Principles, Techniques and Tools*. Addison-Wesley, Reading, MA, 1986.

3. H. Comon, D. Lugiez, and P. Schnoebelen. A rewrite-based type discipline for a subset of computer algebra. *Journal of Symbolic Computation*, 11:349–368, 1991.

4. P. Hudak, S. Peyton Jones, P. Wadler, et al. Report on the programming language Haskell — a non-strict, purely functional language, version 1.2. *ACM SIGPLAN Notices*, 27(5), May 1992.

5. R. D. Jenks and R. S. Sutor. *AXIOM: The Scientific Computation System*. Springer-Verlag, New York, 1992.

6. G. T. Leavens and D. Pigozzi. Typed homomorphic relations extended with subtypes. In S. Brookes, M. Main, A. Melton, M. Mislove, and D. Schmidt, editors, *Mathematical Foundations of Programming Semantics — 7th International Conference*, volume 598 of *Lecture Notes in Computer Science*, pages 144–167, Pittsburgh, PA, Mar. 1991. Springer-Verlag.

7. S. Mac Lane. *Categories for the Working Mathematician*, volume 5 of *Graduate Texts in Mathematics*. Springer-Verlag, New York, 1971.

8. B. C. Pierce. *Basic Category Theory for Computer Scientists*. The MIT Press, Cambridge, MA, 1991.

9. D. L. Rector. Semantics in algebraic computation. In E. Kaltofen and S. M. Watt, editors, *Computers and Mathematics*, pages 299–307, Massachusetts Institute of Technology, June 1989. Springer-Verlag.

10. J. C. Reynolds. Using category theory to design implicit conversions and generic operators. In N. D. Jones, editor, *Semantics-Directed Compiler Generation, Workshop*, volume 94 of *Lecture Notes in Computer Science*, pages 211–258, Aarhus, Denmark, Jan. 1980. Springer-Verlag.

11. B. Stroustrup. *The C++ Programming Language*. Addison-Wesley, Reading, MA, second edition, 1991.

12. A. Weber. A type-coercion problem in computer algebra. In J. Calmet and J. A. Campbell, editors, *Artifical Intelligence and Symbolic Mathematical Computation — International Conference AISMC-1*, volume 737 of *Lecture Notes in Computer Science*, pages 188–194, Karlsruhe, Germany, Aug. 1992. Springer-Verlag.

13. A. Weber. On coherence in computer algebra. In A. Miola, editor, *Design and Implementation of Symbolic Computation Systems — International Symposium DISCO '93*, volume 722 of *Lecture Notes in Computer Science*, pages 95–106, Gmunden, Austria, Sept. 1993. Springer-Verlag.

14. A. Weber. *Type Systems for Computer Algebra*. Dissertation, Fakultät für Informatik, Universität Tübingen, July 1993.

15. A. Weber. Algorithms for type inference with coercions. In *Proc. Symposium on Symbolic and Algebraic Computation (ISSAC '94)*, pages 324–329, Oxford, July 1994. Association for Computing Machinery.

Theories =
Signatures + Propositions Used as Types

Stephan A. Missura*

Institute for Theoretical Computer Science
ETH Zürich, CH-8092 Zürich, Switzerland
E-mail: missura@inf.ethz.ch

Abstract. Languages that distinguish between types and structures use explicit components for the carrier type(s) in structures. Examples are the functional language Standard ML and most algebraic specification systems. Hence, they have to use general sum types or signatures to give types to structures and to be able to build, for instance, the algebraic hierarchy.

Furthermore, in most languages the modelling of properties that the elements of a signature – the structures – should fulfill is not possible or computationally not relevant. This is a major drawback, especially in a computer algebra environment.

This paper presents a calculus for building signatures with explicit type components which are needed for modelling many-sorted structures and structures of the same signature with identical carriers. Additionally, we provide the possibility of using propositions as components needed for modelling the properties of structures. Propositions, which are distinguished from booleans, are inhabited by their proofs and hence, they are types themselves. This idea stems from the "propositions as types" principle known from constructive logic and results in a coherent treatment of carrier types, operations and their properties. It allows us to express theories (specifications) while staying within the framework of signatures.

The framework is used for the construction of function spaces between types *with* equality and the building of some parts of the algebraic hierarchy.

1 Introduction

The functional language Standard ML [17] with its powerful module system and most algebraic specification systems such as OBJ3 [12] distinguish types – called *sorts* in algebraic specification languages – from structures and assign types to the latter in the form of signatures or theories.

---

* Research supported by the Swiss National Science Foundation.

This distinction is analogous to mathematics where an algebraic structure is distinguished from its underlying carrier set(s).[2] Therefore, the languages mentioned can construct

- many-sorted structures such as modules or vectorspaces including their corresponding signatures
- *different* structures of the same signature with *identical* carrier type(s). For instance, the *type* string can be ordered in various ways resulting in *different structures* over the strings. Or the natural numbers can be viewed in two different ways as a monoid using addition and multiplication, respectively, as the binary operation.

Such constructions are especially useful in computer algebra environments. Additionally, functors as in Standard ML and views as offered in [1, 9, 12] allow a style of programming in the large [11].

Typed languages that do not distinguish between types and structures have difficulties in expressing the constructions mentioned. Examples are the language in the computer algebra system Axiom [15], the functional language Haskell [14] or our experimental type system presented in [19]. The carrier type is not *explicit* in these languages but *implicit*. Thus, they do not have to provide type components in signatures[3] and do not need dependent sum types[4], as provided, for instance, by Standard ML's signatures.

We present a signature calculus with the signature operations of adding and removing entries, combining signatures and renaming component names. Because of the advantage of distinguishing types from structures, we allow the use of explicit type components using the kind "Type".

Furthermore, the possibility of modelling properties, which the structures (the "models") of a signature should fulfill, is not possible or computationally not relevant in most languages. This is a major drawback, especially in computer algebra environments. Therefore, the calculus presented allows the use of properties in signatures in the form of propositions. They are inhabited by their proof and hence, they are types themselves. This results in a coherent handling of carrier types, operator and property components, unifying signatures with theories.

Finally, the framework presented will be used to construct function spaces between types with equality and to give some examples of theories appearing in computer algebra.

---

[2] However, not always on the syntactic level: Say in Lang's Algebra [16], a typical algebra textbook, the identifier $\mathbb{Z}$ is used simultaneously for the *set* of integers and the *ring* over the integers.

[3] Signatures correspond to categories in Axiom and type-classes in Haskell.

[4] The elements of a dependent sum type are tuples where the type of a component can depend on preceding components [2, 26].

2   Propositions as Types

Here, we take a constructive standpoint and view propositions in a purely formal way. We are not interested in their truth value under some semantic interpretation but only in whether they are *provable* [22]. Predicates or relations are then functions returning a proposition.

It is important to note that propositions are *not* booleans, the latter being true or false. Only those propositions which are *decidable*, and therefore fulfill the law of the excluded middle [26], can be treated as booleans.

The term "propositions as types" comes from the so called "Curry-Howard" isomorphism [13] between propositions and types in certain type systems [26], resulting in a bijective mapping between proofs and programs. In fact, there are type theories such as Martin-Löf's [22, 23] where both notions are *identified*. In both cases, propositions are inhabited by their proofs, i.e. the type of a proof is its proved proposition. Hence, propositions become types themselves.

We call this property "propositions as types" as opposed to [22, 23, 26] where this term is used exclusively for the Curry-Howard isomorphism and use here only the notion that propositions are inhabited by their proofs, similarly to the calculus of constructions [8], which uses propositions as types but not the Curry-Howard isomorphism.

Because a necessary condition for the Curry-Howard isomorphism is the convergence of all programs since proofs have to terminate, there is no general recursion available in systems using this isomorphism [4]. This seems to be too strong a restriction for a computer algebra environment.

3   Signatures

A signature is a sequence of declarations where identifiers are associated to types. Because the type of an entry in a signature can depend on other preceding entries, signatures are syntactic sugar for dependent sum types [20].

This is an essential difference from the usual presentation in algebraic specification [27], where a signature is a pair (S, Σ) consisting of a set S of sorts and a family Σ of operators indexed by non-empty strings of sorts. Hence, sort and operator declarations are handled differently, as opposed to the definition we use.

3.1   The syntax of signatures

The presented concrete syntax of signatures and their combining operators is influenced by the record type language introduced in [6] and the syntax of the generic theorem prover Isabelle [24].

The language of signatures is given by the grammar in Figure 1, where the non-terminal symbol *Ident* stands for some class of identifiers and the language of *Type* is introduced informally in section 3.3. Signatures are built by

– starting with the empty signature <> as the base case of the grammar,

- adding a declaration as an additional entry, using the operator + (needed for extending a signature by a component),
- combining two signatures with, once again, the operator +,
- deleting an entry using the operator –, and finally,
- renaming the name of an entry and those types which depend on this entry. The operation [$x := y$] should remind the usual substitution operator for terms.

$$
\begin{aligned}
Signature \rightarrow\ & \text{“<>”} \\
|\ & Signature\ \text{“+”}\ Entry \\
|\ & Signature\ \text{“+”}\ Signature \\
|\ & Signature\ \text{“–”}\ Ident \\
|\ & Signature\ \text{“[”}\ Ident\ \text{“:=”}\ Ident\ \text{“]”}
\end{aligned}
$$

$$
Entry\ \ \rightarrow Ident\ \text{“:”}\ Type
$$

Fig. 1. The grammar for signatures

Similarly to [6], we use the form

$$< x_1 : T_1 \ \ \dots \ \ x_n : T_n >$$

as an abbreviation for the signature

$$\text{<>} + x_1 : T_1 + \dots + x_n : T_n$$

Σ, Σ_1 and Σ_2 will denote arbitrary signatures in the remainder of this section. Additonally, we need the following operations on signatures:

$$
\begin{aligned}
\text{Let} \quad \Sigma\ &= <x_1 : T_1, \dots, x_n : T_n> \\
dom(\Sigma) &= \{x_1, \dots, x_n\} \qquad \text{(component names)} \\
ran(\Sigma) &= \{T_1, \dots, T_n\} \qquad \text{(component types)} \\
\Sigma.x_i &= T_i \qquad\qquad\quad \text{(component access)}
\end{aligned}
$$

3.2 Operational semantics

The operational semantics, i.e. how a signature can be "simplified", is given through the definition of the relation "$\Rightarrow$" on signatures. We define it separately for each operation:

- The addition of an entry with a name already existing in Σ is simplified to Σ (the inference rules given later do not allow arbitrary additions of an entry but only well-formed ones):

$$\Sigma + x:T \Rightarrow \Sigma \qquad (x \in dom(\Sigma))$$

- Combination of two signatures:

$$\Sigma + <> \Rightarrow \Sigma$$
$$\Sigma + <x:T> \Rightarrow \Sigma + x:T$$
$$\Sigma + <x_1:T_1, \ldots, x_n:T_n> \Rightarrow \Sigma + x_1:T_1 + $$
$$<x_2:T_2, \ldots, x_n:T_n>$$

- Deleting an entry simply throws out components with the corresponding name:

$$\Sigma - x \Rightarrow \Sigma \qquad (x \notin dom(\Sigma))$$
$$\Sigma + x:T - x \Rightarrow \Sigma - x$$
$$\Sigma + y:T - x \Rightarrow \Sigma - x + y:T \qquad (x \not\equiv y)$$

- Renaming of component names can be defined with the help of the usual substitution operator "$[M/x]$" operating on terms. It replaces all free occurrences of the identifier x with the term M:

$$<> \ [y:=z] \Rightarrow <>$$
$$\Sigma + <x:T> \ [y:=z] \Rightarrow \Sigma \ [y:=z] + x[z/y]:T[z/y]$$

Viewing the relation $\Rightarrow$ as a term rewriting system, we get a normal form for each signature. This normal form uses only the empty signature and the operator + to add a single entry.

3.3 The valid signatures

As already mentioned, signatures are dependent types. Therefore, we have to use inference rules to define the sublanguage of all *valid* (well-formed) signatures. The following judgements are used within the rules where the last two use Σ as a context of type declarations:

- The signature Σ is validly formed:
    Σ *valid*
- Under the assumption of the signature Σ a type declaration can be derived:
    $\Sigma \vdash x : T$
- Using the declarations of Σ the term T is a valid type:
    $\Sigma \vdash T \ type$

$$\overline{\langle\rangle \; valid}$$

$$\frac{\Sigma \; valid \quad \Sigma \; \vdash \; T \; type}{\Sigma + x : T \; valid} \quad (x \in dom(\Sigma) \;\rightarrow\; \Sigma.x \equiv T)$$

Fig. 2. The inference rules for valid signatures

Because every signature can be brought into a normal form, the rule schemes in Figure 2 – defining all valid signatures – are only defined for the empty signature, which is well-formed, and the addition of an entry. Adding an entry is only allowed if we can deduce a valid type declaration (the corresponding rule is given in Figure 3). In case the name of the new entry is already present within the signature, an addition is only allowed if both types of the corresponding entries are identical, hence, we need some equality on types. This behaviour is suitable for combining signatures with the same carrier type component because the relation $\Rightarrow$ throws the second component away.

$$\frac{\Sigma_1 + x : T + \Sigma_2 \; valid}{\Sigma_1 + x : T + \Sigma_2 \; \vdash \; x : T}$$

Fig. 3. The inference rule for type declarations

The language induced by the non-terminal symbol *Type* (in Figure 1) will be given through the type system of the underlying core language and can, in fact, vary. The same is true for an underlying language of terms, including structures and proofs, which depend on a concrete type system. Thus, we don't define a formal type system but for the examples given later we need at least

- the kind of all types [5], written as Type, and the kind of all propositions [8], Prop,
- product types formed by the operator * and function types[5] by ->,
- conjunctions (and), implications (->) and ∀-quantification (forall ...) for building propositions, and
- the minimal set of inference rules given in Figure 4 for forming valid types.

The possibility of having proposition components in our signatures allows the use of a signature immediately as a theory, unifying the two usually distinguished concepts. Therefore, checking if a structure is a model of a theory is simply *type-checking*.

---

[5] Depending on whether the underlying core language also provides partial functions, the logic should be partial [22]. To simplify the presentation we assume we have only total functions available, hence we need not provide propositions in signatures which define the totality of the operations.

$$\frac{\Sigma \ valid}{\Sigma \ \vdash \ \text{Type} \ type}$$

$$\frac{\Sigma \ valid}{\Sigma \ \vdash \ \text{Prop} \ type}$$

$$\frac{\Sigma \ \vdash \ T : \text{Type}}{\Sigma \ \vdash \ T \ type}$$

$$\frac{\Sigma \ \vdash \ P : \text{Prop}}{\Sigma \ \vdash \ P \ type}$$

Fig. 4. The inference rules for valid types

4 Equality: From Types to Sets

In constructive mathematics only those equalities that can be effectively constructed and proven to be an equivalence relation are available [22]. Hence, a type[6] is not equipped with an equivalence (which is senseful because there are theories which don't need an equality) unless we define one and prove it to be really an equivalence relation. Thus, a type can be equipped with more than one equality.

These remarks result in the definition of the theory Set which serves as a first example for a theory. To simplify the reading, we use in this and the coming examples infix syntax for operators such as =, <, *:

```
Set :=
  <
    T      : Type
    =      : T * T -> Prop
    refl   : forall x:T     . x=x
    symm   : forall x,y:T   . x=y -> y=x
    trans  : forall x,y,z:T . x=y and y=z -> x=z
  >
```

This means that a structure having the signature Set as type provides components with a carrier type T, a relation = on this carrier and proofs for reflexivity, symmetry and transitivity of this relation.

The example shows also the dependencies of signature components from preceding components, i.e. the syntactic sugared version of dependent sums.

5 From Functions on Types to Functions on Sets

The function space $\alpha \rightarrow \beta$ is defined for arbitrary *types* α, β and not *sets*. Hence, there are functions which do not respect any available equalities on α and β,

---

[6] Comparable to the term "preset" used in [22].

respectively, and we say that a function f is not *well-defined* on α and β with respect to these equalities:

$$\exists x, y : \alpha \;.\; x =_\alpha y \;\Rightarrow\; f(x) \neq_\beta f(y)$$

Therefore, we define the theory of well-defined functions between two sets and call them morphisms[7]:

```
Mor (A : Set, B : Set) :=
   <
      f          : A.T -> B.T
      well-def : forall x,y:A.T . x A.= y -> f(x) B.= f(y)
   >
```

A model of $\text{Mor}(\alpha, \beta)$ consists of

- a function on the carrier types of α and β
- a proof of well-definedness of the function with respect to the equalities.

An alternative to the above representation of well-definedness of functions are *quotient types* as in Nuprl [7] (as opposed to our treatment of types Nuprl assigns to *every* type a default equality).

6   Examples from computer algebra

Most theories used in computer algebra need an equality for their axiomatisation and therefore build on the theory Set. One of the few examples that does not need one is the theory of pre-orders:

```
PreOrder :=
   <
      T      : Type
      <=     : T * T -> Prop
      refl  : forall x:T      . x<=x
      trans : forall x,y,z:T . x<=y and y<=z -> x<=z
   >
```

The following examples often arise in computer algebra and should justify the signature operations presented:

- Adding an entry is the usual operation for extending a given theory. Say, we create the theory of monoids by extending the theory SemiGroup (assuming a product type constructor * on sets):

---

[7] The dot operator is used as usual to extract a component from a structure.

```
SemiGroup :=
  <
    S      : Set
    *      : Mor(S * S, S)
    assoc : forall x,y,z:S.T .
                          ((x *.f y) *.f z) S.= (x *.f (y *.f z))
  >


Monoid := SemiGroup +
    1      : S.T +
    ident : forall x:S.T . (x *.f 1) S.= (1 *.f x) S.= x
```

Because the components S and * are structures the underlying carrier type and function, respectively, have to be extracted.

– An example for renaming is the definition of abelian monoids which renames the components of Monoid to the usual additive operators + and 0. Then the commutativity property is appended:

```
AbelianMonoid := Monoid [* := +] [1 := 0]  +
    comm : forall x,y:S.T . (x +.f y) S.= (y +.f x)
```

– For the construction of the theory of rings or semi-rings the combination of two signatures is useful (only the distributivity law has to be added):

```
SemiRing := AbelianMonoid + Monoid +
    distr : forall x,y,z:T .
                (z *.f (x +.f y)) S.= (z *.f x +.f z *.f y)   and
                ((x +.f y) *.f z) S.= (x *.f z +.f y *.f z)
```

– Removing fields is normally not used to build hierarchies because the usual way is going from weaker theories – with more models – to stronger ones having less models.

As a final example we can express what homomorphisms between two monoids are: they are structures consisting of a function h between the carriers of the monoids and proofs that h really preserves the identity and the binary operation:

```
Hom(A:Monoid, B:Monoid) : Type :=
  <
    h       : A.S -> B.S
    pres-1 : f(A.1) B.S.= B.1
    pres-* : forall x,y:A.S.T .
                      (h.f(x A.*.f y)) B.S.= (h.f(x) B.*.f h.f(y))
  >
```

7 Comparison to related work

The main difference to Standard ML's module language is that the latter provides no support for expressing properties. Furthermore, the operations required to build new signatures from old ones are rather weak in Standard ML. They do not allow building the algebraic hierarchy in a natural way by extending and combining given signature definitions.

Languages such as Extended ML [25], the work in [21], OBJ3, and Formal [3] provide an algebraic specification style including operations on specifications. This style is usually based on first-order equational logic and distinguishes type declarations, operator declarations, and equational axioms as components in a theory and does not use propositions as types.

Axiom, identifying types with structures (called domains), provides similar signature operations, except for the renaming of components. But properties of domains are computationally irrelevant, as in the case of the polymorphic functional language XFun [9], too. Hence, we can have in these two systems "models" of theories which in fact don't fulfill the axioms in their theory. Furthermore, Axiom builds its algebraic hierarchy on the category SetCategory, which provides a decidable equality. So we can't have domains with undecidable equality (say function spaces) within the hierarchy.

8 Conclusions and further work

The novelty is the use of propositions as types in signatures instead of the usual presentation with dependent sums [7, 23, 26]. This results in more readable specifications and an easier way of building new signatures from older ones through signature operations, as demonstrated by building some parts of the algebraic hierarchy.

The coherent handling of type, operator, and proposition declarations allows the use of signatures as theories, in fact unifying the two concepts, and a coherent component selection mechanism for structures (models of theories). Furthermore, using propositions as types, proofs and propositions become first class objects, similarly to types. This is especially interesting for future computer algebra environments.

Currently, we are investigating partial logics – suitable for a logical core language in a computer algebra environment – such as Lutins of the IMPS system [10] or the logic of partial terms [22]. In the future, support by theorem provers is needed because the creation of proofs for propositions should be done as much as possible mechanically.

Additionally, we want to minimize the shortcomings of distinguishing types from structures, especially the extraction of various operators out of the structures, which makes the specifications difficult to read. The availability of an "open" construct for using a structure as an additional environment such as in Standard ML, and the combination of forgetful functors and coercions [18] seems to be a promising approach to this problem allowing a closer approach to "every day" mathematical notation.

Discussions with P. Baumann, M. Seyfried and A. Weber were very helpful to improve the content of this paper, as were the remarks of D. Basin and P. Jackson about the importance of well-definedness of operations. Furthermore, I would like to thank R. Maeder, N. Mannhart and D. Stotz for reviewing draft versions of this paper.

References

1. K. Abdali, G. Cherry, and N. Soiffer. An Object Oriented Approach to Algebra System Design. In *Symposium on symbolic and algebraic manipulation*, pages 24–30. ACM, 1986.

2. R.M. Burstall and B. Lampson. A Kernel Language for Abstract Data Types and Modules. In *Semantics of Data Types*, volume 173 of *Lecture Notes in Computer Science*, pages 1–50. Springer Verlag, 1988.

3. J. Calmet and I. Tjandra. A Unified-Algebra-based Specification Language for Symbolic Computing. In *Design and Implementation of Symbolic Computation Systems (Proceedings of DISCO '93)*, volume 722 of *Lecture Notes in Computer Science*, pages 122–133. Springer Verlag, 1993.

4. L. Cardelli. A Polymorphic λ-calculus with Type:Type. Technical Report 10, DEC Systems Research Center, 1986.

5. L. Cardelli. Typeful Programming. Technical Report 45, DEC Systems Research Center, 1989.

6. L. Cardelli and J.C. Mitchell. Operations on Records. *Mathematical Structures in Computer Science*, 1(1):3–48, 1991.

7. R.L. Constable et al. *Implementing mathematics with the Nuprl proof development system*. Prentice-Hall, 1986.

8. T. Coquand et al. The Calculus of Constructions. Technical report, INRIA, 1989.

9. S. Dalmas. A polymorphic functional language applied to symbolic computation. In *ISSAC*, 1992.

10. W.M. Farmer, J.D. Guttman, and F.J. Thayer. *IMPS User's Manual*. The MITRE Corporation, September 1993.

11. Joseph A. Goguen. Types as theories. In *Topology and category theory in computer science*, pages 357–390. Oxford science publications, 1991.

12. Joseph A. Goguen et al. Introducing OBJ. Technical report, SRI International, March 1992.

13. W.A. Howard. The formulae-as-types notion of constructions. In *to H.B. Curry: Essays on Combinatory Logic, Lambda Calculus and Formalism*. Academic Press, 1980.

14. P.R. Hudak et al. Report on the programming language Haskell, a non-strict purely functional language, Version 1.2. *ACM SIGPLAN Notices*, 1992.

15. Richard D. Jenks and Robert S. Sutor. *Axiom: The Scientific Computation System*. Springer, 1992.

16. Serge Lang. *Algebra*. Addison-Wesley, 2nd edition, 1984.

17. Robin Milner, Mads Thorpe, and Robert Harper. *The Definition of Standard ML*. The MIT Press, 1990.

18. Stephan A. Missura. Combining Forgetful Functors with Coercions. Unpublished Manuscript, 1993.

19. Stephan A. Missura. Extending AlgBench with a type system. In *Design and Implementation of Symbolic Computation Systems (Proceedings of DISCO '93)*, volume 722 of *Lecture Notes in Computer Science*. Springer Verlag, 1993.

20. J. Mitchell and R. Harper. The Essence of ML. In *Conference Record of ACM Symposium on Principles of Programming Languages*, pages 28–46, 1988.

21. John Mitchell, Sigurd Meldal, and Neel Madhav. An extension of Standard ML modules with subtyping and inheritance. In *Conference Record of the ACM Symposium on Principles of Programming Languages*, pages 270–278, 1991.

22. M.J.Beeson. *Foundations of Constructive Mathematics*. Springer-Verlag, 1985.

23. Bengt Nordström, Kent Petersson, and Jan M. Smith. *Programming in Martin-Löf's Type Theory*. Oxford Science Publications, 1990.

24. Lawrence C. Paulson. Introduction to Isabelle. Technical report, Computer Laboratory, University of Cambridge, 1993.

25. D. Sannella. Formal program development in Extended ML for the working programmer. Technical report, Laboratory for Foundations of Computer Science, University of Edinburgh, December 1989.

26. R. Turner. *Constructive Foundations for Functional Languages*. McGraw-Hill, 1991.

27. Jan van Leeuwen, editor. *Formal Models and Semantics*, volume B of *Handbook of Theoretical Computer Science*. Elsevier, 1990.

The Ideal Structure of Gröbner Base Computations

Stéphane Collart & Daniel Mall

Department of Mathematics
Federal Institute of Technology
CH-8092 Zurich, Switzerland

Abstract. A highly structured algorithm for the computation of Gröbner bases founded on the notion of toric degenerations of a polynomial ideal is described, the "Gröbner Stripping Algorithm". The algorithm relates the complex procedure of Gröbner base computations to the algebraic combinatorial structure of polynomial ideals in a particularly explicit manner, and bears an interesting analogy to the well-known permutation group "Schreier Algorithm" due to Sims.

1 Introduction

The purpose of this note is to show how the notion of *toric degeneration* of a polynomial ideal (cf. [CM94b], [CM94a]) can be used to describe a *structurally iterative* algorithm — the *Gröbner Stripping Algorithm* — for the computation of Gröbner bases (cf. [Buc85]). This algorithm is distinguished by its process of incrementally retrieving a Gröbner base of the ideal under consideration from appropriate initial structures, i.e. from (parts of) toric degenerations, these latter being generated by truncated polynomials.

From a practical viewpoint, the algorithm is structurally iterative in the sense that at any moment of its application to an ideal I it is engaged in imbricated replications of itself on a ladder of sub-ideals of a chain of toric degenerations of I. Moreover, it is always engaged in computations with polynomials which are, loosely speaking, *fragments* of the polynomials of the intended ideal. Another interesting aspect of the Gröbner stripping algorithm is its analogy to the well-known group-theoretic *Schreier Algorithm* introduced by Sims (cf. [Sim70]; [Leo80] for an elaboration).

Within the context of this conference, one should note that since the introduction of Gröbner bases by Buchberger (cf. [Buc65]) schemes and heuristics for computational improvements have been sought and proposed in forms directly applicable to Buchberger's original formulation (see e.g. [Mon92] for a discussion in this conference, 1992). Such heuristics, by and large, revolving around the question of so-called 'S-pair' selection, do not take into consideration the internal structure of the ideal, or more correctly, it is not understood how they do. One may also pursue the question of whether certain AI methods, such as for instance search techniques, can be made of use. Ultimately, by experience one will expect the heuristics and techniques to evolve with further insight to actual algorithms (cf. [CC92]).

Clearly, it is necessary as a prerequisite to recognise first of all which specific information about the ideal can be used to improve heuristics for the selection of S-polynomials, and second, how can the relevant information be efficiently extracted. The Hilbert function, e.g., of the ideal is a celebrated piece of information of this kind. It can be sometimes used as a stop device during the selection of S-polynomials. In this note we are concerned with the combinatorial structure of the ideal expressed in its toric degenerations. To some extent, the Gröbner stripping algorithm explicates the relationship between completion of a polynomial reduction system and the inherent combinatorial structure of the ideal, and the algorithm decribed has an interpretation as a distinguished class of selection strategies. By casting light on the structural relationship between polynomial ideals and Gröbner bases this investigation seeks as a contribution in this place to provide means for a more intrinsic description of the meaning and performance of polynomial completion.

Notation: Let $\Bbbk$ be a field, $A := \Bbbk[x_1, \ldots, x_\ell]$ a polynomial ring, and $\mathcal{J}(A)$ the set of ideals of A. The set of terms $\{x_1^{n_1} \cdots x_\ell^{n_\ell} \mid n_i \in \mathbb{N}, i = 1, \ldots, \ell\}$ is denoted with T^ℓ and the set of admissible term orders with TO_ℓ (see for example [Rob85] and [Wei87]). With 'log' we denote the usual monoid homomorphism between $\Bbbk \cdot T^\ell$ and $\mathbb{N}^\ell$ mapping $c x_1^{n_1} \cdots x_\ell^{n_\ell}$ to $(n_1, \ldots, n_\ell)$. If $f \in A$ then $supp(f)$ denotes the set of *monomials* occuring with non-zero constant coefficient in the distributive normal form of f; for $\prec \in TO_\ell$, $in_\prec f$ denotes the greatest *monomial* in $supp(f)$ with respect to $\prec$ and is called the *initial monomial* of f. For a set of polynomials G, $in_\prec G$ is the set $\{in_\prec f \mid f \in G\}$, and the ideal generated by G is denoted $\langle G \rangle$; for $f, g \in A$, by $f \xrightarrow{(G, \prec)} g$ we mean that f is reducible to g by G with respect to $\prec$, and $R_\prec(I)$ denotes the unique reduced Gröbner base of I with respect to $\prec$ (cf. [Buc85]). The positive real vectors $\{(x_1, \ldots, x_\ell) \in \mathbb{R}^\ell \mid x_i \geq 0, i = 1, \ldots, \ell\}$ are denoted $\mathbb{R}_+^\ell$. As usual, δ_{ab} is the Kronecker symbol.

2 The Combinatorial Structure of Ideals and Gröbner Bases

The purpose of this section is to give a short introduction to the theoretical background of the Gröbner stripping algorithm. For a thorough treatment the reader is referred to [CM94b] and [CM94a]. Our first step is to generalise the notion of admissible term order to that of admissible *partial* term order.

Definition 1. We call a partial order $\lhd$ of T^ℓ admissible if it is a partial semigroup order and if it is well-founded, i.e. there is no term $t \in T^\ell$ with $t \lhd 1$. The set of all admissible partial term orders is denoted $\widetilde{TO}_\ell$.

Definition 2 (Initial Segments). Given an admissible partial order $\lhd \in \widetilde{TO}_\ell$ and a polynomial $g \in A$, the *initial segment* of g *with respect to* $\lhd$ is the polynomial obtained as the sum of all monomials maximal in $supp(g)$ with respect to $\lhd$, and we denote it by extension of notation $in_\lhd g$.

For instance, when $\lhd \in \widetilde{TO}_\ell$ is total, $\lhd$ is admissible as a term order, and $in_\lhd g$ is the initial monomial of g with respect to $\lhd$. We note that every $\omega \in \mathbb{R}_+^\ell$ induces an admissible partial order by means of the prescription

$$t_1 \lhd t_2 : \iff \omega \cdot \log(t_1) < \omega \cdot \log(t_2),$$

and we shall henceforth without loss of generality for our purposes restrict our attention to such $\lhd \in \widetilde{TO}_\ell$. By abuse of language we identify ω with the order it induces, as in the notation $in_\omega f$. Often ω is referred to as a weight vector. We describe next the notion of degree of homogeneity.

Definition 3 (Homogeneity). A polynomial $f \in A$ is called homogeneous with respect to an $\omega \in \mathbb{R}^\ell$ if $\omega \cdot \log(m)$ is constant for $m \in supp(f)$, and this constant value is called the ω-degree of f. Furthermore, f is called homogeneous of degree $l \in \mathbb{N}$ if there exist l linearly independant vectors in $\mathbb{R}^\ell$ with respect to each of which f is homogeneous. A finite family of polynomials is ω-homogeneous (resp. homogeneous of degree l) if all of its members are ω-homogeneous (resp. homogeneous of degree l with respect to the same vectors).

Definition 4. An ideal is homogeneous with respect to an $\omega \in \mathbb{R}^\ell$ (resp. homogeneous of degree l) if it has an ω-homogeneous base (resp. homogeneous of degree l).

For instance, for $\omega = (1, \ldots, 1)$, ω-homogeneity co-incides with homogeneity in the usual sense. Note that if a polynomial or ideal is homogeneous with respect to a family of vectors, it is so with respect to any vector spanned by the family. Note also that ω-homogeneous polynomials are partially ordered by their ω-degree, and that if $f \in A$ is a non-zero polynomial, then f is uniquely decomposable into a strictly ω-decreasing sum of non-zero ω-homogeneous polynomials $f_1, \ldots, f_l$, called its ω-segments.

The concept fundamental to our elaboration, which generalises the notion of initial ideal (cf. [CLO92]; see also [Mac27]), is introduced in the following definition.

Definition 5 (Toric Degeneration). Given an ideal $I \in \mathcal{J}(A)$, and a weight vector $\omega \in \mathbb{R}_+^\ell$, the *toric degeneration* of I with respect to ω is the ideal $\langle \{in_\omega g \mid g \in I\} \rangle$ generated by the ω-initial segments of the elements of I. We denote it $I_{(\omega)}$.

For example, if ω induces an admissible total order then $I_{(\omega)}$ is an initial ideal of I in the usual sense; thus we also write $I_{(\prec)}$ for the initial ideal of I with respect to $\prec$ when $\prec$ is an admissible term order. Noting that the degeneration relation is transitive and reflexive (cf. [CM94a]), we introduce the following notation:

Definition 6 (Degeneration Relation). Given two ideals $I_1, I_2 \in \mathcal{J}(A)$, we write $I_1 < I_2$ if I_1 is a proper toric degeneration of I_2.

Definition 7 (Relative Refinement). Let $\prec$ be an admissible term order in TO_ℓ. One shall say that an $\omega \in \mathbb{R}_+^\ell$ is refined by $\prec$ relatively to a finite family of polynomials F if for all $f \in F$ $in_\prec(in_\omega f) = in_\prec f$ and relatively to an ideal $I \in \mathcal{J}(A)$ if $(I_{(\omega)})_{(\prec)} = I_{(\prec)}$.

For instance, if $\prec$ refines ω in the usual set-theoretic sense, then $\prec$ refines ω relatively to any ideal. The crucial property for the Gröbner stripping algorithm is the one expressed in the following theorem.

Theorem 8. *Let there be given an ideal $I \in \mathcal{J}(A)$ homogeneous of degree $\ell - r$ and an admissible term order $\prec \in TO_\ell$. Then there exists a flag of vector spaces in $\mathbb{R}^\ell$*

$$V_0 \subsetneq V_1 \subsetneq \cdots \subsetneq V_r$$

and a sequence of weight vectors $\omega_1, \ldots, \omega_r \in \mathbb{Q}_+^\ell$ with the following properties:

1. *We have a chain of toric degenerations*

$$I = I_0 > I_{(\omega_1)} > \cdots > I_{(\omega_r)}$$

   with $I_{(\omega_r)} = I_{(\prec)}$.
2. *For all i the vector space V_i has dimension $\ell - r + i$ and $I_{(\omega_i)}$ is homogeneous with respect to V_i, i.e. is homogeneous of degree $\ell - r + i$*

Definition 9 (Toric Resolution). A chain of toric degenerations $I > I_{(\omega_1)} > \cdots > I_{(\omega_r)}$ with the properties of theorem 8 is called a *toric resolution*.

Consider in the following diagram a toric resolution of an ideal I and let F be a base of I.

$$
\begin{array}{ccccc}
I & > & I_{(\omega_1)} & > \ldots > & I_{(\omega_r)} \\
\cup & & \cup & \ldots & \cup \\
F & & in_{\omega_1} F & \ldots & in_{\omega_r} F
\end{array}
$$

(1)

The following 'fundamental lemma' expresses the relationship linking the toric degenerations of an ideal and their Gröbner bases.

Lemma 10 (Fundamental Lemma). *Let $I \in \mathcal{J}(A)$ be an ideal, $\omega \in \mathbb{R}_+^\ell$ a weight vector, and $\prec$ a term order in TO_ℓ which refines ω relatively to I. Then a base F of I is a Gröbner base with respect to $\prec$ if and only if $in_\omega(F)$ is a Gröbner base of $I_{(\omega)}$ with respect to $\prec$. In particular,*

$$R_\prec(I_{(\omega)}) = in_\omega R_\prec(I).$$

As a consequence, we deduce the following further 'polynomial substitution' property (cf. [CKM93] for another application and for a proof):

Lemma 11 (Replacement Lemma). *Let $I \in \mathcal{J}(A)$ be an ideal of A, F a base of I, $\prec \in TO_\ell$ an admissible order, $\omega \in \mathbb{R}_+^\ell$ a weight vector refined by $\prec$ relatively to I, and $\{h_{gf} \mid g \in G, f \in F\}$ a family of ω-homogeneous polynomials such that*

$$\left\{ \sum_{f \in F} h_{gf} \, in_\omega f \mid g \in G \right\}$$

is a reduced Gröbner base of $I_{(\omega)}$; then $\{\sum_{f \in F} h_{gf} f \mid g \in G\}$ is a (not necessarily reduced) Gröbner base of I.

Observe that if in (1) F is not a Gröbner base of I then in general $in_{\omega_i} F, i = 1, \ldots, r$ does not generate $I_{(\omega_i)}$ and in particular $in_{\omega_i} F$ is not a Gröbner base for $I_{(\omega_i)}$. This implies that the ideals generated by the sets $in_{\omega_i} F$ form a 'sub-toric resolution'

$$\langle F \rangle, \langle in_{\omega_1} F \rangle, \ldots, \langle in_{\omega_r} F \rangle$$

of I. The idea of the algorithm will be to

1. find ω_i's to form appropriate 'sub-toric resolutions',
2. extend the generating sets $in_{\omega_i} F$ of the momentary sub-toric resolutions in a coherent way so as to have at the end a toric resolution of I, and hence a Gröbner base G.

3   A Structurally Iterative Algorithm to Compute Gröbner Bases

In this section we present on the basis of the concepts introduced in the preceding section a highly structured algorithm for the computation of Gröbner bases. This algorithm, called *Gröbner Stripping Algorithm*, is doubly iterative in the sense that it iterates itself vertically down a ladder of initial structures and at the same time horizontally along a sequence of sub-structures. Buchberger's S-polynomials are superseded by a notion of O-combinations (to be called null-combinations) of polynomials, which equal the 0-polynomial in distributive normal form (extensionally) but which are recorded (intentionally) as distinct polynomial combinations of a family of base polynomials.

3.1   Description of the Algorithm

First we give an explanation of the progression of the algorithm. The algorithm is most easily explained and perceived to be correct when considered as a recursive algorithm. The input consists of a finite base F of a given ideal I and an admissible term order $\prec \in TO_\ell$. The primary idea, in contrast to Buchberger's algorithm, is to compute not the Gröbner base of I, but of an appropriate proper toric degeneration of I, say $I_{(\omega)}$ for some weight vector $\omega \in \mathbb{R}_+^\ell$ refined with respect to I by $\prec$.

The algorithm proceeds depth first. When active at the bottom-most level, one will in effect be in possession of a ladder of ideals thus:

$$I_0 = I = \langle F_0 \rangle$$
$$I_1 = \langle F_1 = in_{\omega_1}(F_0) \rangle$$

$$\vdots$$

$$I_{\ell-1} = \langle F_{\ell-1} = in_{\omega_{\ell-1}}(F_{\ell-2}) \rangle$$
$$I_\ell = \langle F_\ell = in_{\omega_\ell}(F_{\ell-1}) \rangle$$

accompanied with a sequence of weight vectors $\omega_1 \ldots \omega_\ell \in \mathbb{Q}_+^\ell$ for which for each $f \in F_i$, $in_\prec f = in_\prec(in_{\omega_{i+1}} f)$. Each ideal I_{i+1} is a sub-ideal of the toric degeneration $(I_i)_{(\omega_{i+1})}$ of I_i. The algorithm proceeds by iterating this ladder horizontally, extending each I_{i+1} as an ideal until it is equal to $(I_i)_{(\omega_{i+1})}$. When it has returned to the top-most level with this condition, one is left with I_ℓ equal to the initial ideal $I_{(\prec)}$ and F_0 the Gröbner base with respect to $\prec$ of I sought for.

More in detail now, let us imagine at some level l with $I := I_l$ and $\omega := \omega_{l+1}$ that $I_{(\omega)}$ has a Gröbner base

$$G = \left\{ \sum_{f \in F} h_{gf} in_\omega f \mid g \in G \right\}, \tag{*}$$

for some family F of polynomials of I. Then by the replacement lemma 11 the set $\{\sum_{f \in F} h_{gf} f \mid g \in G\}$ is a Gröbner base of I. In order to make the refinement of this scheme intelligible, the reader should note:

1. we do not know in advance for a given ω whether $\prec$ refines ω with respect to I;
2. we do not have à priori a way of obtaining a Gröbner base G of $I_{(\omega)}$, nor even a family F of polynomials in I such that $\{in_\omega f \mid f \in F\}$ is a base of $I_{(\omega)}$;
3. if it so happened that ω were such that $I_{(\omega)} = I_{(\prec)}$, then (*) would revert to the requirement that F be a Gröbner base of I to begin with, and nothing would be gained.

Observation (3) indicates that we do not want an ω yielding a *maximal* degeneration of I. With respect to (1) and (2) we work iteratively, using the momentarily available 'information' as reflected in the following construction of the recursive procedure.

Inductive construction of recursive procedure GSL *at level* $l < \ell$:

The procedure inputs a set of polynomials F_l and outputs a function

$$H^l : (G_l \cup O_l) \times F_l \longrightarrow A,$$

where G_l is a Gröbner base of F_l with respect to $\prec$ and O_l is the O-combination set of level l.

Given a base F_l for I_l, which is not maximally homogeneous, i.e. is homogeneous of degree $l < \ell$, at first we choose ω such that $F_{l+1} := in_\omega(F_l)$ has one more degree of homogeneity than F_l and for all f in F_l we have $in_\prec f = in_\prec(in_\omega f)$. Observe that in general, $\langle F_{l+1} \rangle = \langle in_\omega(F_l) \rangle \neq I_{l(\omega)}$. Now we assume inductively that GSL is applicable to $F_{l+1} = in_\omega(F_l)$ at level $l + 1$ and yields the function

$$H^{l+1} : (G_{l+1} \cup O_{l+1}) \times F_{l+1} \longrightarrow A$$

with the following properties

1. $\{g = \sum_{f \in F_{l+1}} H^{l+1}(g, f)f \mid g \in G_{l+1}\}$ is a Gröbner base of $\langle F_{l+1} \rangle = \langle in_\omega(F_l) \rangle$ with respect to $\prec$;
2. letting $F_l' := \{\sum_{f \in F_l} H^{l+1}(g, in_\omega f)f \mid g \in G_{l+1}\}$ and letting $O_l' := \{\sum_{f \in F_l} H^{l+1}(o, in_\omega f)f \mid o \in O_{l+1}\}$,
   (a) either $\langle in_\omega(F_l') \rangle = (I_l)_{(\omega)}$
   (b) or else $\langle in_\omega(F_l' \cup O_l') \rangle \supsetneq \langle in_\omega(F_l') \rangle$.

Note: In practice, for $o \in O_{l+1}$, the relation $\sum_{f \in F_{l+1}} H^{l+1}(o, f)f = 0$ will hold (but in general $\sum_{f \in F_l} H^{l+1}(o, in_\omega f)f \neq 0$); hence the designation 'O-combinations' for the elements of the set O_{l+1}.

In case $(2b)$ it is clear that we can iterate this process replacing F_l with $F_l' \cup O_l'$, and that by noetherianity of polynomial rings, this process must come to an end in case $(2a)$ with ourselves in possession of a Gröbner base $in_\omega(\tilde{F}_l)$ of $(I_l)_{(\omega)}$ (but cf. remark 3.3 below). By the fundamental lemma 10, $\tilde{F}_l$ is a Gröbner base of I_l as required. The algorithm assigns $G_l := \tilde{F}_l$.

It remains to construct the function H^l for the current level l: Recording the polynomial combinations expressing the polynomials of F_l' with respect to those of F_l at each step of the horizontal iteration, we obtain by repeated substitution for any $g \in G_l$ a representation $g = \sum_{f \in F_l} \tilde{h}_{gf} f$. The algorithm assigns

$$H^l(g, f) := \tilde{h}_{gf}.$$

The O-combination set O_l is the set of the representations of elements of O_l', equal to 0 in distributive normal form, collected during the iteration process on level l and expressed in terms of the original set F_l. The coefficient $\tilde{h}_{of}$ of $f \in F_l$ in the expression for $o \in O_l$ is assigned as the value:

$$H^l(o, f) := \tilde{h}_{of}.$$

This concludes the formulation of GSL at level $l < \ell$.

Construction of GSL *at level* $l = \ell$:

F_l has degree of homogeneity ℓ and is a monomial base. For monomial ideals, it is clear that no new polynomial combinations are required to produce a Gröbner base, since every base is a Gröbner base. Thus $G_l := \tilde{F}_l = F_l$. The O-combination set simply consists of the expressions

$$o_{mn} := \frac{lcm(m,n)}{m} m - \frac{lcm(m,n)}{n} n$$

for all pairs m, n in F_l.

The function $H^l := (G_l \cup O_l) \times F_l \longrightarrow A$ is defined on $G_l \times F_l$ by

$$H^l(g, f) := \delta_{gf},$$

and on $O_l \times F_l$ by

$$H^l(o_{mn}, m) := n, \, H^l(o_{mn}, n) := -m, \, H^l(o_{mn}, p) := 0 \, (p \neq m, n).$$

This concludes the inductive construction of the recursive procedure GSL.

Finally, to gain an idea of the computational thread of control, one may draw up the following diagram of the bases F_j^i constructed in the course of the algorithm:

$$
\begin{array}{llllllll}
F_0^1 & F_0^2 & F_0^3 & F_0^4 & F_0^5 & \cdots & F_0^{k_0^1} & (K_0 := k_0^1)\\[4pt]
\downarrow & \searrow & \cdots & & & & &\\[4pt]
F_1^1 & \cdots & F_1^{k_1^1} \; F_1^{k_1^1+1} & \cdots & F_1^{k_1^2} & \cdots & F_1^{k_1^{K_0}} & (K_1 := k_1^{K_0})\\[4pt]
\downarrow & \searrow & \cdots & & & & &\\[4pt]
F_2^1 & \cdots & F_2^{k_2^1} \; F_2^{k_2^1+1} & \cdots & F_2^{k_2^2} & \cdots & F_2^{k_2^{K_1}} & (K_2 := k_2^{K_1})\\[4pt]
\downarrow & & & & & & &\\
\vdots & & & & & & &\\
\downarrow & & & & & & &\\[4pt]
F_{\ell-1}^1 \; F_{\ell-1}^2 \; \cdots & F_{\ell-1}^{k_{\ell-1}^1} \; F_{\ell-1}^{k_{\ell-1}^1+1} & \cdots & F_{\ell-1}^{k_{\ell-1}^2} & \cdots & F_{\ell-1}^{k_{\ell-1}^{K_{\ell-2}}} & (K_{\ell-1} := k_{\ell-1}^{K_{\ell-2}})\\[4pt]
\updownarrow \; \updownarrow \; \cdots & \updownarrow \quad \updownarrow & \cdots & \updownarrow & \cdots & \updownarrow & &\\[4pt]
F_\ell^{k_\ell^1} \; F_\ell^{k_\ell^2} \; \cdots & F_\ell^{k_\ell^1-1} \; F_\ell^{k_\ell^1-1+1} & \cdots & F_\ell^{k_\ell^2-1} & \cdots & F_\ell^{k_\ell^{K_{\ell-1}}} & &
\end{array}
$$

In each horizontal row l, $\{F_l^j\}$, $j = 1, \ldots$, is a sequence of base sequences corresponding to the repeated iterations on level l, and the arrows indicate the thread of control. Letting $I_l^j := \langle F_l^j \rangle$, and setting for convenience $k_l^0 := 0$ in each row l one has that

$$F_l^{k_i^{j+1}} \text{ is a Gröbner base of } I_l^{k_i^j+1}, j = 0, \ldots;$$

we note that while

$$I_\ell^{k_i^1} \subsetneqq I_\ell^{k_i^2} \subsetneqq \cdots$$

in general for $l < \ell$ one has

$$I_l^{k_i^j} \neq I_l^{k_i^j+1} \text{ and } I_l^{k_i^j} \not\subset I_l^{k_i^j+1}$$

but

$$\left(I_l^{k_i^j} \right)_{(\prec)} \subsetneqq \left(I_l^{k_i^j+1} \right)_{(\prec)}.$$

3.2   Example

In order to illustrate the stripping procedure, we follow in complete detail its steps on an extremely simple example: we compute a Gröbner base with respect to the so-called total-degree order induced by $y \prec x$ for the ideal I generated by the set $F = \{x^2 + y^2, xy + y\} \subset \Bbbk[x, y]$.

We begin at the top-most level $l = 0$ with the base
$$F_0^1 = \{x^2 + y^2, xy + y\} = F.$$
We select
$$\omega_1^1 = (1, 1)$$
giving us at level $l = 1$ the degenerate base
$$F_1^1 = \{x^2 + y^2, xy\}$$
and again select
$$\omega_2^1 = (1, 0)$$
giving us at level $l = 2$ the degenerate base
$$F_2^1 = \{x^2, xy\}$$
from which we construct the Gröbner base $\tilde{F}_2^1$ and the O-combinations
$$\tilde{F}_2^1 = F_2^1 = \{1 \cdot (x^2) + 0 \cdot (xy), 0 \cdot (x^2) + 1 \cdot (xy)\},$$
$$O_2^1 = \{y \cdot (x^2) - x \cdot (xy)\}.$$
Returning to level $l = 1$ and substituting in these combinations we find
$$F_1^{1'} = \{1 \cdot (x^2 + y^2) + 0 \cdot (xy), 0 \cdot (x^2 + y^2) + 1 \cdot (xy)\} = F_1^1 \text{ and}$$
$$O_1^{1'} = \{y \cdot (x^2 + y^2) - x \cdot (xy) = y^3\}.$$
Thus we extend F_1^1 to
$$F_1^2 = F_1^{1'} \cup O_1^{1'} = \{x^2 + y^2, xy\} \cup \{y^3\}$$
and descend anew to level $l = 2$ with
$$\omega_2^2 = (1, 0) = \omega_2^1$$
to the degeneration
$$F_2^2 = \{x^2, xy\} \cup \{y^3\} = F_2^1 \cup \{y^3\}.$$

Noting that we need not reconsider pairs of monomials previously considered, and that we may apply Buchberger's selection criteria (cf. [Buc85]), we obtain as Gröbner base and new O-combinations

$$\tilde{F}_2^2 = \{1 \cdot (x^2) + 0 \cdot (xy) + 0 \cdot (y^3), 0 \cdot (x^2) + 1 \cdot (xy) + 0 \cdot (y^3), 0 \cdot (x^2) + 0 \cdot (xy) + 1 \cdot (y^3)\},$$
$$O_2^2 = \{y^2 \cdot (xy) - x \cdot (y^3)\}.$$

This yields with substitution in turn at level $l = 1$

$$F_1^{2'} = \{1 \cdot (x^2 + y^2) + 0 \cdot (xy) + 0 \cdot (y^3), 0 \cdot (x^2 + y^2) + 1 \cdot (xy) + 0 \cdot (y^3), 0 \cdot (x^2 + y^2) + 0 \cdot (xy) + 1 \cdot (y^3)\},$$
$$O_1^{2'} = \{y^2 \cdot (xy) - x \cdot (y^3) = 0\}.$$

As the O-combinations yield no new polynomials, the current series of iterations at level $l = 1$ ends and we know that

$$F_1^{2'} = \{x^2 + y^2, xy, y^3\}$$
$$\tilde{F}_1^2 = \{1 \cdot (x^2 + y^2) + 0 \cdot (xy), 0 \cdot (x^2 + y^2) + 1 \cdot (xy), y \cdot (x^2 + y^2) - x \cdot (xy)\}$$

is a Gröbner base of $\langle F_1^1 \rangle$ expressed in polynomial combinations in F_1^1, whereas from $O_1^{2'}$ we obtain O-combinations

$$\tilde{O}_1^2 = \{-xy \cdot (x^2 + y^2) + (x^2 + y^2) \cdot (xy)\}$$

expressed in F_1^1. Rising to level $l = 0$, susbstituting and reducing, we have

$$F_0^{1'} = \{1 \cdot (x^2 + y^2) + 0 \cdot (xy + y), 0 \cdot (x^2 + y^2) + 1 \cdot (xy + y), y \cdot (x^2 + y^2) - x \cdot$$
$$(xy + y)\} = \{x^2 + y^2, xy + y, y^3 - xy \xrightarrow{(F_0^1, \prec)} y^3 - y\} = F_1^0 \cup \{y^3 - y\},$$
$$O_0^{1'} = \{-xy \cdot (x^2 + y^2) + (x^2 + y^2) \cdot (xy + y) = x^2 y + y^3 \xrightarrow{(F_0^{1'}, \prec)} 0\} = \{0\}.$$

As the O-combinations yield no new polynomials, the current series of iterations at level $l = 0$ ends and we know that $\tilde{F}_0^1 = \{x^2 + y^2, xy + y, y^3 - y\}$ is a Gröbner base of $\langle F_0^1 \rangle = \langle F \rangle = I$, as required.

3.3 Concluding Remarks

Remark. The Gröbner stripping algorithm embodies a class of completion 'heuristics' amounting in essence to the *choice* of which of the (many) toric resolutions $I > \cdots > I_{(\prec)}$ of I is computed. This is reflected in the choices of ω_i's made in the course of the algorithm, which in computational terms is an integer programming task.

Remark. After each iteration step, it may well turn out that the degree of homogeneity of F' has increased with respect to F, or it may turn out that $in_\omega(F')$ is more than onefold more homogeneous than F. In these cases, the algorithm in principle may proceed undisturbed. However, it may also turn out that ω is no longer refined by $\prec$ with respect to F'. In this case, ω must be replaced with some ω' such that ω' is refined by $\prec$ relatively to F' and the degree of homogeneity of $in_{\omega'}(F')$ is one more than that of F'. Termination remains guaranteed by the fact that at each iteration step $in_\prec(F')$ has increased.

Remark. It is to be remarked that the algorithm would not be implemented in its expository form above. A detailed analysis would extend beyond the scope of this note. Let it suffice here to give the following indications. It is possible and indeed necessary to perform reduction on the polynomials of O' at each iteration on level l in order to decide between case $(2a)$ and $(2b)$. The usual refinements of Buchberger's scheme, such as inter-reduction of the current base and S- (in our case O-) polynomial selection criteria (cf. [Buc85]) are equally applicable here.

Remark (Analogy to the Schreier Algorithm). The analogy to the permutation group theoretic Schreier algorithm is of a procedural as well as structural nature. In either case, one constructs a chain of distinguished 'sub-structures': there stabilisers, here toric degenerations. In each of these algorithms, it is possible to perform completion at a lower level with respect to partly generated structures higher up. In each case, new generators at lower levels are obtained by 'stripping' away their top levels.

References

[Buc65] B. Buchberger. *Ein Algorithmus zum Auffinden der Basiselemente des Restklassenringes nach einem nulldimensionalen Polynomideal.* PhD thesis, Univ. Innsbruck, Austria, 1965.

[Buc85] B. Buchberger. Gröbner bases: an algorithmic method in polynomial ideal theory. In N. K. Bose, editor, *Recent Trends in Multidimensional System Theory*, chapter 6, Reidel Publ. Comp., 1985.

[CC92] J. Calmet and J. A. Campbell. Artificial intelligence and symbolic mathematical computations. In J. Calmet and J. A. Campbell, editors, *Artificial Intelligence and Symbolic Mathematical Computing*, pages 1–19, Springer, 1992. Proc. Int'l Conf. AISMC-1, Karlsruhe, Germany, Aug. 1992.

[CKM93] S. Collart, M. Kalkbrener, and D. Mall. The Gröbner walk. 1993. Submitted to the J. Symb. Comp.

[CLO92] D. Cox, J. Little, and D. O'Shea. *Ideals, Varieties, and Algorithms: An Introduction to Computational Algebraic Geometry and Commutative Algebra. Undergraduate Texts in Mathematics*, Springer, New York, 1992.

[CM94a] S. Collart and D. Mall. Toric degenerations of polynomial ideals. 1994. In preparation.

[CM94b] S. Collart and D. Mall. Toric degenerations of polynomial ideals and complex duality. In J. Calmet, editor, *Proceedings of the Rhine Workshop on Computer Algebra 1994*, pages 147–154, 1994.

[Leo80] J. S. Leon. On an algorithm for finding a base and a strong generating set for a group given by generating permutations. *Math. Computation*, 35(151):941–974, July 1980.

[Mac27] F. S. Macaulay. Some properties of enumeration in the theory of modular systems. *Proc. London Math. Soc.*, 26:531–555, 1927.

[Mon92] E. Monfroy. Gröbner bases: strategies and applications. In *Proc. Conf. on Artificial Intelligence and Symbolic Mathematical Computations*, pages 64–79, 1992.

[Rob85] L. Robbiano. Term orderings on the polynomial ring. In B. F. Caviness, editor, *Proc. EUROCAL 85*, pages 513–517, Springer, 1985.

[Sim70] C. C. Sims. Computational methods in the study of permutation groups. In John Leech, editor, *Computational Problems in Abstract Algebra*, pages 169–183, Sci. Res. Council Atlas Computer Laboratory, Pergamon Press, Oxford, 1970. Proc. Conf. at Oxford, 29.VIII.-2.IX 1967.

[Wei87] V. Weispfenning. Admissible orders and linear forms. *ACM SIGSAM Bulletin*, 21:16–18, 1987.

Modeling Cooperating Agents Scenarios by Deductive Planning Methods and Logical Fiberings

Jochen Pfalzgraf[1]*, Ute Cornelia Sigmund[2]**, Karel Stokkermans[1]***

[1] RISC-Linz, Austria
[2] TH Darmstadt, Germany

Abstract. We describe a small but non-trivial 3-agent-robotics scenario by two different methods, viz. resource-oriented deductive planning and logical fiberings. The ultimate aim is to find a semantics for planning methods by means of fiberings. To this end, a comparison of the two methods is made and illustrated by the sample scenario, and the correspondences between the basic notions for both methods are clarified. The fiberings method is found to be useful in modeling communication and interaction between cooperating agents, thanks to the local/global distinction that is inherent to this framework. Possible extensions of the framework, like e.g. formulas dependent on space and/or time, are discussed.

1 Introduction

In this paper we set out to provide a semantics for planning methods (in particular resource-oriented disjunctive planning) by logical fiberings, a concept which we have transferred from classical fiber bundle and sheaf theory.

First, a scenario consisting of three robots solving an assembly problem is described. It contains several problems of practical relevance, such as uncertainty about the reason for a failure and disjunctive postconditions of a (part of a) plan, corresponding to non-determinism. Technical details such as precise descriptions of movements have been omitted in order to make the basic problems clear. For a complete description, we refer to [20], where the scenario is presented in much more detail.

After describing the scenario, two approaches to modeling it are given, one based on the method of deductive planning (cf. [4, 8]) and one based on the method of logical fiberings (cf. [18, 19]). The planning model employs conditional planning (because the robots cannot always be sure about the cause of mistakes) and is resource-oriented. The description by logical fiberings is a new attempt to formulate planning problems in a uniform semantical notation. Here, we are proposing a general modeling method which we call the "generic modeling approach". The scenario discussed here is essentially a control problem (the

* jpfalzgr@risc.uni-linz.ac.at
** ute@intellektik.informatik.th-darmsta dt.de
*** kstokker@risc.uni-linz.ac.at

underlying plan is not generated by the robots but prescribed in the specification of the actions), but from the descriptions it will be seen that both approaches are amenable to deal with more complicated robotics scenarios as well. Actually, what we present here is the starting point for a program of work. We fix the notions and test them with a sample scenario. Improvements have to follow as experience grows.

We then describe the connection between the two methods (planning and logical fiberings), and it is shown how actions with their pre- and postconditions correspond to multivariate transjunctions between logical systems, motivated by [18]. It turns out that the fiberings approach allows for a natural way of dealing with conflicting intentions of the robots (cooperating agents). We explicitly model a (logical) state space for each individual agent. This state space is formally presented in terms of a corresponding fiber. The collection of all fibers is the global state space. Specifically, the concepts of local and global sections allow a natural switch of perspective from the individual to the cooperative point of view. In this sense, the mathematical notion of a section gives us a "snapshot" of the current state of the system.

There are various ways to extend the modeling approach presented here. One of them is the introduction of space/time-dependent formulas, on which we touch towards the end of the paper. Another objective is to incorporate the concept of hierarchical planning in the fiberings approach. Another important issue is mirroring actual plan generation (by SLDE-resolution) in the fiberings approach. Further investigations should lead to providing the planning approach with a complete semantic foundation based on logical fiberings.

We would like to thank the referees for suggesting to include some remarks concerning links of our work to the huge field of *distributed AI*. We are aware that distributed AI is an active area of research and that the general "cooperating agents problem" appears in many disguises in various disciplines. We are not experts in distributed AI, but we see that our work has links to it since our approach also attempts to model cooperating agents in a robotics scenario concentrating on logical control and planning issues. Unfortunately we do not feel competent to give here a brief comment on the latest state of the art and mainstream research topics in distributed AI — we are very sorry. Instead we can report that after the talk of the first author on "Logical Fiberings and Polycontextural Systems" at the conference FAIR'91 (Smolenice Castle, Slovakia), Jozef Kelemen discussed with the first author. The question was in which respect the logical fiberings concept can contribute to the mathematical foundations of distributed AI. We conclude by citing the following selected references: [3, 5, 7, 11]. One of our objectives in our future program of work is, among others, to get in to closer contact to those activities which have natural links to our work.

2 Description of the Scenario

In the following description, we do not go into technical details relevant from a technical point of view. We work on a scenario in order to demonstrate the approaches. The scenario consists of three cooperating robots, each carrying out several (up to seven) different actions. They work together in order to assemble workpieces of two different types (a and b) into composite ones (res). However, it can occur that the assembly of two pieces is unsuccessful because of faults in one or both of the pieces. Since the robots have no way to decide which piece is the faulty one, this leads to conditional planning of their respective actions.

The following notations will be used. The work space consists of three robots: r_0, r_1, r_2, working on the following tables: s-a, s-b, s-res (storage of work pieces of type a, b, and final products res, respectively), table (where the actual assembly is carried out), temp (for temporary disposal), and dump (garbage bin for faulty pieces). The control propositions[3]: on(table, a), on(table,b), on(table,res), tried, on(temp, a), on(temp, b), used(dump), empty(s-a), on(s-a,a), empty(s-b), and on(s-b,b). The last four are used in the formulation of the deductive resource-oriented planning approach to determine when the process is finished. The possible actions of the robots: $(r_0$,assemble), $(r_0$,store), $(r_1$,get-a), $(r_1$,get-b), $(r_2$,temp-a), $(r_2$,temp-b), $(r_2$,get-temp), $(r_2$,dump-a), $(r_2$,dump-b), $(r_2$,dump-temp), and $(r_2$,empty-dump). The meaning of the control propositions and robot actions will be clear from the following description of the task and work plan.

The main task of the three robots cooperating in this scenario is to assemble work pieces of two different types, a and b, into an end product, denoted by res. Consequently, we have one robot, r_1, whose task it is to provide work pieces of those two types (i.e. move them from their respective stores to the table) to a robot, r_0, who tries to assemble them and, if successful, stores the product res.

However, it may happen that the assembly is unsuccessful, due to a faulty piece. Robot r_2 takes care of removing faulty pieces. We have prescribed a certain strategy in this process which minimizes the number of pieces dismissed as faulty. This yields a complete description of the multivariate transjunction (cf. Sect. 4) logically controlling the work cell, which corresponds to its work plan.

This strategy consists of first laying one of the two pieces on the temporary dump temp, and replacing it with a piece of the same type at the working table. If the assembly is successful this time, it is assumed that the piece laid apart was indeed faulty, and it is "thrown into the garbage bin", dump. If the assembly is not successful, it is assumed that the piece of the other type (which has been tried twice by now) is faulty. It is dumped, and a new piece of that type is provided for by r_1. If the assembly can be carried out now, the piece laid apart temporarily at temp is tried again in the next cycle, together with a 'fresh' piece of the other type. If, again, the assembly is unsuccessful, the piece which has been tried the last two times is dumped, and the piece lying at temp is taken back again, to be tried for the second time with the new piece of the other type.

---

[3] In the planning approach these correspond to facts that may or may not hold.

On success, r_0 finally can store res and start all over again. On failure, both pieces are thrown away and a fresh start is made.

One could generalize the example by not insisting on any particular strategy from the beginning, and also model the generation of the plan by the fiberings approach (and not just its control, as done here). In Sect. 5, which discusses the links between the deductive planning approach and logical fiberings, we indicate how plan generation works in the fiberings approach.

The task to be carried out by the robots in the cell is supposed to be finished whenever no more pieces are left on s-a or s-b, the tables from which r_1 takes new work pieces. This is checked by additional sensors (instantiating the control propositions empty(s-a), on(s-a,a), empty(s-b), and on(s-b,b)).

3   Planning and Control in a Deductive Resource-Oriented Approach

The construction as well as the control of the execution of robot-plans can be modeled in logic by reasoning about situations, actions and causality. One reason to use logic as the underlying concept is that logic appears to play a fundamental rôle for intelligent behavior. Using logic it is not only possible to reason about actions, change and causality but also to reason about the planning and control process itself, about intentions, knowledge and beliefs of agents, about interactions and dependencies of actions and goals, etc. On the other hand, in deductive approaches to planning and control there also arise some difficulties. In deductive planning a situation, i.e. the state of the world at a certain instant of time, can be modeled as the collection of the facts that (are believed to) hold in the situation. An important property of these facts is that their truth values may change in the course of time. More precisely, an action causes the transition of the world from one situation into a subsequent situation. In this subsequent situation some facts that held in the initial situation may no longer hold. On the other side some additional facts may have been caused by the action. Imagine the situation where two workpieces a and b are on the table and the hand of robot r_0 is empty. If robot r_0 performs the action "picking up a" this will result in a situation where a is no longer on the table and the robot-hand is no longer empty. These facts have to be removed from the description of the situation. On the other hand, we have to add the fact that now the hand holds a. Therefore a straightforward use of classical logic is not practicable.

To model this change and to describe the dependency of facts on situations in logic J.McCarthy and P. Hayes proposed to use the situation calculus [14, 16]. In the situation calculus predicates and functions are given an additional argument representing the situation in which a predicate holds and a function is applicable. Manipulating this additional argument McCarthy and Hayes are able to reason about the change of facts as result of the application of actions. Unfortunately, this solution of associating a particular situation to the truth value of a particular fact brings along the well known technical frame problem, i.e. how to formalize that the truth value of facts not affected by an action is not changed by the execution of this action. For example picking up a will not affect the fact that b is on the table. This requires either to state additional so-called frame axioms [16] or to use a non-monotonic logic and a common-sense law of inertia [12, 15].

3.1 Resource-oriented approaches to deductive planning

A more efficient representation of situations and actions is used in resource-oriented deductive approaches [2, 9, 13]. These approaches are built on the key idea to treat logic formulas as resources which can be produced and consumed through the execution of actions. More precisely, a situation corresponds to a multiset of atomic facts. Facts are consumed (and therefore deleted from the multiset) when the conditions of on action are to be satisfied. They are produced (i.e. added to the multiset) as effects of the application of an action. Hence the change of facts caused by the application of an action can be modeled without the need to state frame axioms explicitly.

In the following we will examine the resource-oriented planning approach based on equational logic [9] more closely. Originally, this approach was restricted to conjunctive planning problems, i.e. problems where situations as well as conditions and effects of actions are restricted to (non-idempotent) conjunctions of atomic facts. Thus, it was only possible to describe transitions from one single situation to one single subsequent situation. Modeling more realistic examples however, the application of an action – like assembling the two workpieces a and b – may have different alternative effects. Therefore, planning and control usually has to deal with a variety of alternative possibilities. In order to be able to describe alternative situations and conditional plans depending on the various alternatives, in [4], [8] and [22] the resource-oriented equational logic approach was substantially enhanced by introducing a concept of non-determinism. In this extension situations are modeled as multisets of resources as before whereas the sets of possible alternatives correspond to sets of situations, i.e. sets of multisets of facts. In the following we will give a short introduction to this approach of disjunctive resource-oriented planning based on equational logic followed by the description of extensions necessary to model our robotics scenario. These extensions overcome some restrictions of former work in so far as they do not only allow for planning and control of actions of single agents but also for reasoning about interactions and dependencies of tasks carried out by several cooperating robots.

One of the basic features of the approach based on equational logic is that situations are described as single terms - so called situation-terms - using a binary function symbol o which is associative (A), commutative (C) and admits a unit element (1), viz. the constant $\emptyset$ denoting the empty situation. The function o connects the various atomic facts which hold in this situation and are represented by elementary terms[4]. For instance, in our robotics scenario the situation where one workpiece of type a and one workpiece of type b are on the worktable, the

---

[4] Elementary terms are non-variable terms that do neither contain the function symbols o and $\ddagger$ nor the constants $\emptyset$ and $\perp$.

temp is empty and the dump is already used can be described by the situation-term[5]

$$\text{on(table,a)} \circ \text{on(table,b)} \circ \text{empty(temp)} \circ \text{used(dump)} \quad .$$

Although, $\circ$ is interpreted as conjunction it is – in contrast to the classical conjunction – not idempotent. For instance the situation-term on(table, a) $\circ$ on(table, a) is not equal to the term on(table, a). It rather represents a situation where two workpieces of type a are on the table. Consequently situation-terms can be adequately interpreted as multisets of facts. To be able to describe a set of alternative situations, situation-terms are connected by the binary function symbol $\ddagger$ which is interpreted as disjunction. The function $\ddagger$ is associative (A) and commutative (C) and admits an unit element $\perp$ (1). In addition $\ddagger$ is also idempotent (I), i.e. $X \ddagger X = X$ holds for all X. For instance the set of alternative situations after assembling two workpieces a and b can be described by the disjunctive situation-term

$$\text{on(table,res)} \ddagger (\text{on(table,a)} \circ \text{on(table,b)} \circ \text{tried}) \quad .$$

If we regard facts as resources and a term of the form $X \ddagger Y$ as having either the resources X or the resources Y but not both, then it is natural to require that $\circ$ distributes over $\ddagger$, i.e. that the law of distributivity

$$X \circ (Y \ddagger Z) = (X \circ Y) \ddagger (X \circ Z) \tag{1}$$

holds[6].

Actions are described specifying their preconditions and their effects as (disjunctive) situation-terms using the ternary predicate $action(c, a, e)$. For instance, a specification of the action "picking up a block X" (which can be performed by all robots in our scenario) can be described by the clause

$$action(\text{on(table, X)} \circ \text{empty(hand)}, \text{ pick\_up(X)}, \text{ holding(hand,X)}) \quad .$$

An action a is applicable if the conditions c of the action are part of the description of the current situation. The application of an action yields a subsequent situation where the conditions c are replaced by the effects e of the action.

Causation is specified using a ternary predicate $causes(s, p, t)$ which is interpreted declaratively as *the execution of the sequence of actions p causes the transformation of the situation s into the situation t*. The following clauses describe the planning and control process. The first clause states that the empty sequence of actions changes nothing. It serves as termination clause for conjunctive planning programs:

$$causes(I, [], G) \leftarrow I =_E G \quad . \tag{2}$$

---

[5] Throughout this chapter, we use a PROLOG-like syntax, i.e. constants and predicates are represented by lower case letters whereas variables are denoted by upper case letters. Free variables are assumed to be universally quantified.

[6] The reader is invited to observe that $\ddagger$ does not distribute over $\circ$ as this contradicts the intended interpretation.

If the goal situation contains alternatives we have to use the clause

$$causes\,(I,[\,],G) \leftarrow G =_E I \mid V \tag{3}$$

instead, which states that a disjunctive goal G is solved if the current situation I is one of the alternatives described in the goal situation G. In addition sometimes we may need a third termination rule, viz.

$$causes(X,[\,],Y) :- X =_E X_1 \mid X_2,\; causes(X_1,[\,],Y),\; causes(X_2,[\,],Y) \tag{4}$$

which states that a problem is solved if it is solved in each alternative separately. Finally, the clause

$$causes\,(I,\; [A\mid P],\; G) \;\leftarrow\; action(C,\; A,\; E), \\ C \circ V =_E I, \tag{5} \\ causes\,(V \circ E,\; P,\; G)$$

defines the entailment of the predicate *causes* for a non-empty list of actions. The predicate $=_E$ denotes the equality under the equational theory for the operators $\circ$ and $\ddagger$[7]. A sequence $[A\mid P]$ of actions causes the transition of the initial situation I into the goal situation G if the following conditions are satisfied. There is an action A such that the conditions C of the action are part of the description of the initial situation I, i.e. the variable V in the body of (5) is bound to the initial situation I without the conditions C. In addition, there is a sequence of actions P which transforms the subsequent situation, viz. V together with the effects E of action A, into the goal situation G. All facts of the current situation which are not part of the conditions of the action are bound to the variable V and thus remain unchanged. This is the way the frame problem is solved without the need of frame axioms.

3.2 Modeling the Scenario

To model some kind of a general controller for the whole working cell we have to extend the deductive planning approach described above in the following way. Because in this example there is more than one robot who is able to perform actions, we will have to label the action-names with the name of the robots which execute them. Furthermore, for the sake of simplicity, we will allow action descriptions to specify not only single actions but also sequences of actions which will be performed successively. These properties in mind we can now give the following formal description of our robotics-scenario.

Starting with a situation where no piece of type a is on table and where no piece a is on temp and where (at least one) piece of type a is at s-a, robot r_1 should perform the action get-a. Afterwards, there will be a piece of type a on the table and one less on s-a. This is described by

$$action(on(\text{s-a,a}) \circ V,\; [(r_1,\text{ get-a})],\; on(\text{table,a}) \circ V) :- \\ \neg\,(V =_E on(\text{table,a}) \circ V'),\; \neg\,(V =_E on(\text{temp,a}) \circ V'') \;. \tag{6}$$

---

[7] An efficient AC1-ACI1-D-Unification procedure for planning problems is presented in [4].

In order to model the conditions that no workpieces of type a are on the table or on temp corresponding literals are added to the body of the clause. Again, with $=_E$ we denote equality under the given equational theory for the functions o and $\ddagger$[8]. $\neg$ denotes negation as failure. A brief discussion of SLDENF-Resolution can be found in [10].

Similarly the action get-b of robot r_1 can be described by the clause

$$action(on(s\text{-}b, b) \circ V, [(r_1, get\text{-}b)], on(table,b) \circ V) :- \\ \neg (V =_E on(table,b) \circ V'), \ \neg (V =_E on(temp,b) \circ V'') \ . \tag{7}$$

If there are workpieces (one of type a and one of type b) available at the table and robot r_0 has not yet tried to assemble them robot r_0 should execute the action assemble. Afterwards either the pieces of type a and b have disappeared and a product of type res is on the table, or both pieces are still there and no result has been produced. In addition we know that robot r_0 has already tried to assemble them:

$$action(on(table, a) \circ on(table, b) \circ V, [(r_0, assemble)], \\ on(table,res) \ \ddagger \ (on(table,a) \circ on(table,b) \circ tried) \circ V) :- \tag{8} \\ \neg (V =_E tried \circ V') \ .$$

As already mentioned above, the result of this action has to be described as a disjunction of alternative situations. In order to model the condition that robot r_0 has not yet tried to assemble the workpieces a corresponding literal is added to the body of the clause.

If assembling of the two workpieces a and b was successful, robot r_0 should store the result res in s-res. Besides, if temp and dump were used they should be cleared by robot r_2 (the piece on temp going back to table to be tried again). This is stated by the following facts:

$$action(on(table, res) \circ on(temp, X) \circ used(dump) \circ V, \\ [(r_0, store), (r_2, get\text{-}temp), (r_2, empty\text{-}dump)], \tag{9} \\ on(s\text{-}res, res) \circ on(table, X) \circ V) \ .$$

If dump was not used, the piece on temp was faulty, i.e. it is necessary to clear temp:

$$action(on(table, res) \circ on(temp, X) \circ V, \\ [(r_0, store),(r_2, dump\text{-}temp), (r_2, empty\text{-}dump)], on(s\text{-}res, res) \circ V) \ . \tag{10}$$

Since the variable X can be bound to the constant a as well as to b it is sufficient to define only one clause to describe these last two actions. If assembling of two workpieces was successful without any faults it is only necessary to store res:

$$action(on(table,res) \circ V, [(r_0, store)], on(s\text{-}res, res) \circ V) \ . \tag{11}$$

---

[8] i.e. the equational theory AC1 of the function o the theory ACI1 of the function $\ddagger$ and the law of distributivity (1).

One problem of this description is that in a situation where (9) is applicable (10) and (11) are also applicable since the multiset of conditions of (9) is a super-multiset of the multisets of conditions (10) and (11). Therefore, we have to force that in each situation only the most specific action is to be used, i.e. the action which has the largest subset of facts (of the description of the current situation) as precondition. A solution to this problem can be found in [10].

If assembling of two workpieces a and b was not successful and there is no workpiece on temp and dump has not been used, robot r_2 can either transport a piece of type a or a piece of type b to temp. This can be described by the following two facts:

$$action(\text{on(table,a)} \circ \text{on(table,b)} \circ \text{tried} \circ V, [(r_2,\text{temp-a})],$$
$$\text{on(table, b)} \circ \text{on(temp,a)} \circ V) \; :- \qquad (12)$$
$$\neg \, (V =_E \text{on(temp, X)} \circ V'), \neg \, (V =_E \text{used(dump)} \circ V'') \; .$$

and

$$action(\text{on(table,a)} \circ \text{on(table,b)} \circ \text{tried} \circ V, [(r_2,\text{temp-b})],$$
$$\text{on(table,a)} \circ \text{on(temp,b)} \circ V) \; :- \qquad (13)$$
$$\neg \, (V =_E \text{on(temp, X)} \circ V'), \neg \, (V =_E \text{used(dump)} \circ V'') \; .$$

In contrast to the three clauses above none of these two clauses is more specific than the other. Hence, in the control process we can arbitrarily choose one of them. In section 5 we will see how this will be reflected in the fiberings approach as a disjunction in the truth-table of the transjunction.

If the assembly-task failed and there is already a workpiece of type a on temp but dump is not used yet, then the workpiece of type b has been tried twice. Therefore, robot r_2 should transport it from table to dump. This is stated by the following fact:

$$action(\text{on(table,a)} \circ \text{on(table,b)} \circ \text{tried} \circ \text{on(temp,a)} \circ V, [(r_2,\text{dump-b})], \qquad (14)$$
$$\text{on(table,a)} \circ \text{on(temp,a)} \circ \text{used(dump)} \circ V) \; .$$

Similar, the action $(r_2,\text{dump-a})$ can be described by the fact

$$action(\text{on(table,a)} \circ \text{on(table,b)} \circ \text{tried} \circ \text{on(temp,b)} \circ V, [(r_2,\text{dump-a})], \qquad (15)$$
$$\text{on(table,b)} \circ \text{on(temp,b)} \circ \text{used(dump)} \circ V) \; .$$

If the assembly-task failed and there is already a workpiece of type a on temp and dump is already used, then the workpiece of type a on table has been tried twice. Now, robot r_2 should transport it from table to dump and then get the piece of type a from temp:

$$action(\text{tried} \circ \text{used(dump)} \circ \text{on(table,a)} \circ \text{on(temp,a)} \circ V,$$
$$[(r_2,\text{dump-a}),(r_2,\text{get-temp})], \text{on(table,a)} \circ \text{used(dump)} \circ V) \; . \qquad (16)$$

Analogously we define

$$action(\text{tried} \circ \text{used(dump)} \circ \text{on(table,b)} \circ \text{on(temp,b)} \circ V,$$
$$[(r_2,\text{dump-b}),(r_2,\text{get-temp})], \text{on(table,b)} \circ \text{used(dump)} \circ V) \; . \qquad (17)$$

If the assembly-task failed, the dump is used and there is no workpiece on temp all pieces have been tried twice. Consequently, the workpieces on the table should be dumped and then the dump should be cleared:

$$action(\text{on(table,a)} \circ \text{on(table,b)} \circ \text{tried} \circ \text{used(dump)} \circ V,$$
$$[(r_2,\text{dump-a}),(r_2,\text{dump-b}), (r_2, \text{empty-dump})], V) :-$$
$$\neg (V =_E \text{on(temp, X)} \circ V') .$$
(18)

Now the planning process can be triggered by formulating the planning problem as the following query to our equational program. We are searching for a plan P such that its execution transforms the initial situation where several workpieces of type a (for instance 5 a) are on stack s-a, several workpieces of type b (for instance 6 b) are on stack s-b and the stack s-res as well as table, dump, and temp are empty to the goal situation where one of the stacks s-a or s-b is empty[9];

$$?- causes(\text{on(s-a,a)}^5 \circ \text{on(s-b,b)}^6, P, (\text{empty(s-a)} \ddagger \text{empty(s-b)}) \circ W) .$$

Connecting the goals with the variable W describes that we want to terminate our assembly-process if one of the stacks s-a or s-b is empty. Besides in this final situation there may hold several other facts which we have not specified and which will be bound to the variable W. As in [10], queries to a planning-program are answered using SLDENF-resolution, where the equational theory of the operators $\circ$ and $\ddagger$ are built into a special unification procedure.

As assembling of two workpieces may result in different alternative situations we have to take care of the fact that our goals are achieved no matter which of these alternatives will occur (this will later be checked by sensors), i.e. we have to solve the problem in both alternatives. This can be done by splitting a set of situations using the following rule:

$$causes(V, cond(V_2, P_2, V'_2, P'_2), W) \leftarrow V =_E V_2 \mid V'_2,$$
$$causes(V_2, P_2, W), causes(V'_2, P'_2, W) .$$

The term $cond(V_2, P_2, V'_2, P'_2)$ is a conditional and should be read as *if the sensor observes V_2 then plan P_2 solves the problem, otherwise, if the sensor observes V'_2 then plan P'_2 solves the problem.* It should be noted, however, that splitting is *not* an action; it is a general rule which is added to any equational logic program to allow for alternative sub-plans.

The planning problem described in the paper is somehow special in so far as in every reachable situation there is no risk to choose an action which would not lead to success of the whole process. Therefore, it is actually not necessary to construct a plan in advance and to carry it out afterwards. The control-process could rather be realized as the planning process itself, i.e. every application of an action-description in our formalism could correspond to the step(s) performed by

---

[9] For abbreviation we denote a situation-term consisting of n facts f connected with the function $\circ$ with f^n.

the robots directly. We would like to point out here, that our approach described above is in general not restricted in that way.

So far we described a general controller which coordinates the actions of all three robots simultaneously. If we want to model an autonomous controller for every robot separately, we have to take care that the actions of the robots are synchronized. This could for example be done by communication via shared variables like in parallel logic programming languages like Concurrent Prolog [21]. Then one of the robots, for instance r_0, could take control over the whole assembly-process by sending commands to the robots r_1 and r_2. After finishing their tasks they would reply with a confirmation.

In the next section we will present an approach to model our scenario by logical fiberings.

4 The Logical Fibering Model

As a base space we have an index system with its own structure like a graph or net or geometric/topological structure. To each element of the base space (index set) there is a state space attached, called the *fiber* over that element. The fibers are "put together" to form the "total space" of the fibering respecting and reflecting the base space structure. In many concrete cases it is natural to cover the base space by certain subsets ("patches") such that we can independently model so-called "local sub-fiberings", which have a comparatively simple local structure. Then this "patchwork" is put together to obtain the whole fibered system. When doing this, we have to model constraints (if they exist) which may arise in the overlap of neighboring patches.

A typical situation in which we want to apply this modeling principle arises in scenarios with cooperating robots. There, two agents may have a common overlap in their work spaces (the "patches"), where they are supposed to cooperate. In such a case specific constraints in the corresponding state spaces of the agents have to be modeled. The whole system is described in a modularized fashion by covering sets of the base space and the corresponding "locally trivial bundles" together with suitable "transition relations" in the overlap of intersecting covering sets.

This modularized modeling principle is characteristic for the classical theory of fiber bundles which has deep applications in fields like geometry, topology, mathematical physics, and systems theory. It is the aim of the following exposition to establish the basic notions from the theory of fiber bundles in order to adapt them to our modeling purposes. This *generic modeling approach* can be applied to various concrete problems; it is always the same principle to which we resort: Given a modeling problem, we have to choose a covering of the corresponding base space (in general a task involving suitable heuristics). The state spaces (fibers) have to be specified and the "putting together" of the local pieces to form the whole fibering amounts to modeling the constraints which arise. In many applications the base space will be a finite (index) set. We wish to stress here that a typical feature of fiberings is that this modeling principle is in some respect recursive or "self-similar". That means that a fiber can again be a fibered structure in itself.

The approach to modeling robotics scenarios presented in this section originates from [18], and was worked out on a small example in [19]. For a brief discussion, see also [17]. For more background on the notions, we refer to [20], where all notions and notations involved are introduced and explained.

4.1 Fiberings and Sections

The classical notion of a fiber bundle consists of a so-called base space B, a typical fiber F, and the total space E which is the collection of all fibers. Both spaces are connected by a projection map $\pi : E \to B$ such that the fiber over a point $b \in B$ is exactly the preimage set $\pi^{-1}(b) = \{x \in E | \pi(x) = b\}$. We now give the formal definition.

Starting from an indexed system $(E_b)_{b \in B}$ (with a given base set B), we define an *abstract fibering* to be a triple $\xi = (E, \pi, B)$, with E the disjoint union (coproduct in categorical terms) of the E_b, denoted by $\coprod_{b \in B} E_b$, and π the canonical projection from E to B defined as $\pi(x) = b$ for all $x \in E_b$. Then, B is called the *base space*, E the *total space*, π the *projection map*, and for every b, we call $E_b = \pi^{-1}(b)$ the *fiber over* b.

The map π respects the corresponding structure of the spaces, for example it has to be a continuous map if we work in the category of topological spaces. The above is the most general notion of a fiber bundle (fibering).

A *global section* $\sigma : B \to E$ is a map such that $\pi \circ \sigma = \mathrm{id}_B$ (so, $\sigma(b) \in \pi^{-1}(b)$ for all b in B). Let $U \subset B$ denote a subset, then a *local section* is a map $\sigma_U : U \to E$ such that $\pi \circ \sigma_U = \mathrm{id}_U$. A *covering* of the base space B consists of a family $\{U_i\}_{i \in I}$ of subsets of B such that B is *covered* by all the U_i. That means that the (set) union of all U_i, $i \in I$, equals B. Note that the local sections play a dominant role in the theory of sheaves and fiber bundles. In our application, they will correspond to descriptions of the situation (state of the scenario) as seen by a particular robot (cooperating agent) or a group of agents—i.e., a local section is a "snapshot" of a part of the current state.

A *logical fibering* is a tuple $\xi = (E, \pi, B, L)$, where E (the total space) and B (the base space, taken to be equal to the indexing set I) are arbitrary sets, and L, denoting the *typical fiber* modeling every fiber $\pi^{-1}(b)$ (for all $b \in B$) of the bundle, is taken to be a classical first order logical space. However, the concept of abstract fiberings is general enough to allow us to "mix logics" in the sense that different logics can occur as local fibers. In the sample scenario, we will mix several many-valued logics. Further development of these aspects is intended for future work.

The simplest form of a fibering is the so called *trivial fibering*, $\xi = (E, \pi, B, F)$, where $E = B \times F$, π is the first projection, and the fiber over $i \in B$ is $\pi^{-1}(i) = \{i\} \times F$.

For logical fiberings, this corresponds to a *parallel system of logics* L_i over an index set I (serving as base space B) for which the typical fiber F is a classical first order logic L. Within each fiber $L_i = \pi^{-1}(i)$, the reasoning processes can run independently and in parallel. Also communication between the fibers can be modeled.

A characteristic feature of a classical fiber bundle is the so-called local triviality property. A locally trivial fiber bundle is composed of parts that locally have a simple structure, in the sense that they are of the type of a product bundle $U_i \times F \to U_i$. Here, the U_i, subsets of the base space B, form a covering of B. The "constraints" arising from forming the entire bundle are modeled by so-called transition functions. They formally describe how the local parts are patched together in all those cases where the covering sets have a non-empty intersection. Each particular fiber $\pi^{-1}(b)$ obtains its structure from the "typical fiber" F. We now give the formal definition of the concept of local triviality.

A fibering is called *locally trivial* with respect to any covering $\{U_i\}_{i \in I}$ of the base space B, if the following diagram is commutative

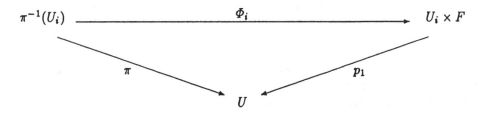

Φ_i is an isomorphism in the corresponding category, where

$$\Phi_i = (\pi, \phi_i), \qquad \phi_i : \pi^{-1}(U_i) \to F \qquad \text{(a morphism)}.$$

For $b \in U_i$, $\phi_{i,b} : \pi^{-1}(b) \xrightarrow{\cong} F$ is the fiber isomorphism induced by Φ_i (this gives π^{-1} its fiber structure).

A product bundle (trivial logical fibering) is given by $E = I \times L$ and $\pi : E \to I$, the projection to the first component. Thus, the fiber over i is $\pi^{-1}(i) = \{i\} \times L =: L_i$, for $i \in I$. We will also call this a *free parallel system*, sometimes denoted by $\mathcal{L}^I$.

The 2-valued subsystems L_i, $i \in I$ are equipped with *local truth values* $\Omega_i = \{T_i, F_i\}$. The set of *(global) truth values* Ω^I for the whole fibering is the disjoint union (coproduct) $\coprod_{i \in I} \Omega_i$.

In such logical systems there are many ways to form logical operations by combining classical logical connectives locally in each L_i and then putting them together in the form of "logical vectors" like $(\phi_i)_{i \in I}$. Furthermore, we can model "system changes" in the sense that we shift logical information (formulas) from one subsystem L_i to another L_j. This corresponds to model communication between fibers (seen as logical state spaces of corresponding agents). For a more formal treatment of such univariate operations and the formation of logical expressions in a logical fibering we refer to [18].

4.2 Transjunctions

Apart from the possibility of forming (bivariate) logical connectives within each subsystem L_i another, more general, non-classical operation arises naturally. Consider a "local" bivariate operation as a mapping $\Theta : L_i \times L_i \rightarrow \mathcal{L}^I$, we can distribute the images of different input pairs $(x_i, y_i) \in L_i \times L_i$ under Θ over different subsystems L_j, L_k, ... in the image space $\mathcal{L}^I$. To be more explicit, semantically, there can be up to four different subsystems for the images of the four possible local input pairs $\Omega_i \times \Omega_i$. For example, as the four image truth values we could have $\Theta(T_i, T_i) = T_\alpha$, $\Theta(T_i, F_i) = F_\beta$, $\Theta(F_i, T_i) = F_\gamma$, $\Theta(F_i, F_i) = F_\delta$, as displayed in the truth-value matrix below. In other words, such bivariate operations can distribute images over several subsystems, a new and basic feature. In [18], a classification of all such bivariate operations, called *transjunctions*, is given. A first, simple demonstration how a transjunction can be applied for the logical control of three cooperating robots was presented in [6, 17, 19]. In the scenario presented here we will use a generalized notion of transjunction which allows multivariate inputs (and outputs). Also, we allow non-classical, many-valued logics in the fibers. The following description for the bivariate and classical case is easily generalized. In order to make a clear distinction between the original notion of transjunction and the generalized version used here, we will denote the latter with *m-transjunction*.

A *transjunction* can be represented by its *truth value matrix*, a mapping from a bivariate truth table within a fixed local system L_i into a bivariate truth table where the T and F values (occurring in the truth table within L_i) are *distributed* over maximally four value sets Ω_α, Ω_β, Ω_γ and Ω_δ, corresponding to the subsystems L_α, L_β, L_γ and L_δ respectively.

Transjunctions can be classified by the type of the truth table under consideration (to which classical connective it corresponds when omitting the indices). As an example, the truth value matrix of a *conjunctional transjunction* looks like:

$$\boxed{\begin{matrix} T_i & F_i \\ F_i & F_i \end{matrix}} \rightarrow \boxed{\begin{matrix} T_\alpha & F_\beta \\ F_\gamma & F_\delta \end{matrix}} .$$

4.3 Modeling the Scenario

We will now model the scenario described above with the machinery of logical fiberings. The individual robots will correspond to local logical systems, and an m-transjunction will be defined which will work as the *logical control function* of the working cell. Finally, the mathematical description of the logical controller will be given by means of a diagram dubbed the *fibered logical controller*.

We will use many-valued logics in the local fibers in order to model the fact that in this scenario the robots have several actions at their disposal (in contrast to the scenario discussed in [19], where each robot had just one possible action to undertake).

The following abbreviations are used for the actions of the robots introduced in Sect. 2: $R_{01}=(r_0,\text{assemble})$, $R_{02}=(r_0,\text{store})$, $R_{11}=(r_1,\text{get-a})$, $R_{12}=(r_1,\text{get-b})$, $R_{21}=(r_2,\text{temp-a})$, $R_{22}=(r_2,\text{temp-b})$, $R_{23}=(r_2,\text{get-temp})$, $R_{24}=(r_2,\text{dump-a})$, $R_{25}=(r_2,\text{dump-b})$, $R_{26}=(r_2,\text{dump-temp})$, and $R_{27}=(r_2,\text{empty-dump})$.

The logical control function will be a heptavariate logical operation; the number of variables of course corresponds to the number of control propositions, giving us the following m-transjunction:

$$\Theta : L_0 \times L_0 \times L_1 \times L_1 \times L_2 \times L_2 \times L_2 \to \mathcal{L} \ .$$

The domain of Θ is forced upon us by the fact that the truth value of the control proposition on(table,res) is checked by a sensor connected to robot r_0, the truth value of tried is determined by the actions of r_0, the truth values of on(table,a) and on(table,b) are determined by the actions of r_1, and the truth values of the last three propositions (on(temp,a), on(temp,b) and used(dump) respectively) are determined by the actions of robot r_2. The codomain (image space) of Θ is determined by the actions to be undertaken upon a given setting of the logical propositions.

We have left the control propositions empty(s-a), on(s-a,a), empty(s-b), and on(s-b,b) out of this discussion. The truth of on(s-a,a) is an extra prerequisite for the action R_{11}, and likewise the truth of on(s-b,b) for R_{12}. For all other actions these propositions are irrelevant. It is therefore a trivial but time and space consuming matter to include them everywhere.

To describe Θ, we will use truth values, expressing it as a function:

$$\Theta : \Omega_0 \times \Omega_0 \times \Omega_1 \times \Omega_1 \times \Omega_2 \times \Omega_2 \times \Omega_2 \to \Omega \ ,$$

where Ω is the set of all (global) truth values.

We will use the following truth values to correspond to the actions:

Truth Value	Logical Subsystem	Action
T_0	L_0	r_0 remains inactive
F_{01}	L_0	r_0 performs R_{01}
F_{02}	L_0	r_0 performs R_{02}
T_1	L_1	r_1 remains inactive
F_{11}	L_1	r_1 performs R_{11}
F_{12}	L_1	r_1 performs R_{12}
T_2	L_2	r_2 remains inactive
F_{21}	L_2	r_2 performs R_{21}
F_{22}	L_2	r_2 performs R_{22}
F_{23}	L_2	r_2 performs R_{23}
F_{24}	L_2	r_2 performs R_{24}
F_{25}	L_2	r_2 performs R_{25}
F_{26}	L_2	r_2 performs R_{26}
F_{27}	L_2	r_2 performs R_{27}

Note that L_0 and L_1 are three-valued logical systems, whereas L_2 turns out to be 8-valued.

We now define the m-transjunction Θ by its truth table, below. In order to render the situation pictorially, we put the values of the first four logical propositions horizontally and the last three vertically. Impossible combinations of the truth values of on(table,res), tried, on(table,a), and on(table,b) have been deleted. The same has been done for impossible combinations of on(temp,a) and on(temp,b), viz. having two pieces on temp. Eliminating these obviously impossible situations, there are only two more combinations of values for the control propositions left that do not occur; these are indicated by blanks in the following diagram, which defines the m-transjunction controlling the working cell. We see that there are several cases in which the resulting truth value is not uniquely determined; this corresponds to a choice of actions for the agents in the scenario.

Θ									
		on(table, res)	0	1	0	0	0		0
		tried	0	1	0	0	0		1
		on(table, a)	0	0	0	1	1		1
		on(table, b)	0	0	1	0	1		1
on(temp, a)	on(temp, b)	used(dump)							
0	0	0	$F_{11} \vee F_{12}$	F_{02}	F_{11}	F_{12}	F_{01}		$F_{21} \vee F_{22}$
0	0	1	F_{27}	F_{02}	F_{25}	F_{24}	F_{01}		$F_{24} \vee F_{25}$
0	1	0	F_{26}	F_{02}		F_{12}	F_{01}		F_{24}
0	1	1	F_{23}	F_{02}	F_{11}	F_{23}	F_{01}		$F_{24} \vee F_{25}$
1	0	0	F_{26}	F_{02}	F_{11}		F_{01}		F_{25}
1	0	1	F_{23}	F_{02}	F_{23}	F_{12}	F_{01}		$F_{24} \vee F_{25}$

Moreover, every individual action R_{ij} should be seen as a morphism on the corresponding local section, thereby regulating the new values of the control propositions except for on(table,res), which is checked by a sensor.

5 Towards a Semantical Foundation for Deductive Planning by Logical Fiberings

In this section we will discuss correspondences between notions from the deductive planning approach and the logical fiberings approach, in order to find a semantic foundation for the planning approach. A complete correspondence will have to await a more rigorous formal statement of all notions involved, but here a list of the most fundamental correspondences is given and illustrated by means of the scenario discussed in the previous sections.

5.1 Situation Versus Global Section

A *situation* in resource-oriented deductive planning simply corresponds to a multiset of facts, where each fact is represented as a term in first order logic. A *global section* is a mapping from the index set connected with a logical fibering to the disjoint union of the indexed subsystems A_i. As we will make the indexed subsystems correspond to the facts (control propositions) to be verified in the scenario, the image of a global section at a given point in time corresponds precisely to the situation at that time ("local snapshot"). For instance, a possible situation in the scenario discussed above could be given by the multiset:[10] {on(table,a), on(temp,b), used(dump)}. This representation of the situation in the planning approach corresponds canonically to the global section mapping $\{0, 1, 2\}$ to $\{\emptyset, \{on(table, a)\}, \{on(temp, b), used(dump)\}\}$. I.e., $\sigma(0) = \emptyset$, $\sigma(1) = \{on(table, a)\}$ and $\sigma(2) = \{on(temp, b), used(dump)\}$.

Taking a *local section* corresponds to viewing the situation from the perspective of one of the agents. As explained in Sect. 4, in this scenario the local section for r_0 only has to take the first four control propositions into account. Likewise, for r_1 only on(table,a) and on(table,b) are relevant. Finally, a local section for r_2 does not have to take into account the truth value of on(table,res).

For instance, from the perspective of r_0 the situation described above looks like {on(table,a)}, implying it has no meaningful action. For r_1, the same situation locally looks like {on(table,a)} as well, which for this robot means that action R_{12}, namely to get a new piece of type b, is invoked. Finally, for r_2, this situation is locally seen as {on(table,a), on(temp,b), used(dump)} and calls for undertaking R_{23} (to move the piece of type b from temp to table).

In this example, possible conflicts can be prevented by looking at r_1 and r_2 together, since taking a local section of r_2 always yields the current section for r_1 as well, as explained in Sect. 5.4 below. A disjunctive situation-term (i.e., a set of multisets) corresponds to a set of sections.

We would like to point out again that the technical machinery of sections works for much more complex applications and scenarios. The example here can only show very elementary principles.

5.2 Action Versus Application of Transjunction

An *action* undertaken in a scenario is described by a specification of its preconditions and its effects, both disjunctive situation-terms. This corresponds to applying the m-transjunction to a global section of the fibering, which gives a mapping from that set of sections to an action evaluation function. The output of this function corresponds to the same action as undertaken on the preconditions (given by a situation-term) in the planning approach. For instance, in the situation described above, the m-transjunction yields as output F_{23} which activates action R_{23}, i.e. robot r_2 will get the piece (of type b) from temp and place it on table. The action also resets some of the control propositions, viz. taking a global section after this action has been carried out would map $\{0, 1, 2\}$ to {on(table,a), on(temp,b), used(dump)}. (In this case sensor interaction is not necessary; a sensor is only needed after carrying out action R_{01}, after which two situations are possible.)

---

[10] Multisets are depicted using the modified curly brackets { and }

In planning terms, this means that (a sufficient) precondition for the same action, namely $(r_2,\text{get-temp})$, is $\{\text{on(table,a)}, \text{on(temp,b)}, \text{used(dump)}\}$, and that the postcondition will contain $\{\text{on(table,a)}, \text{on(table,b)}, \text{used(dump)}\}$ instead of the three terms in the precondition.

Note that the same action $(r_2,\text{get-temp})$ is also evoked on other preconditions, namely $\{\text{on(table,b)}, \text{on(temp,a)}, \text{used(dump)}\}$, with the same postcondition (substituted for these three terms), and also on $\{\text{on(temp,a)}, \text{used(dump)}\}$ and on $\{\text{on(temp,b)}, \text{used(dump)}\}$. In the latter two cases the postconditions are $\{\text{on(table,a)}\}$ and $\{\text{on(table,b)}\}$ respectively.

This corresponds to the fact that in the fiberings approach, the inverse image of the m-transjunction w.r.t. the image F_{23} contains *four* global sections, i.e. $\Theta^{-1}(F_{23}) = \{\{\text{on(table, a)}, \text{on(temp, b)}, \text{used(dump)}\}, \{\text{on(table, b)}, \text{on(temp, a)}, \text{used(dump)}\}, \{\text{on(temp, a)}, \text{used(dump)}\}, \{\text{on(temp, b)}, \text{used(dump)}\}\}$, cf. the table for the logical control function (Sect. 4.3).

5.3 Process Course Versus Sequence of Sections

The *process course* of a scenario in deductive planning is the sequence of all situations that may occur sequentially. This sequence depends on the pre- and postconditions of the actions that are successively applied. This quite naturally corresponds to a (possibly branching) sequence of global sections. However, instead of via pre- and postconditions, the next action is determined via the activation truth value (obtained from the m-transjunction) and the output of the evaluation function *eval*. That output is used as input for the same m-transjunction with which the activation truth value (for the next action to be carried out) was found.

5.4 Plan Generation by Logical Fiberings

In any situation occurring during the process course, the logical controller checks whether there is a m-transjunction (action, in planning terms) which can be applied in the current global section (the complete situation, in planning terms). In the planning approach, this involves matching preconditions of actions with the description of the actual situation. After an action has been selected and carried out, the new situation is constructed by replacing the (instantiated) precondition by the corresponding (instantiated) postcondition. In the fibered approach, the logical controller selects an action (by means of the m-transjunctions) after examining the global section. The m-transjunction almost automatically returns the new situation (only in the case that assembly has been tried, the sensor needs to check whether it was successful or not).

This could be generalized in that one considers (groups of) individual agents as autonomous. By taking local sections according to the control propositions pertinent to each of the three robots, they could determine their next action independently. For r_0, for instance, the only important control propositions are the first four, on(table,res), tried, on(table, a), and on(table,b). Based on these four, the local logical controller can determine the next action (m-transjunction to be applied) for r_0. In the formulation of the scenario as presented above, both r_1 and r_2 need all control propositions to determine their next action. However, one could restrict the local section for r_1 to on(table, a) and on(table,b), and for r_2 to on(table,a), on(table,b), on(table,res), on(temp,a), on(temp,b), and empty(dump). After all, only these are crucial for the actions that these two robots undertake. (For r_1 only needs to "know" whether pieces of type a and b are available on the table, and r_2 in addition needs the information on the temporary and permanent dumps.) The possible ambiguity of actions to undertake (for instance, r_1 might want to supply r_0 with a piece of type a from the store s-a, while r_2 is about to do the same from temp) can be resolved by "checking" the global section. This corresponds to communication between the robots (local fibers).

Summarizing, in the planning approach, the method to generate plans is resolution, as explained in Sect. 3. One tries to match the initial situation with preconditions of actions and searches for a process course that results in the goal situation. (Alternatively, one might start looking backward from the goal situation or mix the two approaches, but search strategy is not our concern here.)

In the fiberings approach things are slightly different. If we take the overall, "global" view, the correspondence is clear. Transjunctions are applied on the initial global section, until the goal situation (corresponding section) is reached. Due to the possibility of failure to assemble pieces of type a and b successfully, the sequence of sections generated will branch (just as in the planning case). Here also, by applying inverse m-transjunctions, backward search can be implemented.

Taking a "local view" however changes things, as robots may be forced to carry out conflicting actions necessitating communication. This communication is modeled by taking global sections in such cases. In the example discussed, conflicts can only arise between r_1 and r_2, suggesting to group the two together as an autonomous unit. What we obtain then is just an optimized version of the "global" approach. For more complicated examples, this concept of localizing the processing of data and parallelizing the decision on which actions to undertake next may greatly improve the efficiency of planning and control. And it is in such more complicated examples that we can exploit the expressive power of fiberings again and again.

6 Prospect: Base Point Dependent Formulas

A major motivation for introducing the handling of logical formulas that are space and/or time dependent comes from the desire to model (logical) state spaces of agents in a scenario varying with space and/or time.

More specifically, we intend to enhance our scenario modeling by the following aspect. In a robotics scenario each (cooperating) agent has its own work space (work cell). The logical formulas corresponding to it are involved in the description of the whole working process. Some actions may depend strongly on the concrete physical position ("coordinates") of an agent, or they can depend on a time parameter. On the other hand, in order to model their specific cooperation tasks, the agents have to fulfil corresponding logical constraints in the overlap of their work spaces. This must be reflected in constraints described in the corresponding local state spaces, since these are technically modeled as fibers over the concrete (local) working areas of the robots. This means that in such applications the physical space of the work cell can serve as base manifold (index system) in our logical fibering. Considerations like these form our natural motivation to model logical formulas which depend on points in space and/or time.

The concepts we introduce to this end can be applied in many ways, since we define them set theoretically, thereby allowing great generality. Specifically, this framework can easily be extended to "fuzzify" logical formulas and information. (In essence, formulas and information is fuzzified by working over $[0, 1]$ instead of $\{0, 1\}$ as the range for the truth values of the formulas involved.)

Here we only cast a first glance at this idea in terms of an example of logical formulas which are space dependent. A more thorough formal treatment is intended for future work.

Let ϕ, ψ, λ, etc., denote logical formulas of our language under consideration which are possible formulas in the logical state spaces, i.e. fibers, over the points $x \in X$ of our base domain, respectively.

We will indicate a formalism of "formula handling" with which we are able to make formulas dependent on the points of the base space. That means that in the state space (i.e. the total space of the fibering) we can work in different ways with logical formulas which can vary from fiber to fiber in the sense that they are "switched on" (evaluable) over a certain subdomain of the base space whereas they are not "switched on" (not evaluable, not present) in the fibers over other subdomains. This allows to model a variety of practical situations by logical means, especially in the case of cooperating agents working in a geometrically modeled environment (base space). A formalism for the handling of such formulas will be subject of future work. Here we only intend to indicate their use by showing a typical example.

For the natural numbers $0, 1$ let $\phi^1 := \phi$, and $\phi^0 :=$ *empty string (empty word)*. That is to say by ϕ^1 we want to express that ϕ is "present" or "switched on", and ϕ^0 is meant to express that ϕ is "not present", "switched off".

Let X denote a domain, for example a geometric domain in the plane in which robots are moving around. $U, V, W, \ldots$ will denote subsets of X. These are to be seen as subareas of the workspace where the robots are allowed to move around under certain working conditions. Let furthermore $\chi_U : X \to \{0, 1\}$ be the *characteristic function* for subsets $U \subset X$ (i.e. $\chi_U(x) = 1$ if and only if $x \in U$ and 0 otherwise). Let now V denote a logical evaluation function which evaluates for every $x \in X$ the local logical formulas over that base point x. (Actually, we would have to define V on the total space of all fibers over the base space X, but we do not go into these technicalities here.)

Now consider expressions of formulas over X like

$$\phi^{\chi_U}, \quad \psi^{1-\chi_U \cup_V}, \quad (\neg^{\chi_U \cap_V})(\phi \wedge \psi), \quad \phi^{\chi_U} \wedge \psi^{\chi_U \cap_V}, \ldots .$$

Then for $x \in X$, we denote by V_x the local evaluation of logical formulas in the state space over x. The following definition

$$V_x(\psi^{1-\chi_{U \cup V}}) := \psi^{1-\chi_{U \cup V}(x)}$$

has the meaning that we insert x in $\chi_{U \cup V}$ for the local evaluation at x. This shall express that ψ appears (and is valid) whenever x is not in U and not in V. Analogously the fourth formula means $\phi \wedge \psi$ does only appear as this conjunction if and only if $x \in U \cap V$. Similarly for the other examples. Many variations are possible using this principle. The same notations can be used for modeling time dependent formulas, namely by extending the base space by time.

We now present an example where we apply this notion of space dependent formulas in the case of three cooperating robots. We deliberately keep this application simple just indicating how we want to exploit the method.

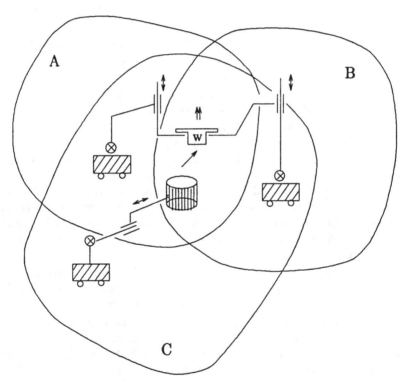

Three robots having workspace areas A, B, and C (we give the corresponding robots the same names A, B, C as well) are able to work independently in all those parts of their workspace where there is no overlap with the workspace of another agent.

The logical control of such (independent) work is modeled by sets of formulas $\{\alpha_A, \ldots\}$, $\{\alpha_B, \ldots\}$, and $\{\alpha_C, \ldots\}$ in the corresponding logical state space of robot A, B, and C, respectively.

In different sets of formulas we could describe the logical control of the scenario in overlaps $A \cap B$, $A \cap C$, $B \cup C$, for example. For reasons of simplicity we only deal with the common overlap of all three robots, namely the workspace area $A \cap B \cap C$.

In this overlap all the three agents should be able to cooperate in order to solve a problem in which the help of all three robots is required. This will correspond to certain logical constraints. Let us consider a simple case of such a situation.

Let W denote a workpiece of weight 62 kg, placed in the area $A \cap B \cap C$ of the work cell. We formulate a task T saying that the workpiece W has to be lifted and placed on a solid basis (a chassis). This chassis can be provided by robot C who can move it in the right position right below the piece W, but of course only in the case that W is lifted. Thus, task T comprises the subtasks "lift W", then "move the chassis right below W".

Now we have the situation that robot A can only lift up to 20 kg ("A can lift up to 20 kg"; this phrase is logically represented by the formula ϕ_A in the logical state space of A). Similarly, robot B can only lift up to 45 kg ("B can lift up to 45 kg" is logically expressed by a corresponding formula ϕ_B in the logical state space of B). Accordingly, neither A alone nor B alone can lift W in their individual workspace area A respectively B. But they can do it jointly in their workspace overlap $A \cap B$. Logically, this will be expressed by the following space dependent formula:

$$(\phi_A^{\chi_A} \wedge \phi_B^{\chi_B}) \Rightarrow (\Box \mathrm{Lift}(W))^{\chi_{A \cap B}} .$$

The box operator $\Box$ in the previous modal formula is used to express that in the workspace overlap $A \cap B$ the robots A and B must fulfil the task of lifting the workpiece W ("Lift(W)") together. The exponents formally express that this formula is only "switched on" or present over $A \cap B$. This formula is contained in the logical state space of both A and B (we interpret it as a logical constraint of the system valid in overlap $A \cap B$).

Now the logical description (control) for performing the whole task T can be modeled as follows. Let [Move$_C$] denote the logical formula corresponding to the subtask of robot C meaning "C can move the chassis right below W". With this, the logical control for the successful performance of task T of the cooperating agents can be expressed by the following space dependent formula:

$$[(\phi_A^{\chi_A} \wedge \phi_B^{\chi_B}) \Rightarrow (\Box \mathrm{Lift}(W))^{\chi_{A \cap B}}] \Rightarrow [\mathrm{Move}_C]^{\chi_{A \cap B \cap C}} .$$

Again, this formula is contained in the logical state space of each robot. We interpret it as a base space dependent constraint of the whole system holding exactly in the common overlap $A \cap B \cap C$. Thus expressing the situation that only in $A \cap B \cap C$ the three robots can fulfil the given task by cooperation.

If we interpret the whole system as a logical fibering with continuous base space (the whole workspace area $A \cup B \cup C$) and discrete total space (the set of all local state spaces of the robots depending on the base points $x \in A \cup B \cup C$), then a fiber over x consists of all the formulas of a local state space of an agent which are "switched on" over that point (that is, which are available or applicable for the corresponding control of the system in that point). A typical feature of this modeling approach is that we can mix discrete and continuous structures.

7 Conclusion

Summarizing, in this paper we have presented two methods to describe robotics scenarios, and indicated how one of them (the logical fiberings) can be used to equip the other (deductive planning) with a semantics. This is part of work in progress. We also indicated how to extend the modeling by logical fiberings with space/time dependent formulas (aiming at modeling space/time logical fiberings), in order to make modeling of very complex scenarios possible and natural. Other extensions, e.g. the incorporation of hierarchical planning in the fibered approach, are intended for future work.

Modeling scenarios with cooperating agents by logical fiberings allows a natural incorporation of communication between the agents, namely by alternating between local and global sections.

Thanks to the modular, fibered approach to modeling an agent's state space, it will be possible to incorporate many different reasoning modules, like modules for geometric reasoning (kinematics problems), modules for communication (both internally, i.e. within the work cell, and externally, with databases located outside), modules for local databases (for data and information to which the agent needs fast and direct access), modules for evaluation of sensor information and the safety thereof, modules for the logical state space description and the reasoning following from that description, modules for planning, and so on.

References

1. J. Allen, J. Hendler, and A. Tate. *Readings in Planning*. Morgan Kaufmann, San Mateo, 1990.
2. W. Bibel. A Deductive Solution for Plan Generation. *New Generation Computing*, 4:115–132, 1986.
3. A.H. Bond and L. Gasser, editors. *Readings in Distributed Artificial Intelligence*. Morgan Kaufmann Publ., San Mateo, California, 1988.
4. S. Brüning, S. Hölldobler, J. Schneeberger, U. C. Sigmund, and M. Thielscher. Disjunction in Resource-Oriented Ddeductive Planning. Technical Report AIDA–94–03, Intellektik, Informatik, TH-Darmstadt, March 1994.
5. E. Csuhaj-Varju and J. Kelemen. On the power of cooperation: a regular representation of recursively enumerable languages. *Theoretical Computer Science*, 81:305–310, 1991.
6. F. Dargam, J. Pfalzgraf, K. Stokkermans, and V. Stahl. Towards a toolkit for benchmarking scenarios in robot multi-tasking. Technical Report 91-45.0, RISC-Linz, J. Kepler University, Linz, Austria, Europe, 1991.

7. L. Gasser and M.N. Huhns, editors. *Distributed Artificial Intelligence.* Morgan Kaufmann Publ., San Mateo, and Pitman, London, 1989.
8. G. Große, S. Hölldobler, J. Schneeberger, U. Sigmund, and M. Thielscher. Equational Logic Programming, Actions, and Change. In *Proc. Joint International Conference and Symposium on Logic Programming JICSLP'92*, 1992.
9. S. Hölldobler and J. Schneeberger. A New Deductive Approach to Planning. *New Generation Computing*, 8:225–244, 1990.
10. S. Hölldobler and M. Thielscher. Actions and Specificity. In D. Miller, editor, *Proceedings of the International Logic Programming Symposium (ILPS)*, pages 164–180, Vancouver, October 1993. MIT Press.
11. J. Kelemen. Syntactical models of distributed cooperative systems. *J. Expt. Theor. Artif. Intell.*, 3:1–10, 1991.
12. V. Lifschitz. Formal theories of action. In *International Joint Conference on Artificial Intelligence*, pages 966–972. Morgan Kaufmann Publishers, Inc., 1987.
13. M. Masseron, C. Tollu, and J. Vauzielles. Generating Plans in Linear Logic. In *Foundations of Software Technology and Theoretical Computer Science*, pages 63–75. Springer, volume 472 of LNCS, 1990.
14. J. McCarthy. Situations and Actions and Causal Laws. Stanford Artificial Intelligence Project, Memo 2, 1963.
15. J. McCarthy. Applications of circumscription to formalizing common-sense knowledge. *Artificial Intelligence*, 28:89–116, 1986.
16. J. McCarthy and P. Hayes. Some philosophical problems from the standpoint of artificial intelligence. In B. Meltzer and D. Michie, editors, *Machine Intelligence, vol. 4*, pages 463–502. Edinburgh University Press, Edinburgh, 1969. Also published in: [1].
17. J. Pfalzgraf. On geometric and topological reasoning in robotics. Submitted to the Special Issue on Artificial Intelligence and Symbolic Mathematical Computation of the Annals of Mathematics and Artificial Intelligence.
18. J. Pfalzgraf. Logical fiberings and polycontextural systems. In Philippe Jorrand and Jozef Kelemen, editors, *Proc. Fundamentals of Artificial Intelligence Research, LNCS 535 (subseries LNAI)*, pages 170–184, 1991.
19. J. Pfalzgraf and K. Stokkermans. Scenario construction continued and extended with a view to test and enhancement of reasoning methods. Technical Report 92-27, RISC-Linz, J. Kepler University, Linz, Austria, Europe, May 1992.
20. J. Pfalzgraf and K. Stokkermans. On robotics scenarios and modeling with fibered structures. In J. Pfalzgraf and D. Wang, editors, *Springer Series Texts and Monographs in Symbolic Computation, Automated Practical Reasoning: Algebraic Approaches.* Springer Verlag, 1994.
21. E. Shapiro and A. Takeuchi. Object oriented programming in Concurrent Prolog. *New Generation Computing*, 1:25–48, 1983.
22. U.C. Sigmund. LLP — Lineare Logische Programmierung. Master's thesis, Intellektik, Informatik, TH Darmstadt, 1992.

Propagation of Mathematical Constraints in Subdefinite Models

Vitaly Telerman

Institute of Informatics Systems,
Russian Academy of Science, Siberian Division
pr. Lavrent'eva 6, Novosibirsk, Russia, 630090
Email: vtel@isi.itfs.nsk.su

Abstract. In the real world we almost always deal only with objects that possess such properties as incompleteness, imprecision, inconsistency, subdefinitess, etc. For this reason, the study of these properties and the nontraditional methods of knowledge processing that are based on them is a very important direction of research.

This paper describes a novel approach to solving problems that is based on *subdefinite calculations.* The use of these methods makes it possible to solve overdetermined and underdetermined problems, as well as problems with uncertain, imprecise and incomplete data. Such problems can be expressed in terms of subdefinite models (SD-models). Propagation of mathematical constraints in all the SD-models is supported by a single data-driven inference algorithm. Several examples are given to show the capabilities of this approach for solving a wide class of constraint satisfaction problems.

Introduction

One of the principal areas of artificial intelligence deals with universal methods and means for working with *imprecise, incomplete* and *uncertain* information. For this reason, the study of these properties and the nontraditional methods of knowledge processing that are based on them is a very important direction of research.

This report is devoted to the description of *subdefiniteness* as one of the most important metacomponents of a knowledge system. Subdefiniteness allows us to use fruitfully all reliable data about real-world simulation objects, including uncertain, noisy, incomplete and imprecise data.

The notion of sudefiniteness has been proposed by A.S.Narin'yani in the beginning of 1980s [Nar80a, Nar80b]. A.S.Narin'yani has developed a formal apparatus which extended set theory and allowed representation of partially known (subdefinite) sets as well as automatic solution of set-theoretic problems incorporating subdefinite sets. This idea was extended to other data types, in particular, to numbers [Nar86a].

Further studies lead to creation of a formal apparatus of data types, which were called active data types (*AcDT*). This apparatus allows us to construct subdefinite extensions of a wider class of objects, concepts, and events [Nar82, Tel88].

The paper is organized as follows. In Section 1 the basic notions of subdefiniteness are presented. The SD-model, which allows us to represent a wide class of computational problems is described in Section 2. In Section 3 we examine a constraint propagation algorithm in the SD-model. The special data type apparatus used for raising the specification level of the SD-model is considered in Section 4. The two examples, that demonstrate the capabilities of the SD-models constraint propagation algorithm are presented in Section 5. In Section 6, four programming systems based on the notion of subdefiniteness are considered. Finally, Section 7 concludes the paper.

1 Basic Notions

The main idea of subdefiniteness can be stated as follows (for a more detailed exposition, see [Nar86a, Nar86b]).

Let us consider an arbitrary multisorted algebra $Q = (A, F)$, where A is a class of objects and F is a set of unary, binary, etc., operations closed on the class A. We associate Q with the algebra: $^*Q = (^*A, ^*F)$ where $^*A$ is the set of all non-empty subsets of A, and all operations $^*F$ are extensions of the corresponding operations of F to $^*A$. The extension of every m-ary operation $f \in F$ is an m-ary operation $^*f$ whose application to an arbitrary m-tuple $^*a_1, ^*a_2, \ldots, ^*a_m$ of the elements of $^*A$ gives the following result:

$$^*f(^*a_1, ^*a_2, \ldots, ^*a_m) = \bigcup_{\substack{a_1 \in {}^*a_1 \\ \vdots \\ a_m \in {}^*a_m}} \{f(a_1, a_2, \ldots, a_m)\}. \tag{1}$$

This means that the result of the operation $^*f$ for an m-tuple $^*a_1, ^*a_2, \ldots, ^*a_m$ is a set of all values obtained by applying the operation f to every m-tuple of the Cartesian $^*a_1 \times ^*a_2 \times \ldots \times ^*a_m$. It is quite obvious that the results of f and $^*f$ are the same for every m-tuple of the elements of A.

Henceforth, we shall call $^*Q$ a subdefinite extension of the Q (or an SD-extension), and each $^*f \in ^*F$, an SD-extension of the corresponding function f. All values in $^*A$ will be called *subdefinite values* (SD-values), the values belonging to A (i.e. singletons) will be called *definite values* and the SD-value equal to A will be called a *complete undefiniteness*.

For example, if Q is a Boolean algebra, where the set A is equal to the set {true, false}, then in the corresponding SD-extension $^*Q$ the set $^*A$ can be equal to the set {{true,false}, {true}, {false}}.

It should be noted, that every object may have several possible SD-extensions. For example, the SD-extension of integers can be defined either as a set of possible values, or as an interval in which this definite value is contained, or as a set of intervals, etc. All these SD-extensions differ in their inference/computation power as well as in the computer resources they require.

Note that some of the properties of the operations of Q may change substantially under SD-extension.

For example, let F contain two operations: the unary $'-'$ and the binary $'+'$ with the identity element $0 \in A$. Let the following conditions be true for these operations for each $a \in A$:

$$-(-(a)) = a,$$

and

$$+(a, (-(a))) = 0. \tag{2}$$

It is clear from (1) that the SD-extension $^*-$ and $^*+$ of these two operations give the following results for the SD-value $^*a = \{3, 4\}$:

$$^*-(^*a) = \{-3, -4\},$$

but

$$^*+(\{3, 4\}, \{-3, -4\}) = \{0, 1, -1\}.$$

Thus, the obtained result only contains the unit element 0, but is not equal to it as in (2).

It the general case the following is true for each $^*a \in {}^*A$:

$$E \in {}^*f(^*a, {}^*a^{-1}),$$

where $^*a^{-1}$ is the inverse element of $^*a$, and E is the identity element of the operation f.

Another property, that is violated under SD-extension is the distributivity law. We can demonstrate that if the operations of Q obey the distributivity law, then their SD-extensions obey the subdistributivity law:

$$^*a \times (^*b + {}^*c) \subseteq {}^*a \times {}^*b + {}^*a \times {}^*c. \tag{3}$$

The violation of some properties of operations under SD-extension $^*Q$, when they are converted from operations on Q, may be explained by the essential difference of interpretation of the notion of value in Q and in $^*Q$. The value of a variable in Q may be either definite (known), or undefinite (unknown). For an SD-variable $^*x$, the assertion $^*x = {}^*a$ has the following meaning: "the definite value of x belongs to $^*a$". Thus, equality of current values of two SD-variables does not imply equality of their definite values, but only the equality of their subdefiniteness.

So, we consider that a typical state of an object of study is subdefiniteness, when we know that its value belongs to a proper subset of its domain of definition. For all elements of the subset, it is equally probable that the value is equal to the specific element. Solving a problem based on subdefiniteness can be organized as a gradual choice of more and more definite values for its parameters. The principle of subdefiniteness treats the state of limited definiteness as a solution of the problem that is available at the present level of knowledge. This subdefinite solution:

- can be practically acceptable;
- can provide a possibility to use methods that are inapplicable in a wider space;
- can stimulate acquisition of additional information.

Note that in everyday practice for the most common problems we cannot find fully definite solutions for lack of necessary information.

2 Subdefinite Models

Let us consider a multisorted algebra Q and its SD-extension ${}^*Q$. Let $X = \{x_1, x_2, \ldots, x_n\}$ be a set of variables with values in A. We associate each $x_i \in X$ with an SD-variable ${}^*x_i$ with the domain of values ${}^*A$. The result of applying each m-ary operation ${}^*f \in {}^*F$ to the values of variables $\{{}^*x_{i1}, {}^*x_{i2}, \ldots, {}^*x_{im}\}$, for ${}^*x_{ij} = {}^*a_j$, $1 \le j \le m$, is defined as the result of the operation ${}^*f({}^*a_1, {}^*a_2, \ldots, {}^*a_m)$, computed by the formula (1), if the values of the variables $\{x_{i1}, x_{i2}, \ldots, x_{im}\}$ are uncorrelated, i.e., no subset of these variables is related by a common relation r that restricts possible combinations of the values of these variables.

If the variables $\{x_1, x_2, \ldots, x_m\}$ are connected by some relation r, then the result of the function can be defined by the following formula:

$$
{}^*f({}^*x_1, {}^*x_2, \ldots, {}^*x_m) = \bigcup_{\substack{a_1 \in {}^*a_1 \cap r_{x_1} \\ \vdots \\ a_m \in {}^*a_m \cap r_{x_m}}} \{f(a_1, a_2, \ldots, a_m)\}. \tag{4}
$$

where r_{x_i} is a projection of r onto variable x_i.

Obviously, the SD-value computed by (4) is contained in the SD-value computed by (1). Thus, in certain cases we can use the operation (1) instead the corresponding function (4) without losing possible solutions. As a consequence, computations may yield results that are significantly wider than the true ones. This can be improved by applying symbolic transformations. For example, given the subdistributivity law, it is more preferably to bring expressions like the right side of (3) to the form of its left one.

To simplify the notation used below, we shall omit the symbol $*$ when referring to SD-extensions of variables and operations if that does not lead to ambiguity.

A problem with subdefinite values of parameters can be represented as a model that is called *SD-model*.

An SD-model M consists of the following four sets:

$$
M = (X, W, C, R),
$$

where

- X is a set of objects,
- W and C are sets of operators of a special kind, called the set of *write operators* and the set of *correctness-check operators*, respectively, and
- R is a set of constraints.

Each object $x \in X$ is associated with a built-in type (or sort) T_x, a SD-value $^*a_x \in {}^*A$, a write operator W_x and a correctness-check operator C_x.

The write operator W_x is a two-argument function which is called for each attempt to assign a new value to the object x and defines the next value of this object as a function of the current and the new values. Thus, depending on the write operator, not every attempt of assignment leads to a change of the current value, and even then the new value is not necessarily equal to the value being assigned. The main requirement to the write operator is that any write operator should ensure that the value of an object changes from a less precise to a more precise one.

The correctness-check operator C_x is an unary function mapping the set of the values of x into the set of Boolean values. This operator examines the SD-value of the object x, and if this value is empty the operator returns the false value.

Each constraint is associated with a set of *interpretation functions*. We consider, that a k-ary function f is a mapping of the k variables values into the set of the n result variables values. Thus, in the function

$$\{y_1, y_2, \ldots, y_n\} = f(x_1, x_2, \ldots, x_k), \quad (k \geq 1, n \geq 1), \tag{5}$$

f is an operation with arity k, and $x_1, x_2, \ldots, x_k, y_1, y_2, \ldots, y_n$ are the variables that represent the k arguments and the n results of operation f, respectively. As was noted above, the same variable can be used instead of different arguments or results of f.

Let us consider the constraint

$$x + y = z, \tag{6}$$

where x, y, and z are three numbers, and $+$ is the addition operation. Then the following three interpretation functions correspond to this constraint:

$$z := x + y; \quad y := z - x; \quad x := z - y. \tag{7}$$

In this case the constraint variables are simultaneously input and output variables. For example, if in the constraint (6) the variables values are

$$x = \{1, 2, 3\}, \quad y = \{2, 3, 4\}, \quad z = \{6, 7, 8, 9\},$$

then after calculation of the interpretation functions (7), we obtain

$$x = \{2, 3\}, \quad y = \{3, 4\}, \quad z = \{6, 7\}.$$

The constraint propagation algorithm will be considered in the Sect. 3.

Traditionally, constraint satisfaction problems are represented as a constraint network. The SD-model also can be represented as a special network, called *SD-network*. An SD-network is a declarative structure that expresses the constraints for parameters. It consists of a number of object nodes connected by constraint interpretation functions.

Each object node is associated with a data type, a value, a write operator and a correctness-check operator.

Each interpretation function is associated with a computational procedure, and an integer number, which is the priority.

The set of constraint interpretation functions determines the computational capabilities of an SD-model inference engine. The functions are data-driven in a certain sense, i.e., every change of an object value leads to a re-computation of all the functions having this object as an argument.

3   Constraint Propagation in SD-Models

The SD-model characterize the following constraint satisfaction problem.

Statement of the Problem. Given is a set R of constraints relating a set X of objects associated with the SD-values $^*A$. Refine the SD-values as far as possible without losing possible exact solutions of R, i.e., determine for each object x the minimal consistent value $^*a_x^{min}$ within its SD-value $^*a_x$. Each SD-value $^*a_x$ should satisfy the following two requirements:

1. its refinement should be consistent with the write operators W_x;
2. the correctness-check operator C_x on $^*a_x$ should return the true value.

In what follows, we shall use the following notation:

- X^t designates a set of variables from X whose values have become more precise at step t;
- $^*a_x^t$ designates a value of the variable x at step t;
- $F|x$ designates a set of interpretation functions having the variable x as one of its arguments;
- $F|X^t$ designates a set of interpretation functions having at least one argument which belongs to X^t.

The inference/computation process may be presented in the form of the following algorithm:

Step 1. $t = 1$; $X^t = \{x_1, x_2, \ldots, x_n\}$. Note that all model variables have values: they are either initialized or their value is the complete undefiniteness.

All the model interpretation functions are placed in set $F \mid X^1$ and all the members of this set form a set of active functions.

Step 2. If the set of active functions $F|X^t$ is empty, go to step 8.

Step 3. Choose and remove a function (5) with the maximal priority from $F|X^t$.

Step 4. Set $F|X^{t+1}$ equal to $F|X^t$.

Step 5. Compute the function (5). Place the results of the computation, i.e., the tuple $\{{}^*a_{y_1}^{new}, {}^*a_{y_2}^{new}, \ldots, {}^*a_{y_n}^{new}\}$ into temporary storage.

Step 6. For each variable $y_i (i = 1, \ldots, n)$ do:

Step 6.1. Compute the write operator

$$*a_{y_i}^{t+1} = W_{y_i}({}^*a_{y_i}^t, {}^*a_{y_i}^{new}).$$

Step 6.2. If the value ${}^*a_{y_i}^{t+1}$ is different from ${}^*a_{y_i}^t$, then execute the correctness check operator $C_{y_i}({}^*a_{y_i}^{t+1})$.

Step 6.3. If $C_{y_i}({}^*a_{y_i}^{t+1})$ is true (i.e. the value is correct), then merge the set of active interpretation functions $F \mid X^{t+1}$ with the set $F \mid y_i$ according to their priority. If the value is incorrect, the model is incompatible. The algorithm terminates.

Step 7. $t = t + 1$ and go to Step 2.

Step 8. Output values of set X.

For all models that have only finite SD-values, this algorithm terminates in a finite number of steps (if all the write operators have been correctly implemented). In the case of infinite sets of values (for examples, intervals for subdefinite real numbers), for a stopping criterion for the algorithm we can take an a priori threshold for the precision of computations. This threshold defines the maximal deviation value for which the new value of the object is still believed to be indiscernible from the previous one.

This constraint propagation algorithm is implemented as a *data-driven virtual processor* (DDVP) [Nar87], which has been used in various software systems based on SD-models [Tel93a].

The above approach to the solution of constraint satisfaction problem has the following features that distinguish it from most of constraint propagation methods the author is aware of:

1. The processing nodes of a computing network contain functions rather than constraints. Thus, a constraint is considered as a special case of the interaction of a group of functions that implement the recalculation of all of its parameters. Such a functional representation of the network is not only more general, but more flexible as well, since it enables one to construct in a regular manner rather complex systems of constraints from a limited set of basic functions. Nevertheless, the end-user always works only with declarative, non-directional constraints.

2. Like constraint propagation, SD-models are based on local computations; however, our approach uses a single data-driven algorithm, which does not depend on the type of the information processed and the complexity of the constraints to be solved. Some algorithms used in constraint propagation also implement a data-driven mechanism, like the Waltz filtering algorithm [Waltz], but they all work with constraints rather than with functions. In our approach, such special algorithms can be used as some of the constraint interpretation functions.

Among the methods of constraint propagation, the approach proposed by Hyvonen is the closest to SD-models [Hyv92]. This approach, which he calls *tolerance propagation (TP)*, has the following characteristics that are common with SD-models:

- representing constraints in the form of a functional network;
- a data-driven algorithm of computations;
- the use of interval arithmetics to implement operations and functions on intervals of real numbers. In addition, the two approaches are related by the common objective, to find sufficiently narrow intervals for the values of the variables, which must contains all the values that satisfy a set of constraints.

Despite this closeness of TP and SD-models, essential differences between them exist.

First, the TP approach is designed to solve only numerical problems in which the values of the variables are represented by intervals of reals. SD-models are oriented to solving problems in various domains. These domains may contain both finite and infinite sets of values. SD-models may include a large number of diverse constraints (linear and nonlinear algebraic equations, inequalities, logical expressions, table relations, set constraints etc.) and solve them as a whole.

Second, a SD-variable always contains a write operator and a correctness check operator which control all modifications of its value. In TP approach, there are no notions that correspond to the write operator and the correctness-check operator. Therefore, the contraction of the ranges of the variables and the consistency checks are ensured by the algorithm rather than by the data themselves (see Sect. 4).

4 Active Data Types

A special data type apparatus (called active data types) can be used to raise the specification level of a SD-model.

AcDT are a hierarchical system of types which is a specific mixture of constraint propagation and object-oriented approach [Nar82, Tel88].

An AcDT T of level t $(t \geq 1)$ includes the following components:

- a write operator and a correctness-check operator;
- a set of *slots* (considered as objects localized in T). Each slot is associated with a data type of level $t'(t > t')$ and an initial value;

- a set of *computational constraints* on the slots;
- a set of *operations* and *constraints* permitted for the objects of the type T.

Consider the field of real numbers $\mathbb{R}$ and the ring of integer numbers $\mathbb{Z}$. Let us define the AcDT *subdefinite number (SD-nuber)*. Since computer number representations are finite, a value of the SD-number can be represented by finite intervals from $\mathbb{R}$ or from $\mathbb{Z}$ which are bounded by some lower and upper boundary numbers Min and Max playing the role of $-\infty$ and $+\infty$, respectively.

Hence, the SD-number can have the folowing representation:

$$\textbf{type } SD\_NUM \ (Wsdnum, Csdnum) = low\_bound : number;$$
$$up\_bound : number;$$
$$\textbf{active } low\_bound := Min;$$
$$up\_bound := Max.$$

Such data types represent approximately known values of a number (integer or real), which belong to the interval between the upper and lower bounds. The write operator $Wsdnum$ is a two-argument function which chooses the minimum of the upper bounds and the maximum of the lower bounds of its arguments. The correctness-check operator $Csdnum$ verifies that the lower bound is less than or equal to the upper bound.

This definition of SD-numbers implies that the interval of a subdefinite object contains the actual value which it represents, and when additional information becomes available, it can contract, reflecting the parameter's value with an increasingly higher accuracy. This means that for this object it is necessary to distinguish its actual value and the current SD-value which is an available estimation of the actual value.

Some of operations for SD_NUM are constructed according to the rules of the interval mathematics [Alef83]. For example, the operation of subtraction of SD_NUM objects must be defined as follows:

$$\textbf{operation} - (a, b : SD\_NUM) : SD\_NUM$$
$$[\ low\_bound, up\_bound \ ] := [\ a.low\_bound - b.up\_bound,$$
$$a.up\_bound - b.low\_bound \ ].$$

Such operations can be utilized to define the constraints for the AcDT SD_NUM.

In the conclusion of this section, we note that the interpretation of a SD-values as a data type offers the following advantages:

- it increases localization of computations and, thus, provides for higher degree of parallelism for the algorithm;
- it enables one to combine naturally in a single network objects of different types;
- it provides an opportunity to use active data types for constructing objects of arbitrary complexity and combining them in a functional network;

– it allows the end-user to deal with non-directional constraints without worrying about the order of execution of functions in the SD-network.

Note also that the model of computations used in our approach is applicable not only to the processing of subdefinite information, since it is the semantics of the write operator that determines how the data can be modified.

5 Solving Problems with SD-Models

In this section, we consider two examples that demonstrate the capabilities of SD-models.

Let us consider the following system of two numerical constraints:

$$x + y = 12; \tag{8}$$
$$2 * x = y; \tag{9}$$

It is also known that x and y are integers between 0 and 100. To represent these variables as subdefinite objects, one can use the type SD-number, defined in Sect. 4, where the types of slots are integer. Hence, the SD-value of these variables can be represented by intervals with the initial state [0,100].

The constraints (8–9) are automatically compiled into the following interpretation functions:

$$f_1 : y := 12 - x;$$
$$f_2 : x := 12 - y;$$
$$f_3 : y := 2 * x;$$
$$f_4 : x := y/2;$$

Taking into account the subdefinite description of the operations $-$, $*$, and $/$, the interpretations of these functions are:

$$*f_1 : [y.low\_bound, y.up\_bound] :=$$
$$[12 - x.up\_bound, 12 - x.low\_bound];$$
$$*f_2 : [x.low\_bound, x.up\_bound] :=$$
$$[12 - y.up\_bound, 12 - y.low\_bound];$$
$$*f_3 : [y.low\_bound, y.up\_bound] :=$$
$$[min(2 * x.low\_bound, 2 * x.up\_bound),$$
$$max(2 * x.low\_bound, 2 * x.up\_bound)];$$
$$*f_4 : [x.low\_bound, x.up\_bound)) :=$$
$$[min(y.low\_bound/2, y.up\_bound/2),$$
$$max(y.low\_bound/2, y.up\_bound/2)];$$

Following the algorithm mentioned in the Sect. 3 we obtain:

$$*X^1 = \{x = [0, 100], y = [0, 100]\};$$
$$F|*X^1 = \{*f_1, *f_2, *f_3, *f_4\}.$$

Table 1 illustrates execution of the algorithm. Initially, all the functions $^*f_1 - {}^*f_4$ are activated. The result of the execution of $^*f_1$ is $y = [-88, 12]$. Then, the write operator for y is called. The old value of y is $[0, 100]$, and the new value is $[-88, 12]$. The write operator returns the value $[0, 12]$ and assigns it to y. Since the value of y changes, the correctness-check operator is called. As the correctness condition is kept ($0 \leq 12$), the functions $^*f_2, {}^*f_4$ are merged with the set of active functions. Then, the next function in the set of active functions is chosen, and so on. The algorithm produces the following results:

$$x = 4; \quad y = 8.$$

It can be verified that further execution of all functions $^*f_1$-$^*f_4$ does not change the values of x or y. Then, the set of active functions becomes empty and the algorithm terminates. It should be stressed that not only numerical objects can

Table 1. Inference steps of the DDVP.

N	Set of Active Functions	Working Function	Old Value	New Value	Change Flag	Functions to be Activated
1	$^*f_1, {}^*f_2, {}^*f_3, {}^*f_4$	$^*f_1$	$y = [0,100]$	$[0,12]$	yes	$^*f_2, {}^*f_4$
2	$^*f_2, {}^*f_3, {}^*f_4$	$^*f_2$	$x = [0,100]$	$[0,12]$	yes	$^*f_1, {}^*f_3$
3	$^*f_3, {}^*f_4, {}^*f_1$	$^*f_3$	$y = [0,12]$	$[0,12]$	no	
4	$^*f_4, {}^*f_1$	$^*f_4$	$x = [0,12]$	$[0,6]$	yes	$^*f_1, {}^*f_3$
5	$^*f_1, {}^*f_3$	$^*f_1$	$y = [0,12]$	$[6,12]$	yes	$^*f_2, {}^*f_4$
6	$^*f_3, {}^*f_2, {}^*f_4$	$^*f_3$	$y = [6,12]$	$[6,12]$	no	
7	$^*f_2, {}^*f_4$	$^*f_2$	$x = [0,6]$	$[0,6]$	no	
8	$^*f_4$	$^*f_4$	$x = [0,6]$	$[3,6]$	yes	$^*f_1, {}^*f_3$
9	$^*f_1, {}^*f_3$	$^*f_1$	$y = [6,12]$	$[6,9]$	yes	$^*f_2, {}^*f_4$
10	$^*f_3, {}^*f_2, {}^*f_4$	$^*f_3$	$y = [6,9]$	$[6,9]$	no	
11	$^*f_2, {}^*f_4$	$^*f_2$	$x = [3,6]$	$[3,6]$	no	
12	$^*f_4$	$^*f_4$	$x = [3,6]$	$[3,4]$	yes	$^*f_1, {}^*f_3$
13	$^*f_1, {}^*f_3$	$^*f_1$	$y = [6,9]$	$[8,9]$	yes	$^*f_2, {}^*f_4$
14	$^*f_3, {}^*f_2, {}^*f_4$	$^*f_3$	$y = [8,9]$	$[8,8]$	yes	$^*f_2, {}^*f_4$
15	$^*f_2, {}^*f_4$	$^*f_2$	$x = [3,4]$	$[4,4]$	yes	$^*f_1, {}^*f_3$
16	$^*f_4, {}^*f_1, {}^*f_3$	$^*f_4$	$x = [4,4]$	$[4,4]$	no	
17	$^*f_1, {}^*f_3$	$^*f_1$	$y = [8,8]$	$[8,8]$	no	
18	$^*f_3$	$^*f_3$	$y = [8,8]$	$[8,8]$	no	

be used in the SD-model. Just as easily we can define the SD-extension of any other data type.

Below, we consider the SD-extension of one of the basic nonnumerical data types, which is a set [Nar80b, Nar86a]. We suppose that there is an universal set (called *the universe* and denoted by U), whose elements can potentially belong to a concrete set. A representation of a subdefinite set (SD-set) is selected according to the pragmatics of the subject domain. For example, if the universe U is infinite, then an SD-set can be represented by the following pair:

$$N = <A^+, M>, \tag{10}$$

where A^+ is the set of elements that belong to the SD-set N, and M is the subdefinite cardinality of N. The slot M can be defined as a subdefinite integer.

The number of elements of the set A^+ can only increase. This number is bounded by the upper bound of its cardinality $(M.up\_bound)$. Any SD-set (10) should satisfy the following condition:

$$|A^+| \le M.low\_bound. \tag{11}$$

If the universe U is finite, then the information that an element of the universe is not in N can also be useful. We denote the set of the elements of a finite universe that do not belong to N by A^-. In this case we can define the SD-set as follows:

$$N = <A^+, A^-, M, U>. \tag{12}$$

Obviously, the SD-set (12) contains additional restrictions besides (11):

$$A^+ \subseteq U;$$
$$A^- \subseteq U;$$
$$M.up\_bound \le |U| - |A^-|;$$
$$A^+ \cap A^- = \emptyset.$$

The number of the elements of the set A^-, as well as the sets A^+, can only increase during the inference/computation process, making N more and more precise.

Consider a simple example of SD-sets (12). Suppose that our universe contains 20 elements. For simplicity, we assume that the elements are the first 20 letters in the alphabet $(\{a-t\})$. Suppose also that we have four SD-sets $(X, Y, Z,$ and $W)$ in the universe U with the following initial values:

$$X = <\{a, b, c, d, e, f, k, l, p\}, \quad \{h, i\},, U>;$$
$$Y = <\{k, l\}, \quad \{g, h, i, j\},, U>;$$
$$Z = <\{c, d, e, f\}, \quad \{a, b, g\},, U>;$$
$$W = <\{o, p, q, r, s, t\}, \quad ,, U>;$$

and the following restrictions:

$$Z = X \setminus Y;$$
$$W \subseteq Z.$$

Suppose that the set X contains at most 14 elements, while Y has more than 5 elements. In this case, the following two constraints are added to the SD-model:

$$X.M \le 14;$$
$$Y.M > 5.$$

After applying the algorithm of constraint propagation to this problem, we obtain the following results:

$$X = < \{a, b, c, d, e, f, k, l, o, p, q, r, s, t\}, \{g, h, i, j, m, n\}, [14, 14], U >;$$
$$Y = < \{a, b, k, l, m, n\}, \{c, d, e, f, g, h, i, j, o, p, q, r, s, t\}, \quad [6, 6], U >;$$
$$Z = < \{c, d, e, f, o, p, q, r, s, t\}, \{a, b, g, h, i, j, k, l, m, n\}, [10, 10], U >;$$
$$W = < \{o, p, q, r, s, t\}, \{a, b, g, h, i, j, k, l, m, n\}, \quad [6, 10], U >;$$

Thus, we have completely determined the values of the sets X, Y, and Z. The set W remains subdefinite. In the current state of the model, it is not clear yet whether the set W contains the elements c, d, e, f.

6 Programming Technology Based on Subdefiniteness

We have been developing not only the theory of subdefiniteness but also a rather powerful software technology, based on a DDVP. The technology has been implemented in the following experimental products:

1. *NeMo-Tec* - an open programming environment for development of models, with the capability of expanding the spectrum of subdefinite data types;
2. *UniCalc* - an end-user system for solving mathematical problems;
3. *ARBOOZ* - an intelligent solver of arithmetic puzzles;
4. *Time-Ex* - an end-user system for calendar scheduling based on a subdefinite temporal model.

All of these systems have exactly the same SD-model-based inference engine (i.e. DDVP), although they use different sets of data types. They are successfully used by the knowledge engineers for experimenting with various test problems. They also can be used as a basis for a wide spectrum of other systems (spreadsheets, CADs, etc.).

6.1 NeMo-Tec Programming Environment

Effective implementation of the end-user systems based on subdefiniteness implies the need for an advanced programming technology. The NeMo-Tec is an open programming environment that enables the knowledge engineer to create new subdefinite data types and operations [Tel93b]. With its help, an experienced knowledge engineer can expand subdefinite technology on to new application areas.

The NeMo-Tec technological environment supports creation of a class of self-defining systems. Working with a system of this kind is reduced to a sequence of steps in which the end-user sets SD-values of some objects, starts the inference process and, when observing its integral results, makes the decision: reject unacceptable consequences of the computations just completed or set the parameter values for the next step. This kind of end-user paradigm is characteristic for such areas as automated design, planning, games, etc.

6.2 UniCalc - an Algebraic Problem Solver

The UniCalc programming system was designed to solve arbitrary systems of algebraic and algebraic-differential equations, inequalities, or logic expressions [UniC1]. Such a system may contain only integer and real variables, or combine both of them.

The following example presents a SD-model with various constraints and data types.

$$x^2 + 6 * x = y - 2^k;$$
$$k * x + 7.7 * y = 2.4;$$
$$(k - 1)^2 < 10;$$
$$(x < 2.5 * y) \rightarrow (k * y \leq 3) \wedge (k > y + 1);$$

In this example x and y are real numbers and k is an integer.

As a result of solving an arbitrary SD-model, the UniCalc inference engine either finds a parallelepiped that contains all solutions of the problem, or issues a message that the problem is inconsistent. If the problem has a unique solution, then the parallelepiped will be reduced, in most cases, to a point (with a given accuracy). If the problem has several solutions, it may be necessary to add appropriate constraints in order to locate each of them, or use the built-in tool for automatic locating of solutions.

For the example above the UniCalc returns the following subdefinite solution:

$$k = [0, 4]; \quad x = [-6.4132, 0.4406]; \quad y = [0.0828, 3.6433].$$

To obtain the exact solution, it is nessesary to impose stronger restrictions on the variables. Thus, if we add to the example the constraint $k < 3$ the solution will be as follows:

$$k = 2; \quad x = -0.6586; \quad y = 0.48261.$$

The more detailed description of the UniCalc system is presented in [UniC2].

6.3 ARBOOZ - Intelligent Solver of Arithmetic Puzzles

The ARBOOZ solver allows the end-user to describe and solve puzzles of the following kind:

$$\begin{array}{r} SEND \\ + MORE \\ \hline MONEY \end{array}$$

This is a well-known type of puzzles: an arithmetic operation on some words for which it is necessary to find a "digit for letter" substitution satisfying this equation [Tel88, Tel93a]. ARBOOZ solve puzzles with sum, subtraction, multiplication and division operations. The senior digits of each number (the leftmost letters) are not equal to zero. Different letters represent different digits. The words in the puzzle can contain letters (Russian or English characters), digits and symbols $*$. The symbol $*$, in contrast with a letter, represents a digit that may coincide with the values of other letters and other symbols $*$.

If the value of a letter is completely unknown, it is equal by default to the whole domain of definition - the set of digits $\{0, 1, \ldots, 9\}$. Therefore, the corresponding subdefinite data type may be defined as a set of integers. Its write-operation calculates the intersection of the current value with the new value and replaces the current value by the result obtained. The correctness check operator checks that the set of permissible digits is not empty. The arithmetic operations are defined with the help of the corresponding standard operations. For example, the sum of the SD-values of two letters is defined as the sum of one pair of digits: the first digit belongs to the SD-value of one letter, the second - to the SD-value of the other letter.

The SD-model is defined as the set of constraints linking the objects in one column. The adjacent columns are related with the help of the carries.

Knowledge engineers can elaborate other and other relations; for example, if the problem contains the following pattern:

$$\begin{array}{r} \ldots \textbf{A B} \ldots \\ + \ldots \textbf{C D} \ldots \\ \hline \ldots \textbf{B A} \ldots \end{array}$$

we infer that $C + D + c_1 + c_2 = 10$, where c_1 is the carry into the first column ($B + D = A$) and c_2 - into the second column ($A + C = B$).

During the inference process in the ARBOOZ system, the user is given explanations of the task solving steps. If the puzzle has a single solution it can be automatically found by the ARBOOZ system. For example, after the puzzle above has been solved, we obtain the following solution:

$$\begin{array}{r} 9\,5\,6\,7 \\ + 1\,0\,8\,5 \\ \hline 1\,0\,6\,5\,2 \end{array}$$

If the ARBOOZ system did not find the solution or if there is more than one solution, the user can apply the exhaustive search method.

6.4 Time-Ex - a Calendar Scheduling System

The situations in which the times for performing certain tasks cannot be specified definitely are typical for practical calendar scheduling. Most often the reason for that is incomplete initial information about the times necessary to perform the tasks. For example, we might know about task A only that it should be started not later than August 3, whereas about task B we know that it has to be completed within 7-10 days. Obviously, such conditions are characterized by subdefiniteness of specific knowledge.

The calendar schedule in the Time-Ex system is described by a SD-model, where objects are subdefinite T-intervals representing tasks to be fulfilled, and constraints relate T-intervals of individual tasks [Borde]. This approach is based on the apparatus of temporal logic [Kan84, Kan86].

An AcDT T-interval has three slots (starting time, finishing time, and duration) which are SD-integers. The slots are bound by the following constraints:

$$starting + duration = finishing;$$
$$starting \leq finishing;$$

The constraints over T-intervals allow a wide class of dependences between tasks to be described. Thus, using the sequence constraint, it is possible to specify a sequence of task execution, whereas using the embedding relation, it is possible to specify the tasks' organization structure. The relation nonsimultaneity of is suitable for simulating conflicts due to the competition between the tasks for recources.

The inference process on the SD-model of the calendar schedule narrows the boundaries of intervals representing subdefinite start and finish points and durations of T-intervals.

Subdefinite calendar scheduling allows one to include time reserves into the calendar schedule and, therefore, to compose calendar schedules that are emergency-stable and easily adaptable to unpredictable reality.

7 Conclusion

A new approach to constraint propagation of mathematical constraints has been presented in this paper. It is based on the notion of subdefiniteness. This approach takes into account the partially known information about values of objects. The task to be solved is represented as an SD-model. A special data type apparatus is used for increasing the specification level of SD-models.

Constraint reasoning based on subdefiniteness has the following features:

- the model can be undetermined or overdetermined, and the parameters of the model may be partially defined;
- no distinction is made between the arguments and the results of the SD-model: the values of all variables are subdefinite to a varying degree and the computations are performed for all variables of the model;
- subdefiniteness of a variable never increases and the computation process converges if the corresponding write operator is implemented correctly;
- this approach determines a parallel, asynchronous indeterministic process with the data-driven control.

The investigation of the subdefiniteness and the use of SD-models in constraint programming is continuing. For the nearest future we have the following objectives:

- further development of the methods for representation of active data types with the corresponding modification of the computational algorithm;

- extending the notion of subdefiniteness to subdefinite functions and subdefinite constraints;
- implementation of symbolic transformations to optimize the problem representation in the SD-network.

References

[Alef83] Alefeld, G., Herzberger, Ju.: *Introduction in Interval Computations*, Academic Press, New York, (1983).

[UniC1] Babichev, A.B., et al.: *UniCalc - an intelligent solver for mathematical problems*, Proceedeings of East-West AI Conference: from theory to practice, Moscow, (1993), 257 – 260.

[Borde] Borde, S.B., et al.: *Subdefiniteness and Calendar Scheduling*, Ibid, 315 – 318.

[Hyv92] Hyvonen, E.: *Constraint reasoning based on interval arithmetic: the tolerance propagation approach*, Artificial Intelligence, **58**, (1992), 71 – 112.

[Kan84] Kandrashina, E.Yu.: *Means for Representing Temporal Information in Knowledge Bases*, Trans. USSR Acad. Sci., Technical Cybernetics, **5**, Moscow, (1984), 15 – 22 (in Russian).

[Kan86] Kandrashina, E.Yu.: *Means for Representing Temporal Information in Knowledge Bases. Event Sequences*, Trans. USSR Acad. Sci., Technical Cybernetics, **5**, Moscow, (1986), 211 – 231 (in Russian).

[Nar80a] Narin'yani, A.S.: *Subdefinite Set - a Formal Model of Uncompletely Specified Aggregate*, Proc. of the Symp. on Fuzzy Sets and Possibility Theory, Acapulco, Mexico, (1980).

[Nar80b] Narin'yani, A.S.: *Subdefinite Sets - New Data Type for Knowledge Representation*, Preprint USSR Acad. Sci., Siberian Division, Computer Center, **232**, Novosibirsk, (1980) (in Russian).

[Nar82] Narin'yani, A.S.: *Active Data Types for Representing and Processing of Subdefinite Information*, In: Actual Problems of the Computer Architecture Development and Computer and Computer System Software, Novosibirsk, (1983), 128 – 141 (In Russian).

[Nar86a] Narin'yani, A.S.: *Subdefiniteness in Knowledge Representation and Processing*, Trans. USSR Acad. Sci., Technical cybernetics, **5**, Moscow, (1986), 3 – 28 (in Russian).

[Nar86b] Narin'yani, A.S.: *Representation of Subdefiniteness, Over-definiteness and Absurdity in Knowledge Bases (Some Formal Aspects)*, Computers and Artificial Intelligence, (1986), **5**, N 6, 479 – 487.

[Nar87] Narin'yani, A.S., Telerman, V.V., Dmitriev, V.E.: *Virtual Data-Flow Machine as Vehicle of Inference/Computations in Knowledge Bases*, In: Ph. Jorrand, V. Sgurev (Eds.) Artificial Intelligence II: Methodology, Systems, Application, North-Holland, (1987), 149 – 154.

[UniC2] Semenov, A.L., Babichev, A.B., Leshchenko A.S.: *Subdefinite computations and symbolic transformations in the UniCalc solver*, This volume.

[Tel88] Telerman, V.V.: *Active Data Types*, - Preprint, USSR Acad. Sci., Siberian Division, Computer Center, **792**, Novosibirsk, (1988) (In Russian).

[Tel93a] Telerman, V.V.: *Constraint propagation in sub-definite models*, In: B.Mayoh, J.Penjam, E.Tyugu (Eds.) NATO Advanced Study Institute, CONSTRAINT PROGRAMMING, Tallinn, (1993), 33 – 45.

[Tel93b]   Telerman, V.V.: *Technological environment for construction and processing
            sub-definite models*, Proc. East-West AI Conference: from theory to practice,
            Moscow, (1993), 356 – 360.

[Waltz]   Waltz, D.L.: *Understanding line drawings of scenes with shadows*, In: P.
            Winston (Ed.) The Psychology of computer Vision, McGraw-Hill, New York,
            (1975).

Combining Computer Algebra and Rule Based Reasoning

Reinhard Bündgen

Wilhelm-Schickard-Institut, Universität Tübingen
Sand 13, D-72076 Tübingen, Fed. Rep. of Germany
phone: x7071/295459 — fax: x7071/295958
e-mail: ⟨buendgen@informatik.uni-tuebingen.de⟩

Abstract: We present extended term rewriting systems as a means to describe a simplification relation for an equational specification with a built-in domain of external objects. Even if the extended term rewriting system is canonical, the combined relation including built-in computations of 'ground terms' needs neither be terminating nor confluent. We investigate restrictions on the extended term rewriting systems and the built-in domains under which these properties hold. A very important property of extended term rewriting systems is decomposition freedom. Among others decomposition free extended term rewriting systems allow for efficient simplifications. Some interesting algebraic applications of canonical simplification relations are presented.

1   Introduction

There has always been mutual interest in the areas of computer algebra and term rewriting systems as can be seen for example from the calls of papers of the main conferences in the two areas (ISSAC and RTA resp.) which each ask for contributions from the other area. Similarly, the major journals for these research fields publish contributions from both areas. However work on actually combining these two areas has been very limited.

- There is a long tradition of research relating the Knuth-Bendix procedure [KB70] and Buchberger's algorithm [Buc65]. The latest results view term completion procedures as a generic Buchberger algorithm which can be parametrised by a specification of the (polynomial) domain in which the completion shall take place [Bün91b, Bün91a, Bün92].
- The discovery of critical pair criteria for Buchberger's algorithm [Buc79] has successfully triggered such research for the Knuth-Bendix procedure too (see e. g. [WB85, Küc85, Bün94a]).
- Knuth-Bendix completion has influenced some methods for computational group theory (see e. g. [BKR87]).
- Some computer algebra systems like Mathematica [Wol91] provide ad-hoc implementations of rewriting as part of their programming language.

Yet there are no combinations of computer algebra systems and term rewriting systems which rigorously combine the algorithmic power of the first with proof strength of the latter systems.

Term rewriting systems appear to be a natural link between computer algebra and artificial intelligence. On the one hand they provide an abstract framework for computing with canonical simplifiers in algebra [BL82] and on the other hand they form a mathematically exact theory for rule based reasoning.

In this article, we will present a discipline of merging rewriting and algebraic computations. Our goal is to obtain a simplification relation that offers both efficient computations of standard operations and the access to equational theorem proving. Therefore termination and confluence (= canonicity) are crucial properties of the simplification relation we want to construct.

Our initial idea is quite simple. We allow to use *mixed terms* consisting of usual first order terms where some subterms may be objects of an *external domain*. These external objects are assumed to be some kind of built-in data type (e. g. infinite precision integers or polynomials) for which the operations occurring in the term can be computed efficiently. Thus each ground term can be computed, resulting in a unique external object. We can view the computation relation as a schematisation of a (possibly infinite) ground term rewriting system. Our simplification relation shall combine a rewrite relation and a computation relation. However it turns out that even if the rewrite relation is canonical, the simplification relation needs neither be terminating nor confluent. We will describe extensions of term rewriting systems with which we can recover the canonicity of the simplification relation. Whether such an extended term rewriting system yields a canonical simplification relation depends among others on the built-in domain. An important rôle play so-called *decomposition free* extended term rewriting systems. These systems never force us to undo a previous computation. In addition we show that decomposition free systems are to some extent robust w. r. t. changes of the external domain. In particular such systems support external domains like fields that cannot be specified by a finite set of equations.

The paper is organised as follows. First we settle the preliminaries for abstract reduction relations and term rewriting systems. Then we introduce extended term rewriting systems to specify simplification relations. Section 4 deals with the termination property of simplification relations and in Section 5, we investigate for which external domains an extended term rewriting system describes a canonical simplification relation. At last some applications of canonical systems for combined rewriting and computations are presented.

2 Preliminaries

We assume the reader is familiar with the theory of term rewriting systems. For a survey on this topic see [DJ90].

Abstract Reduction Relations A *reduction relation* $\rightarrow_A \subseteq \mathcal{D} \times \mathcal{D}$ is an asymmetric binary relation. Then $\leftarrow_A$ is the inverse relation of $\rightarrow_A$. $\leftrightarrow_A$, $\rightarrow_A^*$ and $\leftrightarrow_A^*$ are the symmetric -, transitive and reflexive -, and the symmetric, transitive and reflexive closures of $\to_{\mathcal{A}}$ respectively. The relation $\to_{\mathcal{A}}$ is *terminating* if there is no infinite chain $a_1 \to_{\mathcal{A}} a_2 \to_{\mathcal{A}} \dots$. An object which is irreducible w. r. t. $\to_{\mathcal{A}}$ is called a *normal form*. The relation $\to_{\mathcal{A}}$ is *confluent* if for all $a, b, c \in \mathcal{D}$ such that $b \leftarrow_{\mathcal{A}}^* a \to_{\mathcal{A}}^* c$ there is a $d \in \mathcal{D}$ with $b \to_{\mathcal{A}}^* d \leftarrow_{\mathcal{A}}^* c$. We then write $b \downarrow_{\mathcal{A}} c$. Confluent and terminating reduction relations are called *canonical*. They compute a unique normal form for each object. To prove the confluence of a relation it often suffices to show a weaker condition. $\to_{\mathcal{A}}$ is *locally confluent* if for all $a, b, c \in \mathcal{D}$ with $b \leftarrow_{\mathcal{A}} a \to_{\mathcal{A}} c$, $b \downarrow_{\mathcal{A}} c$ follows. In [New42] Newman showed that for all terminating relations, confluence is equivalent to local confluence.

Terms Throughout this paper, we denote by $\mathcal{F} = \bigcup_i \mathcal{F}_i$ the finite set of ranked *function symbols* (the $\mathcal{F}_n$ are the n-ary function symbols) and by $\mathcal{X}$ the set of *variables*. Each function symbol is implicitly accompanied by a sort description $f : s_1 \times \dots \times s_n \to s$ for $f \in \mathcal{F}_n$ and the s, s_i are elements of a fixed set of sorts. In the same way each variable is assigned a fixed sort. Then $T(\mathcal{F}, \mathcal{X})$ denotes the set of all sort-correct terms freely generated by $\mathcal{F}$ and $\mathcal{X}$. Terms without variables are called *ground terms*. For a term t, $\mathcal{X}(t)$ is the set of variables occurring in t. Let p be a *position* (or *occurrence*) in a term t. Then $t|_p$ denotes the subterm of t at position p, $t(p)$ is the symbol labelling position p and for $s \in T(\mathcal{F}, \mathcal{X})$, $t[s]_p$ is the result of replacing in t the subterm at position p by s. A subterm $t|_p$ of t is called *proper* if $t|_p \neq t$.

A *substitution* $\sigma : \mathcal{X} \to T(\mathcal{F}, \mathcal{X})$ is a mapping from variables to terms. If we extend the application of a substitution σ to a term t, we write $t\sigma$ meaning that all variables x in t are simultaneously replaced by $\sigma(x)$. $t\sigma$ is an *instance* of t and we say that t is *more general than* s if s is an instance of t. If there is a substitution μ such that $s\mu = t\mu$ then s and t *unify* and μ is a *unifier* of s and t.

Rewrite Systems A *term rewriting system* $\mathcal{R}$ is a set of *rewrite rules* $l \to r$, where l and r are terms. l is called the *left-hand side* of the rule and r is its *right-hand side*. A rule is *left-linear* if in its left-hand side every variable occurs at most once. The term s reduces in one step to t, $s \to_{\mathcal{R}} t$, if $s = s[l\sigma]_p$, $l \to r \in \mathcal{R}$ and $t = s[r\sigma]_p$. $s|_p$ is then called a *redex*.

In computer algebra applications associative and commutative operations play an important rôle. Therefore we also consider rewriting modulo associativity and commutativity (AC). I. e. certain operators in $\mathcal{F}_{AC} \subseteq \mathcal{F}_2$ are known to be both associative and commutative. We assume the basic algorithms (equality-test, match, unification) do have built-in knowledge of the associativity and commutativity of the operators in $\mathcal{F}_{AC}$ [Hul79, Sti81].

Equational Specifications A set $\mathcal{E}$ of equations $s = t$ where s and t are terms induces a relation $\leftrightarrow_{\mathcal{E}} = \to_{\overline{\mathcal{E}}}$ where $\overline{\mathcal{E}} = \{s \to t, t \to s \mid s = t \in \mathcal{E}\}$. A term rewriting system $\mathcal{R}$ is *equivalent* to a set of equations $\mathcal{E}$ if $\leftrightarrow_{\mathcal{R}}^* = \leftrightarrow_{\mathcal{E}}^*$. The Knuth-Bendix completion procedure [KB70] transforms on success a set of equations into an equivalent canonical term rewriting system. Let $l \to r$ and $l' \to r'$ be two rules where l contains a subterm s which unifies with l' such that the most general unifier of s and l' is μ. Then $l\mu$ can be reduced by each of the two rules and the two terms resulting from the two different one-step reductions are called a *critical pair*. Knuth and Bendix [KB70] showed that a terminating term rewriting system is confluent iff all critical pairs are confluent. The Knuth-Bendix completion procedure can be extended to term rewriting systems with built-in AC-operators (e. g. [PS81, JK86]).

An *equational specification* is a pair $(\mathcal{F}, \mathcal{E})$ of a signature $\mathcal{F}$ and a set of equations $\mathcal{E}$ of terms in $T(\mathcal{F}, \mathcal{X})$. If we speak of the equational specification $\mathcal{E}$ it is understood that $\mathcal{F}$ is the set of functions occurring in $\mathcal{E}$.

3   Extended Term Rewriting Systems

In this section, we want to describe a simplification relation for mixed terms based on extended term rewriting systems with a built-in computation relation.

By D we will denote a set of *external objects* disjoint from the signature and variables. A *mixed term* is a structure obtained by replacing some leaf nodes of a term (variable or constant nodes) by external objects. In a many sorted term algebra D should be partitioned into $\biguplus_{s \in S} D_s$ where S is the set of sorts occurring in the term algebra. Then elements of D_s are considered as mixed terms of sort s. E. g. let $x \in \mathcal{X}$ be of sort *Nat*, $+ : Nat \times Nat \to Nat \in \mathcal{F}$ and $57 \in D_{Nat} = \mathbf{N}$ then $x + 57$ is a well-formed mixed term. The set of mixed terms generated by a signature $\mathcal{F}$, the variables in $\mathcal{X}$ and the external objects in D will be denoted by $T(\mathcal{F}, \mathcal{X}, D)$. External objects will be denoted by d, d_1, d_2, d_3 and so on.

Computations map a (mixed) ground term to an external object. This can be described by an $\mathcal{F}$-algebra $\mathcal{D} = (D, \mathcal{F}_D)$ where $\mathcal{F}_D = \{ f_D : D_{s_1} \times \cdots \times D_{s_n} \to D_s \mid f : s_1 \times \cdots \times s_n \to s \in \mathcal{F} \}$ is a set of interpretation functions for the operators in $\mathcal{F}$. $\mathcal{D}$ will be called the *external domain*. The *computation relation* $\to_{\mathcal{D}} \subseteq T(\mathcal{F}, \mathcal{X}, D) \times T(\mathcal{F}, \mathcal{X}, D)$ can now be defined as $s \to_{\mathcal{D}} t$ if $s|_p = f(d_1, \ldots, d_n)$ and $t = s[f_D(d_1, \ldots, d_n)]_p$. Since the $f_D \in \mathcal{F}_D$ are total functions, $\to_{\mathcal{D}}$ is a canonical relation. Let $\phi : T(\mathcal{F}, \mathcal{X}, D) \to T(\mathcal{F}, \mathcal{X}, D)$ be the function that computes the $\to_{\mathcal{D}}$-normal form for each mixed term. By ϕ^{-1} we will denote some inverse of ϕ. However ϕ^{-1} is not yet determined because so far we did not require that ϕ be injective or surjective.

A simplification relation on mixed terms shall allow for computations, rewrites by a term rewriting system (where variables may match mixed terms), and combined rewrites and computations. Extended term rewriting systems describe the latter two kinds of reductions.

Definition 1. An *extended term rewriting system (XTRS)* is a set of rules of the following form:

- Each rule of a pure term rewriting system is also a rule of an extended term rewriting system. The variables in an extended term rewriting system match any mixed term.

- The left-hand sides of an extended term rewriting system may contain *external variables* which match only external objects. As with ordinary variables, each external variable occurring in a right-hand side of a rule must also occur in its left-hand side. External variables will be denoted by barred variables (e.g. $\bar{x}, \bar{y}, \bar{z} \ldots$).

- The right-hand sides of rules in an extended term rewriting system may contain subterms *marked for immediate computation*. These subterms may not contain ordinary variables. We will denote these marks by an exclamation mark (!) in front of the subterm.

- The left-hand sides of rules in an extended term rewriting system may contain proper subterms *marked for decomposition*. These subterms may not contain ordinary variables. We will denote these marks by a question mark (?) in front of the subterm. A subterm $?t$ marked for decomposition matches an external object d if t matches $\phi^{-1}(d)$ for some fixed decomposition function ϕ^{-1}.

To distinguish the rewrite systems defined in Section 2 from extended term rewriting systems we will call the former ones *pure* term rewriting systems. In the sequel we will adopt the following conventions. If $\mathcal{R}^x$ is an extended term rewriting system then $\mathcal{R}$ is the set of pure rewrite rules in $\mathcal{R}^x$. By default $\mathcal{F}$ is the set of function symbols occurring in $\mathcal{R}$.

Example 1. The rule $(\bar{x} \cdot y) + y \rightarrow !(\bar{x} + 1) \cdot y$ can be interpreted as a scheme for a possibly infinite set of rules $(d_1 \cdot y) + y \rightarrow d_2 \cdot y$ where $d_2 = d_1 +_D 1_D$. □

Example 2. Immediate computation is sometimes required to ensure the termination property of an extended term rewriting system. The system

$$\mathcal{R}_1^x = \{(x + 1) \cdot y \rightarrow (x \cdot y) + y, \ (\bar{x} \cdot y) + y \rightarrow (\bar{x} + 1) \cdot y\}$$

is certainly not terminating whereas

$$\mathcal{R}_2^x = \{(x + 1) \cdot y \rightarrow (x \cdot y) + y, \ (\bar{x} \cdot y) + y \rightarrow !(\bar{x} + 1) \cdot y\}$$

is terminating. □

Example 3. Let $D = \mathbf{N}$ and for $n \in \mathbf{N}$, $\phi^{-1}(n) = s(\cdots s(0) \cdots)$ where s is stacked n times on a 0. Then the rule $fib(?s(s(\bar{x}))) \rightarrow fib(s(\bar{x})) + fib(\bar{x})$ describes the recursive evaluation of the Fibonacci function. □

Example 4. The computation relation $\rightarrow_D$ can be described by the set of rules

$$\mathcal{R}_D = \{f(\bar{x}_1, \ldots, \bar{x}_n) \rightarrow !f(\bar{x}_1, \ldots, \bar{x}_n) \mid f \in \mathcal{F}\}.$$

□

Even though the computation relation can be captured by an extended term rewriting system we do not want to denote it explicitly. The computation relation will be treated as built-in, and extended term rewriting systems do not generally contain the rules $\mathcal{R}_D$ of Example 4. However, Example 4 shows that computations fit smoothly into the framework of extended term rewriting systems.

Definition 2. Let $\mathcal{D} = (D, \mathcal{F}_D)$ be an external domain and $\mathcal{R}^x$ be an extended term rewriting system. The *simplification relation* of $\mathcal{R}^x$ and $\mathcal{D}$ is the relation $(\to_{\mathcal{R}^x} \cup \to_D)$. If $\mathcal{R}^x$ and $\mathcal{D}$ are understood we write $\to_S$ for this relation.

An important property of extended term rewriting systems is the lack of marks for decomposition.

Definition 3. An extended term rewriting system $\mathcal{R}^x$ is *decomposition free* if none of its left-hand sides contains a mark for decomposition.

Provided we can decide the equality of two external objects, decomposition free extended term rewriting systems allow us to apply reductions without looking into the internal structure of external objects. Thus the simplification relation is decidable. This is not necessarily the case for non-decomposition free systems. The applicability of decomposing rules depends strongly on the choice of ϕ^{-1}.

Decidability of the simplification relation is one aspect of decomposition free extended term rewriting systems. Another important aspect of decomposition freedom is *economy of computation*: When normalising a mixed term w.r.t. a decomposition free extended term rewriting system, we are never forced to undo a previous computation.

4 Termination of Combined Rewriting and Computation

Termination is a crucial aspect of reductional systems. Here we want to look at the combination of a pure term rewriting system $\mathcal{R}$ and a computation relation. The computation relation is by definition (uniquely) terminating and we will assume that $\mathcal{R}$ is terminating too. However termination is not a modular property of arbitrary combinations of rewrite systems. Here it helps to consider computation as the rewrite system presented in Example 4.

Modularity of termination could be shown for some restricted classes of rewrite systems, see [Ohl94] for a survey and [KR94] for recent results on hierarchical combinations of term rewriting systems. Unfortunately, combined rewriting and computation does not fit into a class for which the modularity of the termination property has been proved. Although $\mathcal{R}_D$ has a very simple structure (all rules are ground, no right-hand sides contain a symbol occurring in $\mathcal{R}$), the combination of $\mathcal{R}$ and $\mathcal{R}_D$ does not terminate in general.

Example 5. Let $\mathcal{R} = \{x + x \to 1 + 0\}$ which is clearly terminating. Further let $1 \to_\mathcal{D} d$ and $0 \to_\mathcal{D} d$ for some $d \in D$. Then

$$d + d \to_\mathcal{R} 1 + 0 \to_\mathcal{D}^* d + d \to_\mathcal{R} 1 + 0 \to_\mathcal{D}^* \cdots$$

is an infinite reduction sequence. □

Let us give some criteria for the termination of combined term rewriting and computation systems.

Lemma 4. *Let $\mathcal{R}$ be a terminating left-linear term rewriting system over a signature $\mathcal{F}$ and let $\mathcal{D} = (D, \mathcal{F}_D)$ be an external domain. Then the combined rewriting and computation relation $\to_\mathcal{R} \cup \to_\mathcal{D}$ is terminating.* □

Unfortunately, many interesting term rewriting systems are not left-linear. We can give another non-syntactical criterion for the termination of the combined system. This criterion relies on the ordering used for proving the termination of the rewrite system. See [Der87] for a survey on termination orderings.

Lemma 5. *Let $\mathcal{R}$ be a term rewriting system over a signature $\mathcal{F}$ and $\mathcal{D} = (D, \mathcal{F}_D)$ be an external domain. If the termination of $\mathcal{R}$ can be shown by a (recursive/lexicographic) path ordering, a polynomial interpretation ordering or a Knuth-Bendix ordering then the combined rewriting and computation relation $\to_\mathcal{R} \cup \to_\mathcal{D}$ is terminating.*

Proof: It is easy to extend these orderings in a way such that all elements in D are minimal. □

The termination of a combination of a decomposition free extended term rewriting system with a computation relation can be proved in a similar way.

Lemma 6. *Let $\mathcal{R}^x$ be an extended term rewriting system and $\mathcal{D} = (D, \mathcal{F}_D)$ be an external domain. Let the termination of $\mathcal{R}^x$ be provable by a (recursive or lexicographic) path ordering, a polynomial interpretation ordering or a Knuth-Bendix ordering such that external variables and terms marked for decomposition or immediate computation are treated as minimal objects in the ordering. Then the simplification relation $\to_{\mathcal{R}^x} \cup \to_\mathcal{D}$ is terminating.* □

In the sequel we will always assume that our simplification relation described by an (extended) term rewriting and a computation relation is terminating.

5 Confluent Simplifications

In this section, we want to investigate confluence aspects of extended term rewriting systems and their associated simplification relations. Since we always assume termination of the rewrite systems this implies that the reduction relation is canonical. Confluence considerations are important for two reasons. First they allow us to define a semantics for a specification (cf. [AB94]) and second canonical systems are decision procedures for the equality defined by an equational specification and a computation relation.

In general combining a pure canonical term rewriting system with a computation relation does not yield a canonical simplification relation.

Example 6. Consider the following canonical term rewriting system for monoids

$$\mathcal{R} = \{1 \cdot x \to x, \ x \cdot 1 \to x, \ (x \cdot y) \cdot z \to x \cdot (y \cdot z)\}$$

and 2×2 matrices over $\mathbf{Q}$ as external domain. Then the rewrite ambiguity

$$\begin{pmatrix} 1 & 2 \\ 2 & 1 \end{pmatrix} \cdot \left(\begin{pmatrix} 2 & 0 \\ 0 & 1 \end{pmatrix} \cdot x \right) \leftarrow_{\mathcal{R}} \left(\begin{pmatrix} 1 & 2 \\ 2 & 1 \end{pmatrix} \cdot \begin{pmatrix} 2 & 0 \\ 0 & 1 \end{pmatrix} \right) \cdot x \to_{\mathcal{D}} \begin{pmatrix} 2 & 2 \\ 4 & 1 \end{pmatrix} \cdot x$$

does not have a common $\to_S$-normal form. □

We will investigate in how far extended term rewriting systems can solve this problem. The rôle of the extended term rewriting systems iwill be to link the simplification relation with an equational specification.

Definition 7. Let $\mathcal{D} = (D, \mathcal{F}_D)$ be an external domain and $\mathcal{E}$ be an equational specification. An extended term rewriting system $\mathcal{R}^x$ is $\mathcal{E}$-*equivalent for* $\mathcal{D}$ if

$$(\leftrightarrow_{\mathcal{R}^x} \cup \leftrightarrow_{\mathcal{D}})^* = (\leftrightarrow_{\mathcal{E}} \cup \leftrightarrow_{\mathcal{D}})^*.$$

If $\mathcal{R}^x$ is $\mathcal{E}$-equivalent for the initial model of $\mathcal{E}$ we just say $\mathcal{R}^x$ is $\mathcal{E}$-*equivalent*.

By abuse of notation we say an algebra $\mathcal{D}$ is a *model of an extended term rewriting system* $\mathcal{R}^x$ if $\mathcal{R}^x$ is $\mathcal{E}$-equivalent and $\mathcal{D}$ is a model of $\mathcal{E}$.

Next we define confluence for the simplification relations in terms of their defining extended rewrite systems.

Definition 8. Let $\to_S$ be the simplification relation specified by an extended term rewriting system $\mathcal{R}^x$ and an external domain $\mathcal{D}$. Then $\mathcal{R}^x$ is *(ground) confluent for* $\mathcal{D}$ if for all (ground) terms $s, t_1, t_2 \in T(\mathcal{F}, \mathcal{X}, D)$ with $t_1 \leftarrow_S^* s \to_S^* t_2$ there is a term $\hat{s} \in T(\mathcal{F}, \mathcal{X}, D)$ with $t_1 \to_S^* \hat{s} \leftarrow_S^* t_2$.

We say $\mathcal{R}^x$ is *(ground) confluent* if $\mathcal{R}^x$ is (ground) confluent for the initial model of $\mathcal{R}^x$. If $\mathcal{R}^x$ is terminating, (ground) confluence for $\mathcal{D}$ implies that $\mathcal{R}^x$ is *(ground) canonical for* $\mathcal{D}$.

Lemma 9. *Let $\mathcal{R}$ be a ground canonical term rewriting system and $\mathcal{D}$ be an external domain that is a model of $\mathcal{R}$ such that $\to_{\mathcal{R}} \cup \to_{\mathcal{D}}$ is terminating. Then the simplification relation $\to_S$ is ground canonical.* □

Given an equational specification $\mathcal{E}$ and an external domain $\mathcal{D}$, we want to find an extended term rewriting system $\mathcal{R}$ which is both $\mathcal{E}$-equivalent for $\mathcal{D}$ and canonical for $\mathcal{D}$. The existence of such extended term rewriting systems was shown in [Bün94b] in case that $\mathcal{E}$ allows for a canonical term rewriting system and $\mathcal{D}$ is the initial model of $\mathcal{E}$.

Theorem 10. *Let $\mathcal{R}$ be a canonical term rewriting system that is equivalent to a set of equations $\mathcal{E}$. Then there exists an $\mathcal{E}$-equivalent, canonical extended term rewriting system.* □

The proof of the theorem relies on a simulation argument where the built-in domain is presented by a (renamed) copy of the original system. Using this technique different canonical systems can be searched with the help of the Knuth-Bendix completion procedure.

Example 7. The following two extended term rewriting systems are canonical (for the integers modulo 3). Let $+$ and $\cdot$ be both associative and commutative. We also identify the term constants 0 and 1 with the associated built-ins.

$$
\begin{aligned}
\mathcal{R}_1^x = \{\, & 0 + x && \to x, & 0 \cdot x && \to 0, \\
& 1 \cdot x && \to x, & x + x + x && \to 0, \\
& x \cdot (y + z) \to (x \cdot y) + (x \cdot z), & x \cdot ?(\bar{y} + \bar{z}) \to (x \cdot \bar{y}) + (x \cdot \bar{z}) \,\}
\end{aligned}
$$

Given ϕ^{-1} with $\{\phi^{-1}(0) \mapsto 0, \phi^{-1}(1) \mapsto 1, \phi^{-1}(2) \mapsto 1+1\}$, $\mathcal{R}_1^x$ is canonical but not decomposition free. There exists a decomposition free alternative:

$$
\begin{aligned}
\mathcal{R}_2^x = \{\, & 0 + x && \to x, & 0 \cdot x && \to 0, \\
& 1 \cdot x && \to x, & x + x + x && \to 0, \\
& x \cdot (y + z) && \to (x \cdot y) + (x \cdot z), & (x \cdot \bar{y}) + x && \to x \cdot !(\bar{y} + 1), \\
& (x \cdot \bar{y}) + (x \cdot \bar{z}) && \to x \cdot !(\bar{y} + \bar{z}), & (x + x) && \to x \cdot 2 \,\}.
\end{aligned}
$$

Note that because of the last rule in $\mathcal{R}_2^x$ we can eliminate the fourth rule. □

Before we use an arbitrary extended term rewriting system $\mathcal{R}^x$, we must fix the decomposition function ϕ^{-1} for the domains we are interested in. Given that $\mathcal{R}$ is canonical, this decomposition function should specialise to a mapping from each element of the initial model of $\mathcal{R}$ to the corresponding ground term in $\mathcal{R}$-normal form.

Definition 11. Let $\mathcal{D}$ be an initial model of $(\mathcal{F}, \mathcal{E})$. Then $\mathcal{D}^*$ is a *free $\mathcal{E}$-enrichment of $\mathcal{D}$* if $\mathcal{D}^*$ is an initial model for $(\mathcal{F} \cup \mathcal{F}_0^*, \mathcal{E})$ where $\mathcal{F}_0^*$ is a set of new constants. The objects in $\mathcal{D}^*$ that correspond to $T(\mathcal{F}_0^*, \emptyset)/\mathcal{E}$ are the *generators* of $\mathcal{D}^*$ w.r.t. $(\mathcal{F}, \mathcal{E})$.

Definition 12. A function $\phi^{-1} : \mathcal{D} \to T(\mathcal{F}, \emptyset, D)$ is called a *standard decomposition for $\mathcal{R}^x$* if $\phi(\phi^{-1}(d)) = d$ and $\phi^{-1}(d) = t$ is a $\mathcal{R}$-irreducible mixed term in $T(\mathcal{F}, \emptyset, D')$ where D' is a set of generators of $\mathcal{D}$ w.r.t. $(\mathcal{F}, \mathcal{R})$.

We can now slightly extend the set of domains for which a canonical extended term rewriting system specifies a canonical simplification relation.

Lemma 13. *Let $\mathcal{R}^x$ be a canonical extended term rewriting system that is $\mathcal{E}$-equivalent. Then $\mathcal{R}^x$ is canonical for every free enrichment of D if the standard decomposition for $\mathcal{R}^x$ is used.* □

The proof of the lemma is a direct consequence of the simulation argument used to prove Theorem 10.

Example 8. By Lemma 13, $\mathcal{R}_1^x$ from Example 7 is also canonical for (multivariate) polynomials over $\mathbf{Z}/(3\mathbf{Z})$. □

The next theorem proves a further advantage of decomposition free extended rewrite systems. Decomposition free systems are very versatile with regard to the external domains they are applicable to.

Definition 14. Let $\mathcal{D} = (D, \mathcal{F}_D)$ be an algebra. Then an algebra isomorphic to $\mathcal{D}^* = (D^*, \mathcal{F}_{D^*})$ is a *consistent enrichment* of $\mathcal{D}$ if $D^* \supseteq D$ and $\mathcal{F}_{D^*}$ restricted to D is equal to $\mathcal{F}_D$.

Theorem 15. *Let $\mathcal{R}^x$ be a canonical decomposition free extended term rewriting system whose termination can be proved by Lemma 6. Then $\mathcal{R}^x$ is canonical for any consistent enrichment of the initial model of $\mathcal{R}^x$.*

Proof: To prove the theorem, we apply the following transformation. Assume that all external objects in $\mathcal{D}_s$ belong to a new sort $\bar{s}$. In the same way $\to_{\mathcal{D}}$ is represented by rewrite rules $\bar{f}(d_1, \ldots, d_n) \to d$ where $\bar{f} : \bar{s}_1 \times \cdots \times \bar{s}_n \to \bar{s}$ for $f : s_1 \times \cdots \times s_n \to s \in \mathcal{F}$. To obtain well-sorted terms in this setting, external objects must be introduced into mixed terms via unary coercion operators $c_s : \bar{s} \to s$ for every sort s. Using this transformation, we see that for critical pair computation the rules describing $\to_{\mathcal{D}}$ need only be superposed with external objects occurring in the left-hand sides of $\mathcal{R}^x$. A consistent enrichment can be described by adding rules of the form $\bar{f}(d_1, \ldots, d_n) \to d$ only. The consistency requirement inhibits the addition of rules of the form $d_1 \to d_2$. Therefore no new critical pairs can be created if $\mathcal{D}$ is extended consistently. □

Remark: The proof of Theorem 15 exhibits that we can further relax the restrictions on the external domains. Namely $\mathcal{R}^x$ is canonical for every model $\mathcal{D}$ of $\mathcal{R}^x$ if all external objects occurring in the left-hand sides of $\mathcal{R}^x$ are minimal in the equivalence classes introduced by $\mathcal{D}$.

So far canonical decomposition free extended term rewriting systems have been found for (Abelian) monoids, (Abelian) groups, commutative rings, modular commutative rings, Boolean rings and lists.

Example 9. By Theorem 15, $\mathcal{R}_2^x$ from Example 7 is canonical for

- Galois fields $GF(3^q)$,

- vectors over $\mathbf{Z}/(3\mathbf{Z})$ where the operations are performed component wise and

- all quotients $(\mathbf{Z}/(3\mathbf{Z}))[X_1,\ldots,X_n]/I$ for ideals I where $[0]_I$, $[1]_I$ and $[2]_I$ are distinct. □

Example 10. A decomposition free canonical extended term rewriting system for commutative rings is

$$\mathcal{R}^x = \{\,0 + x \quad\quad \to x, \quad\quad\quad 0 \cdot x \quad\quad \to 0,$$
$$1 \cdot x \quad\quad\quad \to x, \quad\quad\quad -(x) \quad\quad \to x \cdot! - 1,$$
$$x \cdot (y + z) \quad \to (x \cdot y) + (x \cdot z), (x \cdot \bar{y}) + x \to x \cdot!(\bar{y} + 1),$$
$$(x \cdot \bar{y}) + (x \cdot \bar{z}) \to x \cdot!(\bar{y} + \bar{z}), \quad\quad x + x \quad\quad \to x \cdot 2 \,\}$$

where $+$ and $\cdot$ are associative and commutative. By Theorem 15, $\mathcal{R}^x$ is canonical for the Gaussian integers, integral polynomials, but also for infinite fields like the rationals, algebraic numbers, reals, complex numbers and polynomials and vectors over these fields.

By the remark following the proof of Theorem 15, $\mathcal{R}^x$ is actually canonical for all non-trivial commutative rings. □

6 Applications

In this section, we want to briefly high-light the benefits of canonical simplification relations as combined term rewriting and built-in computations.

6.1 Canonical Simplification

First of all, a canonical system allows us to effectively decide the equality of two terms. Note that we allow for terms with variables which range over all objects of the external domain. E.g. let $\mathbf{Z}[X]$ be an external domain and $p \in \mathcal{X}$ then $p + 2X^3 - p \cdot X$ represents elements of $\mathbf{Z}[X]$ where p stands for any integral polynomial. For commutative algebra, computer algebra systems often simulate this situation using polynomials with additional indeterminates (in our example $\mathbf{Z}[X,p]$). But note that the primary operation for terms (and variables) is substitution application (and its inverse) and this is a rather complex operation when simulated on data structures for polynomials.

6.2 Functional Programming

Given an external domain that is the initial model of a base rewrite system, we can write functional programs with built-in data types (see also [KC89]).

Example 11. Given a canonical extended term rewriting system $\mathcal{P}^x$ for a Peano-axiomatisation of the natural numbers, the following ground confluent system describes a functional program for the Fibonacci function.

$$\{\, fib(0) \quad \to 1,$$
$$fib(1) \quad \to 1,$$
$$fib(s(s(x))) \to fib(s(x)) + fib(x),$$
$$fib(s(?s(\bar{x}))) \to fib(s(\bar{x})) + fib(\bar{x}),$$
$$fib(?s(s(\bar{x}))) \to fib(s(\bar{x})) + fib(\bar{x}) \,\}$$

□

6.3 Deduction

Next to reductions, critical pair computation is the main deduction mechanism for term rewriting systems. This can be carried over to extended term rewriting systems. E. g. we can deduce

$$y + 100 = X^2$$

from

$$z + (y + 100) = z + X^2$$

for every group by computing the critical pair between $z + (y + 100) \to z + X^2$ and the rule $-x + (x + y) \to y$ from the canonical system describing groups (see Example 12). Note that unification of external objects does not pose any problems since in our setting they can be treated as constants.

6.4 Generic Equation Solving

If there is a canonical term rewriting system that is equivalent to a set of equations $\mathcal{E}$, narrowing [Hul80] provides a semi-decision procedure for unification problems modulo $\mathcal{E}$. Thus we can use narrowing based on a canonical simplification relation as an approach to a generic procedure to solve equations in the built-in domain. Again decomposition freedom turns out to be important because with a decomposition free canonical extended term rewriting system we will never be forced to perform unifications modulo $\mathcal{D}$ during the narrowing process.

Example 12. Given a canonical extended term rewriting system for groups

$$\mathcal{R}^x = \{\, x \cdot 1 \quad \to x, \qquad\qquad 1 \cdot x \quad \to x,$$
$$x \cdot x^{-1} \quad \to 1, \qquad\qquad x^{-1} \cdot x \quad \to 1,$$
$$(x \cdot y) \cdot z \quad \to x \cdot (y \cdot z), \quad 1^{-1} \quad \to 1,$$
$$(x^{-1})^{-1} \quad \to x, \qquad\qquad x^{-1} \cdot (x \cdot y) \to y,$$
$$x \cdot (x^{-1} \cdot y) \to y, \qquad\qquad (x \cdot y)^{-1} \quad \to y^{-1} \cdot x^{-1},$$
$$\bar{x} \cdot (\bar{y} \cdot z) \quad \to !(\bar{x} \cdot \bar{y}) \cdot z \,\}.$$

Let the external domain $\mathcal{D} = \langle t, r; t^2, r^3, tr^2 = rt \rangle$ be the group S_3. The elements of $\mathcal{D}$ are represented by strings over the alphabet $\{t, t^{-1}, r, r^{-1}\}$. In order to find a solution for the equation

$$x \cdot \text{``$tr$''} = \text{``t''} \cdot (y \cdot \text{``t''})$$

we can narrow both $x \cdot \text{``$tr$''}$ and $\text{``$t$''} \cdot (y \cdot \text{``$t$''})$ to 1 using rules of $\mathcal{R}^x$ only. This results in the solution $x = (\text{``tr''})^{-1}$ and $y = \text{``$t$''}^{-1} \cdot \text{``$t$''}^{-1}$. The solution can then be further simplified (e.g. to $\text{``$r^{-1}t$''}$ and 1). □

7 Conclusion

We have presented an extension of term rewriting systems to describe canonical simplification relations defined by rewrite relations and built-in computations. An important property of such systems is decomposition freedom. The impact of this property is manifold. First it allows for efficient simplification. Second a decomposition free canonical extended term rewriting system is canonical for every consistent model of the system. Thus we are allowed to use built-in domains like the rationals or complex numbers that cannot be specified as equational varieties. Third decomposition free systems permit deductions by critical pair computations and narrowing without the necessity to unify modulo the built-in domain. Decomposition free canonical extended term rewriting systems exist for many important algebraic domains.

8 Acknowledgements

I would like to thank Prof. R. Loos for supporting this research and Andreas Weber and Wolfgang Küchlin for some helpful discussions.

References

[AB94]   J. Avenhaus and K. Becker. Operational specifications with built-ins. In P. Enlbert, E. W. Mayr, and K. W. Wagner, editors, *STACS 94 (LNCS 775)*, pages 263–274. Springer-Verlag, 1994. (Proc. STACS'94, Caen, France, February 1994).

[BKR87] B. Benninghofen, S. Kemmerich, and M. M. Richter. *Systems of Reductions*. Springer-Verlag, Berlin, 1987.

[BL82]   Bruno Buchberger and Rüdiger Loos. Algebraic simplification. In *Computer Algebra*, pages 14–43. Springer-Verlag, 1982.

[Buc65] Bruno Buchberger. *Ein Algorithmus zum Auffinden der Basiselemente des Restklassenringes nach einem nulldimensionalen Polynomideal*. PhD thesis, Universität Innsbruck, 1965.

[Buc79]   Bruno Buchberger. A criterion for detecting unnecessary reductions in the construction of Gröbner-Bases. In E. Ng, editor, *Symbolic and Algebraic Computing (LNCS 72)*, pages 3–21. Springer-Verlag, 1979. (Proc. EUROSAM'79, Marseille, France).

[Bün91a] Reinhard Bündgen. Completion of integral polynomials by AC-term completion. In Stephen M. Watt, editor, *International Symposium on Symbolic and Algebraic Computation*, pages 70 – 78, 1991. (Proc. ISSAC'91, Bonn, Germany, July 1991).

[Bün91b] Reinhard Bündgen. Simulating Buchberger's algorithm by Knuth-Bendix completion. In Ronald V. Book, editor, *Rewriting Techniques and Applications (LNCS 488)*, pages 386–397. Springer-Verlag, 1991. (Proc. RTA'91, Como, Italy, April 1991).

[Bün92]   Reinhard Bündgen. Buchberger's algorithm: The term rewriter's point of view. In G. Kuich, editor, *Automata, Languages and Programming (LNCS 623)*, pages 380 –391, 1992. (Proc. ICALP'92, Vienna, Austria, July 1992).

[Bün94a] Reinhard Bündgen. On pots, pans and pudding or how to discover generalized critical pairs. In *12th International Conference on Automated Deduction, (LNCS)*. Springer-Verlag, 1994. (Proc. CADE'94, Nancy, France, July 1994).

[Bün94b] Reinhard Bündgen. Preserving confluence for rewrite systems with built-in operations. In *Workshop on Conditional and (Typed) Term Rewriting Systems*, 1994. (also to appear in LNCS).

[Der87]   Nachum Dershowitz. Termination of rewriting. *Journal of Symbolic Computation*, 3:69–115, 1987.

[DJ90]   Nachum Dershowitz and Jean-Pierre Jouannaud. Rewrite systems. In Jan van Leeuven, editor, *Formal Models and Semantics*, volume B of *Handbook of Theoretical Computer Science*, chapter 6. Elsevier, 1990.

[Hul79]   Jean-Marie Hullot. Associative-commutative pattern matching. In *Fifth IJCAI*, Tokyo, Japan, 1979.

[Hul80]   Jean-Marie Hullot. Canonical forms and unification. In *Proc. Fifth International Conference on Automated Deduction (LNCS 87)*, pages 318–334. Springer-Verlag, 1980.

[JK86]   Jean-Pierre Jouannaud and Hélène Kirchner. Completion of a set of rules modulo a set of equations. *SIAM J. on Computing*, 14(4):1155–1194, 1986.

[KB70]   Donald E. Knuth and Peter B. Bendix. Simple word problems in universal algebra. In J. Leech, editor, *Computational Problems in Abstract Algebra*. Pergamon Press, 1970. (Proc. of a conference held in Oxford, England, 1967).

[KC89]   Stéphane Kaplan and Christine Choppy. Abstract rewriting with concrete operators. In Nachum Dershowitz, editor, *Rewriting Techniques and Applications (LNCS 355)*, pages 178–186. Springer-Verlag, 1989. (Proc. RTA'89, Chapel Hill, NC, USA, April 1989).

[KR94]   M. R. K. Krishna Rao. Simple termination of hierarchical combinations of term rewriting systems. In P. Enlbert, E. W. Mayr, and K. W. Wagner, editors, *STACS 94 (LNCS 775)*, pages 203–223. Springer-Verlag, 1994. (Proc. STACS'94, Caen, France, February 1994).

[Küc85]   Wolfgang Küchlin. A confluence criterion based on the generalised Knuth-Bendix algorithm. In B. F. Caviness, editor, *Eurocal'85 (LNCS 204)*, pages 390–399. Springer-Verlag, 1985. (Proc. Eurocal'85, Linz, Austria, April 1985).

[New42]   M. H. A. Newman. On theories with a combinatorial definition of "equivalence". *Annals of Mathematics*, 43(2):223–243, 1942.

[Ohl94]   Enno Ohlebusch. *Modular Properties of Composable Term Rewriting Systems*. PhD thesis, Universität Bielefeld, D-33501 Bielefeld, Germany, Mai 1994.

[PS81]   G. Peterson and M. Stickel. Complete sets of reductions for some equational theories. *Journal of the ACM*, 28:223–264, 1981.

[Sti81]   Mark E. Stickel. A unification algorithm for associative-commutative functions. *JACM*, 28(3):423–434, July 1981.

[WB85]   Franz Winkler and Bruno Buchberger. A criterion for eliminating unnecessary reductions in the Knuth-Bendix algorithm. In *Proc. Colloquium on Algebra, Combinatorics and Logic in Computer Science*. J. Bolyai Math. Soc., J. Bolyai Math. Soc. and North-Holland, 1985. (Colloquium Mathematicum Societatis J. Bolyai, Györ, Hungary, 1983).

[Wol91]   Stephen Wolfram. *Mathematica: a system for doing mathematics by computer*. Addison-Wesley, Redwood City, CA, 1991.

Algebraic Specification of Empirical Inductive Learning Methods based on Rough Sets and Matroid Theory

Shusaku Tsumoto and Hiroshi Tanaka

Department of Information Medicine, Medical Research Institute,
Tokyo Medical and Dental University
1-5-45 Yushima, Bunkyo-ku Tokyo 113 Japan
E-mail:{tsumoto, tanaka}@tmd.ac.jp

Abstract. In order to acquire knowledge from databases, there have been proposed several methods of inductive learning, such as ID3 family and AQ family. These methods are applied to discover meaningful knowledge from large databases, and their usefulness is ensured. However, since there has been no formal approach proposed to treat these methods, efficiency of each method is only compared empirically. In this paper, we introduce matroid theory and rough sets to construct a common framework for empirical machine learning methods which induce the combination of attribute-value pairs from databases. Combination of the concepts of rough sets and matroid theory gives us an excellent framework and enables us to understand the differences and the similarities between these methods clearly. In this paper, we compare three classical methods, AQ, Pawlak's Consistent Rules and ID3. The results show that there exists the differences in algebraic structure between the former two and the latter and that this causes the differences between AQ and ID3.

1 Introduction

1.1 Motivation

In order to acquire knowledge from databases, there have been proposed several methods of inductive learning, such as ID3 family[2, 13] and AQ family[1, 4, 5]. These methods are applied to discover meaningful knowledge from large database, and their usefulness is ensured. However, since there has been no formal approach proposed to treat these methods, efficiency of each method is compared by using real-world databases[1, 2, 5, 7, 19, 13], such as medical databases. These results suggest some differences between these methods. However, sometimes these differences may depend on applied domains, so general discussion is left unsolved.

In this paper, we introduce matroid theory[20, 21, 22] and rough sets[10] to construct a common framework for empirical machine learning methods which induce knowledge from attribute-value pattern databases. Combination of the concepts of rough sets and matroid theory gives us an excellent framework and enables us to understand the differences of these methods clearly. Using this framework, we compare three classical methods: AQ, Pawlak's Consistent Rules[10] and ID3 and we obtain six interesting conclusions from our approach. First, while AQ and Pawlak's method are equivalent to the greedy algorithm for finding bases of Matroid from space spanned by attribute-value pairs, ID3 method calculates ordered greedoids, which are defined by weaker axioms than matroids. Second, a matroid defined by AQ method(AQ matroid) is a dual matroid of one defined by Pawlak's method[10](Pawlak's matroid). Third, according to the computational complexity of the greedy algorithm, the efficiency of both methods depends on the number of attributes, especially, independent variables. Fourth, the induced results are optimal to the training samples if and only if the conditions on independence are hold. So if adding some new examples make independent attributes change their nature into dependent ones, the condition of deriving optimal solution is violated. Fifth, in addition to the fourth conclusion, since a greedoid of ID3 has weaker structure than the other two methods, ID3 method is the most sensitive to training samples, although its computational complexity is the lowest. Finally, it is shown that pruning and truncation methods do not always correct overfitting phenomenon. Furthermore, for pruning and truncation methods, we assume that future accuracy is related to the complexities of induced rules. If this assumption is violated, we cannot obtain optimal rules or trees by applying pruning and truncation.

This work is a preliminary work towards development of a multistrategy machine learning system [6]. Recently, it has been pointed out that each machine learning method is a little powerless to extract sufficient knowledge from data inputs and that it is more effective to integrate some machine learning methods with other machine learning methods or other subsymbolic methods, such as neural networks, statistics. And we stress that technique of symbolic mathematical computing is indispensable to realize such multistrategic integration, especially when domain-knowledge can be represented as algebraic equations.

The paper is organized as follows: section in Section 2 and 3, we briefly discuss original rough set model and AQ method respectively. In Section 4, the elementary concepts of matroid theory are introduced, and several characteristics are discussed. Section 5 presents comparison between AQ matroid and Pawlak's matroid. Section 6 gives ID3 greedoid, which is algebraic structure of decision tree induction. In section 7, we discuss about optimal solutions given by the greedy algorithm. Section 8 presents discussion on pruning and truncation in terms of rough sets and matroid theory, and in Section 9, we discuss why we need technique of computer algebra to proceed our work. Finally, in Section 10, we conclude the results of this paper.

1.2 Notation and Some Assumptions

In this paper, we focus on algebraic specification of domain-independent aspects of classical empirical learning methods, AQ, Pawlak's method and ID3. So we do not consider about constructive generalization[4], since this methods explicitly needs domain-specific knowledge. And, moreover, we omit the proofs of the theorems, since all of the proofs in Section 4 are originated from matroid theory, which readers could refer to [20, 21], and since most of the theorem in Section 5 and Section6 are trivial. For further information on rough sets and matroid theory, readers could refer to [10, 20, 21].

Below in this subsection, we mention about the following four notations used in this paper. First, for simplicity, we deal with classification of two classes, one of which are supported by a set of positive examples, denoted by D and the other of which are by a set of negative examples, $U - D$. And the former class is assumed to be composed of some small clusters, denoted by D_j, that is , $D = \cup_j D_j$. Second, we regard an attribute-value pair as an **elementary equivalence relation** as defined in rough sets[10]. We denote the combination of an attribute-value pairs, which is called *the complex of selectors* in terms of AQ theory, by an equivalence relation, R. A set of elements which supports R, which is called a *partial star* in AQ, is referred to as an indiscernible set, denoted by $[x]_R$. For example, let $\{1, 2, 3\}$ be a set of samples which supports an equivalence relation R. Then, we say that a partial star of R is equal to $\{1, 2, 3\}$ in terms of AQ. This notion can be represented as $[x]_R = \{1, 2, 3\}$ in terms of rough sets. Third, when we describe a conjunctive formula, we use the ordinary logical notation. Furthermore, when an equivalence relation is described as attribute-value pair, we denote this pair by $[attribute = value]$. For example, if an equivalence relation R means "a=1 and b=0", then we write it as $R = [a = 1] \bigwedge [b = 0]$. Finally, we define partial order of equivalence as follows:

Definition 1. Let $A(R_i)$ denote the set whose elements are the attribute-value pairs included in R_i. If $A(R_i) \subseteq A(R_j)$, then we represent this relation as:

$$R_i \preceq R_j.$$

For example, let R_i represent a conjunctive formula, such as $a \wedge b \wedge c$, where a, b, c are elementary equivalence relations. Then $A(R_i)$ is equal to $\{a, b, c\}$. If we use the notation of Michalski's APC(Annotated Predicate Calculus)[4], R_i can be represented as, say $[a = 1]\&[b = 1]\&[c = 1]$, then $A(R_i)$ is equal to a set of selectors, $\{[a = 1], [b = 1], [c = 1]\}$.

2 Rough Set Theory

2.1 Elementary Concepts of Rough Set Theory

Rough set theory is one of the most important approach in inductive learning, developed and rigorously formulated by Pawlak[10]. This theory can be used to acquire certain sets of attributes for classification and can also evaluate how precisely the attributes of database are able to classify data. In this paper, we only mention what we need in relation to our reasoning strategy, since including whole discussion on rough sets is too lengthy and redundant. For further information, readers could refer to [10].

Table 1 is a small example of database which collects the patients who complained of headache. First,let us consider how an attribute "location" classify

Table 1. a Small Example of Database

No.	location	nature	history	prodrome	jolt	nausea	M1	M2	class
1	occular	persistent	persistent	0	0	0	1	0	m.c.h.
2	whole	persistent	persistent	0	0	0	1	1	m.c.h.
3	lateral	throbbing	paroxysmal	0	1	1	0	0	migraine
4	lateral	throbbing	paroxysmal	1	1	1	0	0	migraine
5	occular	persistent	persistent	0	0	0	1	0	psycho.
6	occular	persistent	subacute	0	1	1	0	0	i.m.l.
7	occular	persistent	acute	0	1	1	0	0	psycho.
8	whole	persistent	chronic	0	1	0	0	0	i.m.l.
9	lateral	throbbing	persistent	0	1	1	0	0	common
10	whole	persistent	persistent	0	0	0	1	1	m.c.h.

The above abbreviations stand for the following meanings: jolt: Jolt headache, M1, M2: tenderness of M1 and M2, 1: Yes, 0: No, m.c.h.: muscle contraction headache, psycho: psychogenic pain, and i.m.l.: intracranial mass lesion.

the headache patients' set of the table. The set whose value of the attribute "location" is equal to "whole" is $\{2,8,10\}$(In the following,the numbers represent each record number). This set means that we cannot classify $\{2,8,10\}$ further solely by using the constraint $R = [location = whole]$. This set is defined as indiscernible set over relation R and described as follows: $[x]_R=U/R= \{2,8,10\}$ (U denotes the total set of database). In this set, $\{2,10\}$ suffer from muscle contraction headache("m.c.h."), $\{8\}$ suffers from intracranial mass lesion("i.m.l."). Hence we need other additional attributes to classify this set of patients as to their disease. Using this concept, we can evaluate the classification power of each attribute. For example, $[prodrome = 1]$ is specific to the case of classic migraine ("classic"). We can also extend this indiscernible relation to multivariate cases, such as $[x]_{[location=whole]\wedge[M2=1]} = \{2, 10\}$. Moreover, we can take not only an attribute-value set as a relation, but also an attribute itself as a relation. For example, $[x]_{location} = \{[x]_{[location=whole]}, [x]_{[location=lateral]}, [x]_{[location=occular]}\} = \{\{2, 8, 10\}, \{3, 4, 9\}, \{1, 5, 6, 7\}\}$. Using these basic concepts, several topological sets and measures for these sets can be defined as follows:

Definition 2. Let R be an equivalence relation and X be the subset of U.

positive region $\quad Posi_R(X) = \bigcup\{Y \in U/R : Y \subseteq X\}$

possible region $\quad Poss_R(X) = \bigcup\{Y \in U/R : Y \bigcap X \neq \phi\}$

boundary region $\quad Bound_R(X) = Poss_R(X) - Posi_R(X)$

accuracy measure $\quad \alpha_R(X) = \dfrac{card \ Posi_R(X)}{card \ Poss_R(X)}.$

(In [10], Pawlak does not use the above term "possible region". He refers to our possible region as *upper* approximation of X, and he rarely use it. In this paper, we focus on this region in order to deal with probabilistic domain. Compared with the word "positive", we use "possible" which reflects the intuitional meaning of an *upper* approximation.)

For example, the set whose class is "m.c.h." is composed of $\{1,2,5,10\}$. Let us take this set as X. We take the following relations R_i such that $U/R_i \bigcap X \neq \phi$:

$R_1 = [location = occular] \wedge [nature = persistent] \wedge [history = persistent]$
$\quad \wedge [prodrome = 0] \wedge [jolt = 0] \wedge [nausea = 0] \wedge [M1 = 1] \wedge [M2 = 1]$, and

$R_2 = [location = whole] \wedge [nature = persistent] \wedge [history = persistent]$
$\quad \wedge [prodrome = 0] \wedge [jolt = 0] \wedge [nausea = 0] \wedge [M1 = 1] \wedge [M2 = 1].$

Let R denote $R_1 \cup R_2$. That is, $U/R = \{[x]_{R_1}, [x]_{R_2}\} = \{\{1,5\}, \{2,10\}\}$. The positive region of "m.c.h." over the relation R is $U/R - \{1,5\} = \{2,10\}$. And then its possible region is $\{1,2,5,10\}$, which includes one case for "psycho": $\{5\}$. Furthermore, we can derive $\{1,5\}$ as the boundary region, and the accuracy measure is $2/4$.

2.2 Pawlak's Consistent Rules

Based on the concepts of rough sets, Pawlak[10] introduces *Reduction of Knowledge*, which is a method to examine the independencies of the attributes iteratively and extract the minimum indispensable part of equivalence relations. Here we only mention about the definition of *consistent rules* and their knowledge reduction. For further details, readers could refer to [10].

Definition 3. Let R_j be an equivalence relation and D be a set of samples which belongs to a target concept. $R_j \Rightarrow D$ is called a *consistent rule* when $Posi_{R_j}(D)$ is given by:

$$Posi_{R_j}(D) = D_k = [x]_{R_j} \subseteq D,$$

where $Posi_{R_j}(D)$ denotes the positive region of D in terms of R_j. □

Definition 4. Let R_0 be equal to $R \wedge [a = v]$, where $[a = v]$ denotes a certain attribute-value pair. If an attribute-value pair $[a = v]$ is satisfied with the following equation:

$$Posi_{R_0}(D) = Posi_{R \wedge [a=v]}(D) = Posi_R(D),$$

then we say that $[a = v]$ is *dispensable* in R_0, and can be deleted from R. □

Intuitively, reduction procedure removes redundant variables which do not contribute to classification of a class.

For the above example, since $Posi_{R_2}(X) = \{2,10\}$, $R_2 \Rightarrow X$ is a consistent rule of X. Here, when we delete an attribute "prod", that is, we decompose R_2 into $R_3 \wedge [prod = 0]$, R_3 is equal to $[location = whole] \wedge [nature = persistent] \wedge [history = persistent] \wedge [jolt = 0] \wedge [nausea = 0] \wedge [M1 = 1] \wedge [M2 = 1]$ and $[x]_{R_3} = [x]_{R_2}$. Therefore, $Posi_{R_3 \wedge [prodrome=0]}(X) = Posi_{R_3}(X)$. We can delete this attribute. Applying this method iteratively, we derive the following minimum equivalent relations: $[location = whole] \wedge [M1 = 1]$ and $[history = persistent] \wedge [M1 = 1]$.

If we use some weight function for efficiency, this algorithm is also a kind of greedy algorithm which finds independent variables. However, while AQ is based on incremental addition of equivalence relations, Pawlak's method is based on incremental removal of dependent equivalence relations. This characteristic is also discussed in Section 5.

3 AQ method

3.1 Bounded Star as Positive Region

AQ is an inductive learning system based on incremental STAR algorithm[4]. This algorithm selects one "seed" from positive examples and starts from one "selector"(attribute-value pair) contained in this "seed" example. It adds selectors incrementally until the "complexes" (conjunction of attributes) explain only positive examples, called a **bounded star**. Since many complexes can satisfy these positive examples, AQ finds the most preferred one, according to a flexible extra-logical criterion.

It would be worth noting that the complexes supported only by positive examples corresponds to the lower approximation, or the positive region in rough set theory. That is, the rules induced by AQ is equivalent to consistent rules defined by Pawlak when constructive generalization rules[4] are not used. As a matter of fact, AQ's star algorithm without constructive generalization can be re-formulated by the concepts of rough sets. For example, a bounded star denoted by $G(e|U - D, m)$ in Michalski's notation is equal to $G = \{R_i | [x]_{R_i} = D_j\}$, such that $|G| = m$ where $|G|$ denotes the cardinality of G. This star is composed of many complexes, which is ordered by LEF_i, lexicographic evaluation functional, which is defined as the following pair:$< (-negcov, \tau_1), (poscov, \tau_2) >$ where $negcov$ and $poscov$ are numbers of negative and positive examples, respectively, covered by an expression in the star, and where τ_1 and τ_2 are tolerance threshold for criterion $poscov, negcov$ ($\tau \in [0..100\%]$). This algorithm shows that AQ method is a kind of greedy algorithm which finds independent variables using selectors which are equivalent to equivalence relations in terms of rough sets. We will discuss this characteristic later in Section 5.

3.2 INDUCE method of AQ algorithm

An algorithm to derive a bounded star is called INDUCE method in AQ algorithm [4]. Here we illustrate how this INDUCE method works. For example, let us consider the above example of database shown in Table 1, which collects the patients whole complained of headache. Then an attribute value pair, $[location = whole]$ can be regarded as *selector*. And if we choose a seed, whose record number is 2, then a partial star which includes this seed and supports $[location = whole]$ is $\{2,8,10\}$.

This is not a bounded star for a class "m.c.h." (muscle contraction headache), since the class of "8" is "i.m.l." (intracranial mass lesion), that is, this star includes a *negative example* as to "m.c.h". So we have to add selectors to remove "8" from a star. This means that we have to add some selectors in order to get a bounded star. In AQ method, we choose a selector from the selectors which is supported by the seed, sample "2". For example, if we choose a selector [*history* = *persistent*], then a star of [*location* = *whole*]&[*history* = *persistent*] is equal to {2,10}, which only consists of positive samples as to "m.c.h."

Therefore [*location* = *whole*] ∧ [*history* = *persistent*] can be regarded as a premise of a rule for classification of "m.c.h.", that is, if a sample satisfies [*location* = *whole*] ∧ [*history* = *persistent*], then a class of this sample is "m.c.h.". It is also notable that [*location* = *whole*] ∧ [*M1* = 1] and [*location* = *whole*] ∧ [*M2* = 1] generate a bounded star whose element is only "2". In order to choose a suitable selector from possible selectors, we have to apply some extra-logical criterion, such as aforementioned *LEF* criterion, which includes domain-specific knowledge. That is, in AQ algorithm, domain knowledge is applied in order to select attribute-value pairs, or premises of rules from possible combination of those pairs, which are suitable to describe the structure of domain-knowledge.

4 Matroid Theory

4.1 Definition of a Matroid

Matroid theory abstracts the important characteristics of matrix theory and graph theory, firstly developed by Whitney[22] in the thirties of this century. The advantages of introducing matroid theory is the following: 1) Since matroid theory abstracts graphical structure, this shows the characteristics of formal structure in graph clearly. 2) Since a matroid is defined by the axioms of independent sets, it makes the definition of independent structure clear. 3) Duality is one of the most important structure in matroid theory, and by this definition we can treat relation between dependency and independency rigorously. 4) The greedy algorithm is one of the algorithms for acquiring an optimal base of a matroid. This algorithm is studied in detail, so we can use well-established results in our problem.

Although there are many interesting and attractive characteristics of matroid theory, we only discuss about duality, and the greedy algorithm, both of which are enough for our algebraic specification. For further information on matroid theory, readers might refer to [20].

First, we begin with definition of a matroid. A matroid is defined as an independent space which satisfies the following axioms:

Definition 5. The pair $M(E, \mathcal{J})$ is called a matroid, if

1) E is a finite set,
2) $\emptyset \in \mathcal{J} \subset 2^E$,
3) $X \in \mathcal{J}, Y \subset X \Rightarrow Y \in \mathcal{J}$,
4) $X, Y \in \mathcal{J}, card(X) = card(Y) + 1 \Rightarrow (\exists a \in X - Y)(Y \cup \{a\}) \in \mathcal{J}$.

If $X \in \mathcal{J}$, it is called **independent**, otherwise X is called **dependent**. □

One of the most important characteristic of matroid theory is that this theory refers to the notion of independence using the set-theoretical scheme. As shown in [10], we also consider the independence of the attributes in terms of rough sets, which uses the set-theoretical framework. Therefore our definition of independence can be also partially discussed in terms of matroid theory.

4.2 Duality

Another important characteristic is duality. While this concept was firstly introduced in graph theory, a deeper understanding of the notion of the duality in graph theory can be obtained by examining matroid structure. Definition of duality is as follows:

Definition 6. If $M = (E, \mathcal{J})$, is a matroid with a set of bases β, then the matroid with a set of elements E, and a set of bases $\beta^* = \{E - B | B \in \beta\}$ is termed the **dual** of M and is denoted by M^*. ☐

¿From this definition, it can be easily shown that $(M^*)^* = M$, and M and thus M^* are referred to a **dual matroid pair**. And we have the following theorem:

Theorem 7. *If M is a matroid, then M^* is a matroid.* ☐

4.3 the Greedy Algorithm

Since it is important to calculate a base of a matroid in practice, several methods are proposed. In these methods, we focus on the greedy algorithm. This algorithm can be formulated as follows:

Definition 8. Let B be a variable to store the calculated base of a matroid, and E denote the whole set of attributes. We define the Greedy Algorithm to calculate a base of a matroid as follows:

1. $B \leftarrow \phi$
2. Calculate "priority queue" Q using weight function of E.
3. If B is a base of $M(E, \mathcal{J})$ then stop. Else go to 4.
4. $e \leftarrow first(Q)$, which has a minimum weight in Q.
5. If $B \cup \{e\} \in \mathcal{J}$ then $B \leftarrow B \cup \{e\}$. goto 2. ☐

This algorithm searches one solution which is optimal in terms of one weight function. Note that a matroid may have many bases. The base derived by the greedy algorithm is optimal to some **predefined** weight function. Hence if we cannot derive a suitable weight function we cannot get such an optimal base. In the following, we assume that we can define a good weight function for the greedy algorithm. For example, we can use *information gain* as defined in [2, 13] for such function. When information gain is used as a weight function, the greedy algorithm with this weight function gives a solution optimal to apparent accuracy. since this gain is closely related with apparent accuracy or apparent accuracy. In other words, the solution is optimal to apparent rate, that is, in the language of statistics, the algorithm calculates the best class allocation of training samples. Under this assumption, this algorithm has the following characteristics:

Theorem 9. *The complexity of the greedy algorithm is*

$$\mathcal{O}(mf(\rho(M)) + m \log m),$$

where $\rho(M)$ is equal to a rank of matroid M, m is equal to the number of the elements in the matroid, $|E|$, f represents a function of computational complexity of an independent test, which is the procedure to test whether the obtained set is independent, and is called independent test oracle. □

Theorem 10. *The optimal solution is derived by this algorithm if and only if a subset of the attributes satisfies the axioms of the matroid.* ⊔

This theorem is very important when we discuss about optimal solution of learning algorithms. This point is discussed in section 7.

5 AQ matroid and Pawlak's Matroid

Here we show that our "rough sets" formalization of AQ algorithm is equivalent to the greedy algorithm for calculating bases of a matroid and that the derived bases are dual to those derived by Pawlak's reduction method.

5.1 AQ matroid

Under the above assumption we can constitute a matroid of AQ method, which we call *AQ matroid* as follows:

Theorem 11. *Let B denote the base of a matroid such that $[x]_B = D_k$. If we define an independent set $\mathcal{J}(D_k)$ as $\{A(R_j)\}$ which satisfies the following conditions:*

1) $R_j \preceq B$,
2) $[x]_B \subseteq [x]_{R_j}$,
3) $\forall R_i$ s.t. $R_i \prec R_j \preceq B$, $D_j = [x]_B \subseteq [x]_{R_j} \subset [x]_{R_i}$,

where the equality holds only if $R_j = B$. Then this set satisfies the definition of a matroid. We call this type of matroid, $M(E, \mathcal{J}(D_k))$, AQ matroid. □

The first condition means that a base is a maximal independent set and each relation forms a subset of this base. And the second condition is the characteristic which satisfies all of these equivalence relations. Finally, the third condition denotes the relationship between the equivalence relations: Any relation R_i which forms a subset of $A(R_j)$ must satisfy $[x]_{R_j} \subset [x]_{R_i}$. Note that these conditions reflects the conditional part of AQ algorithm. For example, let us consider the example shown in Table 1. Let us take two equivalence relations, $[location = whole]$ and $[M1 = 1]$. $[x]_{[location=whole]}$ and $[x]_{[M1=1]}$ are equal to $\{2,8,10\}$, and $\{1,2,5,10\}$. Since, these two sets are supersets of $D = [x]_{[location=whole]\wedge[M1=1]} = \{2, 10\}$, which is a positive region of class "m.c.h.", we derive the following relations: $D \subset [x]_{[location=whole]}$ and $D \subset [x]_{[M1=1]}$. Therefore, $\{[location = whole]\}$, $\{[M1=1]\}$, and $\{[location = whole] \wedge [M1 = 1]\}$ belong to the independent sets of the target concept, classification of a class "m.c.h.". It is also notable that each D_k has exactly one independent set $J(D_k)$. Therefore the whole AQ algorithm is equivalent to the greedy algorithm for acquiring a set of bases of AQ matroid, denoted by $\{J(D_k)\}$. Furthermore, since the independent test depends on the calculus of indiscernible sets, is less than $\mathcal{O}(\rho(M) * n^2)$ where n denotes a sample size, the computational complexity is given as follows:

Theorem 12. *Assume that we do not use constructive generalization. Then the complexity of AQ algorithm is less than*

$$\mathcal{O}(mn^2\rho(M)) + m\log m),$$

where $\rho(M)$ is equal to a rank of matroid M, m is equal to the number of the elements in the matroid, $|E|$. □

Hence the computational complexity of AQ depends mainly on the number of the elements of a matroid, since it increases exponentially as the number of the attribute-value pairs grows large.

5.2 Pawlak's Matroid

On the other hand, since $\rho(M)$ is the number of independent variables, $m-\rho(M)$ is equal to the number of dependent variables. From the concepts of the matroid theory, if we define an dependent set $\mathcal{I}$ as shown below, then $M(E,\mathcal{I})$ satisfies the condition of the dual matroid of $M(E, \mathcal{J})$.

Theorem 13. *Let B denote the base of a matroid such that $[x]_B = D_k$. If we define an independent set $\mathcal{I}(D_k)$ as $\{A(R_j)\}$ which satisfies the following conditions:*

1) $B \prec R_j$,
2) $[x]_B = [x]_{R_j}$,
3) $\forall R_i$ s.t. $B \prec R_i \preceq R_j$, $D_k = [x]_B = [x]_{R_j} = [x]_{R_i}$,

then $M(E,\mathcal{I}(D_k))$ is a dual matroid of $M(E, \mathcal{J}(D_k))$, and we call $M(E,\mathcal{I}(D_k))$ Pawlak's matroid. □

The first condition means that a base is a maximal independent set and each relation forms a superset of this base. And the second condition is the characteristic which satisfies all of these equivalence relations. Finally, the third condition denotes the relationship between the equivalence relations: Any relation R_i which forms a subset of $A(R_j)$ must satisfy $[x]_{R_i} \subset [x]_{R_j}$. Note that these conditions reflects the conditional part of reduction method. For example, let us take R_2 in Section 2.1. as an example. In this case, R_2 is equal to a positive region of a class "m.c.h.". If we describe R_2 as the conjunction of R_3, which is equal to $[location = whole] \wedge [nature = persistent] \wedge [history = persistent] \wedge [prodrome = 0] \wedge [M1 = 1] \wedge [M2 = 1]$, $[jolt = 0]$, and $[nausea = 0]$ that is , $R_2 = R_3 \wedge [jolt = 0] \wedge [nausea = 0]$, then we get the following result: $\{2, 10\} = [x]_{R_2} = [x]_{R_3 \wedge [jolt=0]} = [x]_{R_3}$. Therefore $[jolt = 0]$, $[nausea = 0]$, and $[jolt = 0] \wedge [nausea = 0]$ are the elements of a Pawlak's matroid.

As shown above, the algorithm of Pawlak's method is formally equivalent to the algorithm for the dual matroid of AQ matroid, and the computational complexity of Pawlak's method is less than $\mathcal{O}((p - \rho(M)) * (n^2 + 2n) + m \log m)$. Hence, we get the following theorem.

Theorem 14. *The complexity of the Pawlak's method is less than*

$$\mathcal{O}(mn^2(p - \rho(M))) + m \log m),$$

where p is a total number of attributes, $\rho(M)$ is equal to a rank of matroid M, and m is equal to the number of the elements in the matroid, $|E|$. □

¿From these consideration, if $\rho(M)$ is small, AQ algorithm performs better than Pawlak's one under our assumption.

6   ID3 Greedoid

Induction of decision trees, such as CART[2] and ID3[13] is another inductive learning method based on the ordering of variables using information entropy measure or other similar measures. This method splits training samples into smaller ones in a top-down manner until it cannot split the samples, and then prunes the overfitting leaves.

As to pruning methods, we discuss independently later, so here we briefly illustrate splitting procedures. For simplicity, let consider classification of "m.c.h." in the example shown in Table 1. Then positive samples consists of $\{1,2,10\}$. For each attributes, we calculate information gain, which is defined as the difference between the value of entropy measure (or other similar measures) before splitting and the averaged value after splitting. And we select the attribute which gives the maximum information gain.

For example, positive examples consist of three elements, the root entropy measure is equal to: $-\frac{3}{10} \log_2 \frac{3}{10} = 0.5211$. In the case of an attribute "M1", the number of positive examples in a sample which satisfy $[M1 = 1]$ is three of four, and that of positive examples in a sample which satisfy $[M1 = 0]$ is zero of six. So, the expected entropy measure is equal to: $-\frac{4}{10}(\frac{3}{4} \log_2 \frac{3}{4}) - \frac{6}{10}(0 \log_2 0) = 0.1245$. Since $0 \log_2 0$ is defined as 0, information gain is derived as: $\frac{3}{10} \log_2 \frac{3}{10} - \frac{4}{10}(\frac{3}{4} \log_2 \frac{3}{4}) = 0.5211 - 0.1245 = 0.3966$. On the other hand, in the case of an attribute "location", since the excepted entropy measure is equal to: $-\frac{4}{10}(\frac{1}{4} \log_2 \frac{1}{4}) - \frac{3}{10}(\frac{2}{3} \log_2 \frac{2}{3}) - \frac{3}{10}(0 \log_2 0) = 0.3170$, information gain of this attribute is equal to: $0.5211 - 0.3170 = 0.2041$. Therefore, "M1" is better for classification at the root. In fact, "M1" is the best attribute for information gain, and we split training samples into two subsamples, one of which satisfies "M1=1" and the other of which satisfies "M1=0". Then we recursively apply these procedures to subsamples. In this case, we get the following small tree for classification of "m.c.h.":

$$
\begin{cases}
[\text{M1}=1](m.c.h. : 3 \quad non - m.c.h. : 1) \\
\quad \begin{cases} [\text{M2}=1]\cdots\cdots(m.c.h. : 2) \\ [\text{M2}=0]\cdots\cdots(m.c.h. : 1 \quad non - m.c.h. : 1) \end{cases} \\
[\text{M1}=0](m.c.h. : 0 \quad non - m.c.h. : 6)
\end{cases}
$$

"Non-m.c.h." denotes the negative samples as to "m.c.h." For each node, (m.c.h.: p non-m.c.h.: n) denotes the number of elements of positive examples and that of negative examples. In other words, p "m.c.h." samples belong to that node, while n negative samples are also included.

The main characteristic of the bases derived by ID3 is the following. First, the attribute-value pairs are totally ordered, and in each branch, which corresponds to each base for D_j, subsets of each branch have to preserve this order. For example, let a base be composed of binary attributes, say, $\{a, b, c\}$, in which ID3 algorithm chooses these attributes from the left to the right. Then the allowable subsets are: $\{a\}$, $\{a, b\}$, and $\{a, b, c\}$. Second, each base have the common attribute at least in the first element. For example, if one base is composed of $\{a, b, c\}$, then another base is like $\{a, b, \bar{c}\}$, or $\{\bar{a}, d, c\}$, where $\bar{a}$ denotes a complement of a.

Although these global constraints, especially the first one, decreases the search space spanned by attribute-value pairs, those makes a family of subsets lose the characteristics of a matroid. In fact, a set of the subsets derived by ID3 method does not satisfy the axiom of a matroid. It satisfies the axiom of a greedoid[21], which is a weaker form of a matroid, defined as follows.

Definition 15. The pair $M(E, \mathcal{J})$ is called a Greedoid, if

1) E is a finite set,
2) $\emptyset \in \mathcal{F} \subset 2^E$,
3) $X \in \mathcal{F}$, there is an $x \in X$ such that $X - x \in \mathcal{F}$,
4) $X, Y \in \mathcal{F}, card(X) = card(Y) + 1 \Rightarrow (\exists a \in X - Y)(Y \cup \{a\}) \in \mathcal{F}$.

If $X \in \mathcal{J}$, it is called **feasible**, otherwise X is called **infeasible**. □

Note that the third condition becomes a weaker form, which allows for the total ordering of elements. Because of this weakness, some important characteristics of matroids, such as duality, are no longer preserved. Hence ID3 has no dual method like AQ and Pawlak's method. However, since the optimality of the greedy algorithm is preserved, so we can discuss about this characteristic. Using the above formulation, we can define the search space for ID3 as a ordered greedoid.

Definition 16. Let B denote the base of a matroid such that $[x]_B = D_k$. If we define a feasible set $\mathcal{K}(D_k)$ as: $\{A(R_j)\}$ which satisfies the following conditions:

1) $R_j \preceq B$,
2) $[x]_B \subseteq [x]_{R_j}$,
3) $\forall R_i$ s.t. $R_i \preceq R_j \preceq B$, $D_j = [x]_B \subseteq [x]_{R_j} \subset [x]_{R_i}$,

where the equality holds only if $R_j = B$, and if we demand that the each $K(D_k)$ should satisfy the following conditions:

(1) for all R_i and R_l, $R_i \preceq R_l$ or $R_l \preceq R_j$ holds ,
(2) $\forall\ \mathcal{K}(D_q)$ and $\mathcal{K}(D_p)$, For all $R_j \in \mathcal{K}(D_q)$ and $R_i \in \mathcal{K}(D_p)$,
 if $[x]_{R_i} \cap [x]_{R_j} \neq \phi$, then $R_j \preceq R_i$,

then this set satisfies the definition of a greedoid. We call this type of greedoid, $G(E, \mathcal{K}(D_k))$, *ID3 greedoid*. □

Note that each D_k has exactly one feasible set $\mathcal{K}(D_k)$. Therefore the whole ID3 algorithm is equivalent to the greedy algorithm for acquiring a set of bases of ID3 greedoid, denoted by $\{\mathcal{K}(D_k)\}$.

6.1 Computational Complexity of ID3

As shown above, ID3 algorithm is also the greedy algorithm for deriving a base of a greedoid. However, the main feature of this algorithm is that two constraints to independent sets are given. This reduces the search space of independent sets, because the sets which satisfy the above two constraints are not so many. Here, we obtain the following theorem:

Theorem 17. *The complexity of ID3 algorithm is less than*

$$\mathcal{O}(mn^2 \rho(M)) + m \log m),$$

where $\rho(M)$ is equal to a rank of greedoid M, m is equal to the number of the elements in the greedoid, $|F|$. □

The difference in computational complexity between AQ and ID3 is the value of m. This difference is illustrated as follows. Let all attributes be binary and the total number of attributes be p. Then, for AQ, since the search space is spanned by the whole combination of attribute-value pairs, so $|E|$ is almost equal to 2^p. On the other hand, the search space for ID3 is equal to $2^{\rho(M)+1} - 1$. Therefore, if $\rho(M) \ll p$, then the computational complexity of ID3 is much lower than AQ.

Hence, in many cases, this ID3 method is faster than the other two methods. However, it does not mean that ID3 performs well, because some optimal solutions will never be found by the two constraints to independent sets. They will not appear in the space of ID3 greedoid. This phenomenon sometimes make ID3 performs worse. For example, when training samples do not reflect the importance of variable, that is, when some less important variables are given more weight than important ones, some relations between important ones can be never found. Therefore this fact explains one aspect that ID3 is more sensitive to training samples.

7 Optimal Solution

If we adopt a weight function which is described as a monotonic function of apparent accuracy, we obtain an optimal solution which is the best for apparent accuracy. So, in this case, Theorem 10 tells us that an optimal solution is obtained only when relations between training samples and attributes-value pairs satisfy the conditions of AQ matroid. So we obtain the following corollary:

Corollary 18. *The optimal solution is derived by AQ method and Pawlak's method if and only if a subset of the attributes satisfies the axioms of the matroid.* □

However, this assumption is very strict, since apparent accuracy depends on only given training samples. In practice, it is often violated by new additional training samples. For example, when in the old training samples, $R_i \prec R_j$ implies $[x]_{R_j} \subset [x]_{R_i}$, additional samples cause the latter relation to be $[x]_{R_j} = [x]_{R_i}$. In other words, additional samples cause independent variables to be dependent. In this case, the former derived solution is no longer optimal to this weight function. On the other hand, since ID3 greedoid is totally ordered, the situation is much severer. Even if an attribute in an feasible set remains feasible, if the above two constraints are violated, this set is no longer ID3 greedoid. So the optimal solution is no longer optimal. This suggests that this method is more sensitive to training samples than AQ and Pawlak's method.

Note that these results is obtained by weight function which is described as a function of apparent accuracy. Therefore these results suggest that, if we have a weight function which is a monotonic function of true accuracy, then we derive a base optimal to true accuracy. Unfortunately, it is impossible to derive such function, since we can only estimate true accuracy. Some approaches discuss about these functions, which are known as MDL(Minimum Description Length) function[9, 15], and their usefulness is ensured in [9, 15].

In the next section, we consider the relation between the solutions by MDL function and true accuracy.

8 Pruning, MDL and True Error Rate

8.1 Our Model of Pruning and Truncation

It was pointed out that the results induced by splitting (ID3) and INDUCE method (AQ) are overfitted to original training samples. In order to avoid overfitting, pruning and truncation are introduced [1, 2, 13]. These two methods remove some attribute-value pair from the induced trees or rules according to some heuristic criterion. For example, in the case of CART, cross-validation method and minimal-cost complexity measure are applied as follows. First, cross-validation method estimates predictive optimal tree size. Then we choose tree size whose minimal-cost complexity is within 1-standard deviation of the complexity of tree size determined by cross-validation (1-SE rule).

For simplicity, we use the following simple model to discuss the above mechanism. First, we describe true accuracy in the language of rough sets.

Definition 19. Let D^c or $[x]_{R_i}^c$ denote unobserved future cases of a class D or those which satisfies R_i, respectively. Then true accuracy $\hat{\alpha}_{R_i}(D)$ can be defined as:

$$\hat{\alpha}_{R_i}(D) = \frac{card \ \{([x]_{R_i} \cap D) \bigcup ([x]_{R_i}^c \cap D^c)\}}{card \ \{[x]_{R_i} \bigcup [x]_{R_i}^c\}}$$
$$= \varepsilon_{R_i} \alpha_{R_i}(D) + (1 - \varepsilon_{R_i}) \alpha_{R_i}^c(D),$$

where ε_{R_i} is equal to the ratio of training samples to total population,

$$\varepsilon_{R_i}(D) = \frac{card \ [x]_{R_i}}{card \ ([x]_{R_i} \cup [x]_{R_i}^c)},$$

$\alpha_{R_i}(D)$ denotes an apparent accuracy, and $\alpha_{R_i}^c(D)$ denotes the accuracy of classification for unobserved cases. □

This is a fundamental formula of accuracy. Resampling methods, MDL principle, and pruning methods focus on how to estimate ε_{R_i} and $\alpha_{R_i}^c$, and makes some assumption about these parameters.

Next, we define the following sequence of equivalence relations derived by the above greedy algorithm.

Definition 20. Let $R_1, \cdots, R_n$ be a track of equivalence relations derived by the greedy algorithm. Then, using the defined partial order relation, we obtain the following sequence, which we call **partial order sequence**:

$$\phi \prec R_1 \prec R_2 \prec \cdots \prec R_{n-1} \prec R_n,$$

where ϕ denotes null equivalence relation which contains no equivalence relation. R_n is equal to a optimal base calculated by the greedy algorithm. □

Let $\epsilon_{R_i}(D) = 1 - \alpha_{R_i}(D)$ denote an apparent error rate of R_i to classify a class which is supported by a set of samples, D. Here we assume that the weight function of the above greedy algorithm is a monotonic function of apparent accuracy. Then the following sequence of apparent error rate is derived:

$$1 > \epsilon_{R_1}(D) > \epsilon_{R_2}(D) > \cdots > \epsilon_{R_{n-1}}(D) > \epsilon_{R_n}(D) \geq 0.$$

Next we consider a complexity of an induced equivalence relation R_i, which we denote by l_i. If we assume the complexity of R_i, l_i to be proportional to $card \ A(R_i)$, that is, $l_i \propto card \ A(R_i)$, then we also obtain the following sequence according to partial order sequence:

$$0 < l_1 < l_2 < \cdots < l_{n-1} < l_n.$$

Using these notation, we define total cost of R_i, denoted by c_i which is a simplified version of cost complexity of CART as follows:

$$c_i = \epsilon_{R_i}(D) + kl_i,$$

where k is a constant for normalization of l_i. Furthermore, we can define the sequence of c_i. In CART, we choose a tree whose cost is minimal over the c_i sequence. For example, if a c_i sequence is derived as follows:

$$1 > c_1 > c_2 > \cdots > c_k < c_{k+1} < \cdots < c_n,$$

then we choose a tree whose complexity is equal to l_k, that is, R_k. In this strategy, we assume that this total cost reflects predictive behavior of induced rules or trees. So if the cost is not related to true error rate, then it is not guaranteed that this procedure chooses a solution optimal to true error rate $\hat{\alpha}_{R_i}(D)$. In the next subsection, we show this assumption more formally.

Note also that this result is not confined to the pruning method in CART, called *minimal cost complexity pruning* [2]. Other pruning methods or truncation methods also use similar measures or some heuristics which can be reformulated in terms of cost function. So the discussion in the next subsection can be applied to other pruning methods.

8.2 Pruning and True Error Rate

Simple calculation gives us a formula of true error rate as follows:

$$\hat{\epsilon}_{R_i}(D) = \epsilon_{R_i}(\epsilon_{R_i}(D) + \frac{1 - \epsilon_{R_i}}{\epsilon_{R_i}}\epsilon_{R_i}^c(D)),$$

where $\epsilon_{R_i}(D)$ denotes an apparent accuracy, and $\epsilon_{R_i}^c(D)$ denotes the accuracy of classification for unobserved cases. Therefore, compared this function with the formula of c_i, the following result is obtained:

Theorem 21. *If*

$$l_i \propto \frac{1 - \epsilon_{R_i}}{\epsilon_{R_i}}\epsilon_{R_i}^c(D),$$

then the above pruning method is equal to a method to choose a model optimal to true error rate, under the condition that an optimal solution should be in the partial order sequence. □

So if the above relation does not hold, we cannot determine whether results induced by pruning method or truncation is optimal to prediction of future cases. Unfortunately, it is impossible to determine whether the above relation satisfies or not, since it includes two terms related to future observations, ϵ_{R_i}, and $\epsilon_{R_i}^c(D) = 1 - \alpha_{R_i}^c(D)$.

The former term, as shown in the definition, denotes the ratio of training samples to total population. So it indicates whether the training samples is large enough for estimation of true error rate or not. On the other hand, the latter one shows predictive behavior of induced rule, R_i for future classification. Therefore it indicates whether the training samples have little sampling bias, compared with total population. Therefore the term, $((1 - \varepsilon_{R_i})/\varepsilon_{R_i})\varepsilon_{R_i}^c(D)$, represents two kinds of effect of future unobserved case, one of which reflects sampling size, and the other of which is related to sampling bias. According to this discussion, the premise of the above definition intuitively means that the complexity of a rule (an equivalence relation) is closely related to suitable sampling size and sampling bias.

However, this assumption is *ad hoc*, and it seems that this relation is highly domain-dependent. These results are the same as qualitative ones which Schaffer discusses as "a form of bias" in [18].

8.3 MDL Principle and True Error Rate

The minimum description length principle (MDL principle) [15, 16, 17, 11, 12] states that the best "theory" to infer from a set of data is the one which minimizes the sum of 1) the length of the theory and 2) the length of the data when encoded using the theory as a predictor for the data. Expressed mathematically according to [11], the best theory T is the one that minimizes

$$l(T) + l(z_1, z_2, ..., z_n | T),$$

where $l(T)$ is the length in bits to describe T, $z_1, z_2, ..., z_n$ are the observations, and $l(z_1, z_2, ..., z_n, |T)$ is the number of bits needed to encode the observations with respect to T. The quantity $l(T)$ effectively measures the complexity of T and $l(z_1, z_2, ..., z_n)$ measures the degree to which T accounts for the observations, with fewer bits indicating a better fit. The sum of these two quantities defines the number of bits needed to exactly represent the observations.

In this case, since the description length of apparent error rate, $\varepsilon_{R_i}(D)$ is equal to $-\log_2 \alpha_{R_i}(D)$, so we obtain: $l(z_1, z_2, ..., z_n | T) = -log_2 \varepsilon_{R_i}(D)$. Therefore, total cost is described as: $l(T) - log_2 \varepsilon_{R_i}(D)$. On the other hand, the logarithm of the formula of true error rate is rewritten as follows:

$$-\log_2 \varepsilon_{R_i}(D) - \log_2 \varepsilon_{R_i}(1 + \frac{1 - \varepsilon_{R_i}}{\varepsilon_{R_i}} \frac{\varepsilon_{R_i}^c(D)}{\varepsilon_{R_i}(D)}).$$

Therefore choice of a model under MDL principle is almost equivalent to the pruning method discussed in subsection 8.2, which is based on cost function. So in the similar way of the above subsection, we obtain the following result:

Theorem 22. *If*

$$l(T) \simeq -\log_2 \varepsilon_{R_i}(1 + \frac{1 - \varepsilon_{R_i}}{\varepsilon_{R_i}} \frac{\varepsilon_{R_i}^c(D)}{\varepsilon_{R_i}(D)})),$$

then the above pruning methods is equal to a method to choose a model optimal to true error rate, under the condition that an optimal solution should be in the partial order sequence. □

Note that the conditional of the above term is much complicated. However, in this case, the premise of the above definition also intuitively means that the complexity of a rule (an equivalence relation) is closely related to suitable sampling size and sampling bias.

9 Towards a Multistrategic Machine Learning System

This work is a preliminary work towards development of multistrategy machine learning system [6]. Recently, it has been pointed out that each machine learning method is a little powerless to extract sufficient knowledge from data inputs and that combination of some machine learning methods with other machine learning methods or other methods, such as neural networks, statistics, and other symbolic reasoning methods is more effective. One of the promising study on multistrategic approach is goal-driving learning [3]. In this approach, a meta-cognitive agent controls machine learning methods by domain knowledge, mainly algebraic constraints, evaluate the characteristics of each method and chooses suitable machine learning methods for a goal.

As shown in the above sections, some effective strategies, constraints are based on some algebraic equations. So if we would like to estimate these constraints automatically and to extract when we can apply these constraints, it is indispensable to apply methods of symbolic mathematical computing. For example, pruning methods for decision tree are based on the algebraic relation between true error rate function and complexity measure. It seems effective to apply algebraic reasoning to the validity of these pruning equations, since some domain knowledge can be transformed into algebraic constraints: $l_i \simeq \epsilon_{R_i}^c(D)$. Then, a equation: $\epsilon_{R_i}^c(D) \simeq ((1 - \varepsilon_{R_i})/\varepsilon_{R_i})\epsilon_{R_i}^c(D)$ is derived and we get a constraints on ε_{R_i}. The same discussion can be applied to algebraic constraints on statistical measures, such as MDL principle.

Although this is a simple application of computer algebra technique, some domain knowledge or learning strategy may represent a form of couple of some complex algebraic equations. In this case, we need a powerful technique of symbolic mathematical computing.

The above is a brief discussion mainly on what symbolic mathematical computing gives to machine learning methods. Finally, we briefly discuss what our works request to future works of computer algebra. Most of our work discusses that empirical learning methods can be viewed as calculating the basis of a space which is spanned by some equivalence relations (attribute-value pairs). So if we have an effective method to obtain basis, we can apply these techniques to empirical learning methods, although data transformation may be needed. Furthermore, it becomes easier to discuss about the characteristics of problem spaces of machine learning methods. And we also suppose that the calculus of basis can be applied not only to empirical learning methods, but also to other AI methods, such as qualitative reasoning, nonmonotonic reasoning.

10 Conclusion

In this paper, we combine the concepts of matroid theory with those of rough sets, which gives us an excellent framework and enables us to understand the differences between AQ, Pawlak's method and ID3 clearly. Using this framework, we obtain six interesting conclusions from our approach. First, while AQ and Pawlak's method are equivalent to the greedy algorithm for finding bases of matroids, ID3 method calculates those of ordered greedoids. Second, AQ matroids are dual to Pawlak matroids. Third, the efficiency of AQ and Pawlak's method depends on the number of attributes, especially, independent variables. Fourth, the induced results are optimal to the training samples. Fifth, in addition to the fourth conclusion, since a greedoid of ID3 has weaker structure than the other two methods, ID3 method is the most sensitive to training samples, although its computational complexity is the lowest. Finally, it is shown that pruning and truncation methods do not always correct overfitting phenomenon. Furthermore, for pruning and truncation methods, we assume that future error rate is related to the complexities of induced rules. If this assumption is violated, we cannot obtain optimal rules or trees by applying pruning and truncation.

Although these results were observed by some experimental results[1, 2, 5, 7, 13, 19, 25], they have not yet been explained by formal theory. We feel that the extension of these approach can be applied to the extension of the above three original methods, such as POSEIDON(AQ16)[1], VPRS[24, 25] and C4[14]. And it is our future work to formalize these methods and to analyze the relationship between these existing algorithms by using our framework.

Acknowledgements

The authors would like to thank Zdzislaw Pawlak, Andrej Skowron and Nitin Indurkhya for giving insightful comments on the manuscript. This research is supported by Grants-in-Aid for Scientific Research No.04229105 from the Ministry of Education, Science and Culture, Japan.

References

1. Bergadano, F., Matwin, S., Michalski, R.S. and Zhang, J. Learning Two-Tiered Descriptions of Flexible Concepts: The POSEIDON System, *Machine Learning*, **8**, 5-43, 1992.
2. Breiman, L., Freidman, J., Olshen, R. and Stone, C. *Classification And Regression Trees*. Belmont, CA: Wadsworth International Group, 1984.
3. Hunter, L.(eds). *Proceedings of AAAI-94 Spring Workshop on Goal-Driven Learning*, AAAI Press, 1994.
4. Michalski, R.S. A Theory and Methodology of Machine Learning. Michalski, R.S., Carbonell, J.G. and Mitchell, T.M., *Machine Learning - An Artificial Intelligence Approach*, 83-134, Morgan Kaufmann, CA, 1983.
5. Michalski, R.S., et al. The Multi-Purpose Incremental Learning System AQ15 and its Testing Application to Three Medical Domains, *Proc. of AAAI-86*, 1041-1045, Morgan Kaufmann, CA, 1986.

6. Michalski, R.S., and Tecuci, G.(eds) Machine Learning vol.4 - A Multistrategy Approach -, Morgan Kaufmann, CA, 1994.

7. Mingers, J. An Empirical Comparison of Selection Measures for Decision Tree Induction. *Machine Learning*, **3**, 319-342, 1989.

8. Mingers, J. An Empirical Comparison of Pruning Methods for Decision Tree Induction. *Machine Learning*, 4, 227-243, 1989.

9. Nakakuki,Y., Koseki, Y., and Tanaka, M. Inductive Learning in Probabilistic Domain in *Proc. of AAAI-90*, 809-814, 1990.

10. Pawlak, Z. *Rough Sets*, Kluwer Academic Publishers, Dordrecht, 1991.

11. Pendnault, E.P.D. Some Experiments in Applying Inductive Inference Principles to Surface Reconstruction, *Proceedings of IJCAI-89*, 1603-1609, 1989.

12. Pendnault, E.P.D. Inferring probabilistic theories from data, *Proceedings of AAAI-88*, 1988.

13. Quinlan, J.R. Induction of decision trees, *Machine Learning*, **1**, 81-106, 1986.

14. Quinlan, J.R. Simplifying Decision Trees. *International Journal of Man-Machine Studies*, **27**, 221-234, 1987.

15. Quinlan, J.R. and Rivest, R.L. Inferring Decision Trees Using the Minimum Description Length Principle, *Information and Computation*, **80**, 227-248, 1989.

16. Rissanen, J. Stochastic complexity and modeling, *Ann. of Statist.*,14, 1080-1100, 1986.

17. Rissanen, J. Universal Coding,Information, Prediction, and Estimation, *IEEE. Trans. Inform. Theory*, **IT-30**, 629-636, 1984.

18. Schaffer, C. Overfitting Avoidance as Bias. *Machine Learning*, **10**, 153-178, 1993.

19. Tsumoto, S. and Tanaka, H. PRIMEROSE: Probabilistic Rule Induction Method based on Rough Sets. in: Ziarko, W.(eds) *Rough Sets, Fuzzy Sets, and Knowledge Discovery*, Springer, London, 1994.

20. Welsh, D.J.A. *Matroid Theory*, Academic Press, London, 1976.

21. White, N.(ed.) *Matroid Applications*, Cambridge University Press, 1991.

22. Whitney, H. On the abstract properties of linear dependence, *Am. J. Math.*, **57**, 509-533, 1935.

23. Ziarko, W. The Discovery,Analysis, and Representation of Data Dependencies in Databases, in:*Knowledge Discovery in Database*, Morgan Kaufmann, 1991.

24. Ziarko, W. Variable Precision Rough Set Model, *Journal of Computer and System Sciences*, **46**, 39-59, 1993.

25. Ziarko, W. Analysis of Uncertain Information in the Framework of Variable Precision Rough Sets, *Foundation of Computing and Decision Science*, **18**, 381-396, 1993.

Subsymbolic Processing using Adaptive Algorithms

David J. Nettleton* and Roberto Garigliano

Laboratory for Natural Language Engineering,
Department of Computer Science,
University of Durham Science Site,
Stockton Road,
Durham. DH1 3LE

Abstract. Subsymbolic approaches have been adopted in attempting to solve many AI problems. In order to find a near optimal solution to the problem a procedure is needed by which the subsymbolic components can be manipulated. In searching all but the simplest of solution spaces algorithms such as hill climbing will often result in only suboptimal solutions being found. Often search algorithms do not make sufficient use of information acquired from previous evaluations of possible solutions. Several forms of adaptive algorithm have been developed in an attempt to overcome this problem and produce robust search mechanisms, *e.g.*, evolutionary algorithms, classifier systems. This paper discusses some adaptive algorithms and presents initial work on a novel form of adaptive algorithm.

1 Introduction

Subsymbolic approaches have been adopted in attempting to solve many AI problems. Such a strategy is appropriate when it is necessary to operate at a level below that of traditional logical-symbolic approaches. In situations such as these a procedure is needed by which the subsymbolic components can be manipulated in order to find a near optimal solution to the problem. This procedure must not only be capable of producing near optimal solutions, it must also be able to do so in an efficient manner. Occasionally some specialised algorithm exists which can carry out this manipulation in the optimal or near optimal number of steps. However, it is often the case that no such procedure is available and some other approach needs to be adopted.

Solutions which are coded by some underlying structure need to be tested in the environment and some measure of performance returned. The measure of performance allows for a 'fitness landscape' to be envisaged and the aim of the optimisation process is often intuitively considered as finding the solution which corresponds to the 'highest peak' of the landscape. The process of optimisation is often carried out my manipulating the coding of the structure and testing new

---

* e-mail: D.J.Nettleton@durham.ac.uk

versions in the environment. Many alternative methods for this manipulation have been suggested. The remainder of this section discusses some of the simplest methods and examines the concept of a fitness landscape.

It is often the case that the search space, in which solutions to a problem exist, is extremely large and complex. For any search algorithm in such a space there exists a fundamental trade-off between exploration and exploitation. One example of an exploratory search would involve the successive enumeration of all possibilities. Though ensuring that the optimal is found such algorithms are practical in only the smallest of search spaces.

In order for an algorithm to search a space efficiently it must be able to exploit opportunities for improved performance. This often involves making use of information acquired from previous evaluations of possible solutions. If such information is not used then the search could degenerate to the point at which it is little better than a random sampling of the solution space. An example of a search strategy which makes use of current information is a hill-climbing algorithm.

Hill climbing algorithms concentrate the search effort around the best solution discovered. Intuitive concepts such as 'peaks' and 'valleys' are often used to describe how such algorithms traverse the fitness landscape. However, such analogies need to be made with care, as Fig. 1 and the following discussion demonstrates. The object of the hill climber is to maximise the (local) fitness, *i.e.*, climb one of the 'peaks'. The normal procedure for this involves making small changes to the solution's representation and accepting as the new best a solution which outperforms the current one. Following this method and using an integer representation, climbs such as those from X to X_1, and from Y to Y_1, would intuitively be how the search would be expected to proceed. With a binary representation a climb from A to B would be expected, but what about one from A to C? In fact both of these are equally likely since $B = 00111$ and $C = 00010$ are both exactly one 'small change away' from $A = 00110$. This would appear to contradict the idea of a 'hill' climber since a 'valley' has been traversed. The concept of what constitutes a hill climb is, therefore, dependent on the representation used [12].

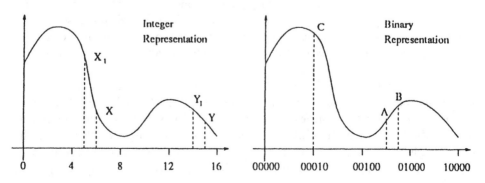

Fig. 1. Hill-climbing and the underlying representation.

Simulated annealing [13] is a search algorithm with a natural metaphor. Inspired by the process of annealing crystalline solids the algorithm models the behaviour of thermodynamic state transitions. One of the important features of simulated annealing is a theory which provides sufficient conditions for asymptotic convergence to the global optimum. Natural processes have also inspired a class of search algorithms known as evolutionary algorithms. These are based on the principles of natural evolution and are discussed in the next section.

2   Evolutionary Algorithms

A feature of both random search and hill climbing is that they discard much of the information which is presented to them during the course of a search. In order to try and retain some of this information a class of algorithms have been developed that maintain a population of solutions. Inspired by natural evolution these algorithms are collectively known as evolutionary algorithms and include genetic algorithms, evolutionary programming, evolution strategies and genetic programming.

Evolutionary algorithms (EAs) are loosely based upon the Darwinian principles of biological evolution and in fact many of the strategies and operators used in their application bear similar names to their biological counterparts, *e.g.*, survival of the fittest, crossover and mutation. An EA creates a set of possible (usually randomly generated) solutions to the problem under consideration and calls this the initial generation. Successive generations are produced via a series of operators which act on the previous generation. The operators produce new solutions which are variations of those that have survived to that point and probabilistically culls the worst using a "survival of the fittest strategy". This process continues until the desired number of generations has been completed. The solutions in the final generation are in effect the 'answers' to the search problem. The main advantage of EAs is their ability to be able to consider many solutions 'at the same time' and from these solutions produce other solutions that converge to the optimal.

2.1   Genetic Algorithms

Genetic Algorithms (GAs) are perhaps the most well known form of evolutionary algorithm and have been applied to a wide range of problems [1] [9] [10]. While there is much experimental evidence to support their success a comprehensive theoretical basis is lacking and research into this continues [2] [7] [17].

It would appear, at first, that a particular generation of a GA possesses only a selection of possible solutions. However, each of these solutions is made up of a string of data and within each string there are many smaller strings of data, usually called schemata. A schema is a set of individuals which share common attributes. In the case of a binary alphabet a schema is denoted by a string consisting of elements taken from the set $\{0, 1, \square\}$ where $\square$ means "don't care".

For example, 00101 and 00111 are both elements of the schema 000□1, but 10101 is not. Each binary string of length k will be an instance of 2^k schemata.

The *Schema Theorem* of Holland [11] relates the number of instances of a schema ξ in two successive generations. If the expected proportion of schema in one generation is $P(\xi, t)$ then:

$$P(\xi, t+1) \geq P(\xi, t) \frac{\hat{\mu}_\xi(t)}{\hat{\mu}(t)} [1 - P_D(\xi)]$$

where $\hat{\mu}_\xi(t)$ is the observed average performance of ξ, $\hat{\mu}(t)$ the average fitness of the schema ξ and $P_D(\xi)$ is the probability of the loss of ξ through the effect of operators such as crossover and mutation.

Throughout a single generation the solutions present will contain many schemata which can be combined (via crossover) to represent other individuals that are not present in the population at that time. The power of GAs is believed to stem from regarding the performance of a single solution as a test on the large number of schemata of which it is an instance. Thus a test on a solution of length k will simultaneously sample instances of 2^k schemata. In a population consisting of M solutions of length k there are between 2^k and $M2^k$ schemata with instances contained within the population. The proportion of each schema which survives to the subsequent generation is largely dependent only on its own observed fitness $\hat{\mu}_\xi(t)$ and is largely independent of what is happening to the other schemata in the population.

An approach using a GA allows for the highly fit, short schemata to be propagated quickly through a population whilst at the same time considering other less fit possibilities. This is termed by Holland [11] as intrinsic parallelism (more recently known implicit parallelism). The proportion of schema in a population is in part dependent on its past performance and so this serves as a record of the performance.

In order to maximise the implicit parallelism, a binary approach to encoding is often advocated [8]. Such an approach allows for the maximum number of schemata to be represented and thus processed. The success of GAs is described by Goldberg [8] in terms of the building block hypothesis — short highly fit segments of each binary string can be combined to form larger, fitter segments of each binary string. The GA, therefore, constitutes a 'bottom up' approach to solution construction.

Criticisms of GAs range from matters of detail to more fundamental questions regarding the underlying paradigm. Radcliffe [17] and Mason [15] state that there is an overemphasis on a binary representation. While Fogel [3] argues that the GA's modelling of the evolutionary process is flawed.

2.2 Evolutionary Programming

Evolutionary programming (EP) originated in the early 1960s, and was initially used on a population of algorithms in order to study the possibilities of evolving artificial intelligence [6]. Since then EP has been extended to cope with real valued functions [4] and has been applied to a wide range of problems [5] [16].

The EP perspective of the evolutionary process is very different to the bottom-up building block approach of GAs. By determining how well solutions are performing in the current environment, improvements are made via a flow of information from the environment back to the underlying genotypic representation. Solutions are probabilistically culled in such a way that those which perform worst are more likely to be removed from the population. EP adopts a top-down approach to solution improvement, as opposed to the bottom-up approach of GAs.

Unlike GAs, EP does not use a crossover operator. Instead solutions are mutated by an amount that is dependent on their fitness, the fitter solutions being less likely to be mutated to the same degree as less fit solutions.

Initial criticisms of the EP approach included those by Solomonoff [18] and Lindsay [14]. Solomonoff argued that the method was only applicable to the simplest of problems and expressed concern at the lack of a crossover operator. Lindsay states that "...such a strategy [EP] amounts to random search...". Criticisms of EP and other evolutionary methods were often based on a misunderstanding of the work, but were enough to lead to the AI community largely rejecting research into evolutionary systems during the 1970s. However, since 1985 there has been a resurgence of interest in the theory and application of evolutionary based algorithms.

3 Classifier Systems

Classifier systems are used in the study of machine learning. A rule set governing a system's behaviour is manipulated so as to optimise the behaviour with respect to its environment. The system has a set of detectors associated with it, each of which interacts with the environment and outputs some message. These messages are then processed by the system and used to direct the system's effectors which determine how the system interacts with the environment. Occasionally the environment provides a measure of performance of the system and it is this performance that is to be optimised.

The processing of the input messages is performed by a set of conditional rules or classifiers. An example of a classifier is given by Holland as:

IF there is (a message from the detectors indicating) an object left of center in the field of vision
THEN (by issuing a message to the effectors) cause the eyes to look left.

In the event of the effectors being required to carry out mutually exclusive actions a competition is held to resolve which is selected. Each classifier, therefore, has associated with it a strength. The rules can be considered as hypotheses for actions in particular situations with the strengths corresponding to the degree of past usefulness —- those with higher strengths are more likely to have been more useful to the system in the past than those with lower strengths. By modifying the strengths over many interactions of the system with the environment the system adapts to it. The modification is carried out by credit assignment and this constitutes part of the learning process. Figure 2 shows how a classifier system interacts with an environment.

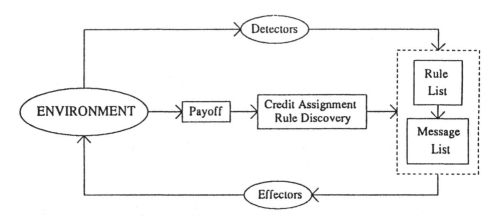

Fig. 2. A classifier system.

The two basic problems which are encountered by classifier systems are those of credit assignment and rule discovery. Credit assignment must be such that rules which result in favourable actions have their strengths increased. If the actions of the rules are independent of each other then the assignment is trivial. However, this is often not the case and there are complex interactions between rules. A bucket brigade algorithm is often used in assignment of credit. The problem of how to create new rules suitable for consideration by the system can also be a difficult task. Genetic algorithms have been suggested as a means of solving this problem.

4 A Voting Algorithm

The algorithms discussed in the previous sections are a selection of the paradigms that have been suggested as a means of producing an adaptive algorithm. Genetic algorithms and evolutionary programming act on a population of solutions in an attempt to find a solution which performs as well as possible within the environment. In the case of Classifier systems a set of classifiers compete to provide responses to the environment in which they are tested. This section discusses a means of selecting between a few structures with regard to many different evaluation functions, each of which may act on only a small amount of the total information contained within a particular structures. Such information is termed 'local' information and the algorithm called a voting algorithm.

A voting algorithm is composed of three basic components; a set of structures, an environment and a set of individuals. Figure 3 demonstrates how these components interact.

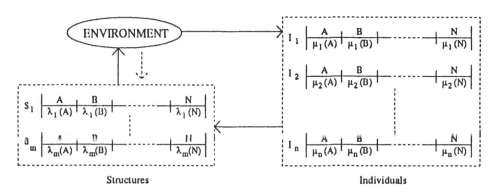

Fig. 3. Components of a voting algorithm. Typically $n \gg m$.

Each of the structures $\{S_1, ..., S_m\}$ is composed of a set of policies $\subseteq \{A, ..., N\}$. It will be assumed that the policies are linearly independent and that each structure contains all of them (this need not be the case in general). If a structure i, contains a policy K, then associated with that is information $\lambda_i(K)$. This information may simply be 1 (yes) or -1 (no) which indicates the structure's agreement (or not) of the policy. In general > 0 indicates agreement, < 0 disagreement and 0 no opinion. The collection of policies contained in each structure is a strategy which can be used within the environment.

In order to find the optimal strategy some measure of its performance is needed. In contrast to GAs and EP the voting algorithm does not assign a measure of performance directly to the structure as a whole. Instead a set of individuals use local evaluation functions to determine how a strategy performs on different aspects of the environment.

In particular for the set of policies $\{A, ..., N\}$ there is a set of fitness functions $\{f_A, ..., f_N\}$ which determines the performance of each policy within the environment E. For example if $f_A(E) \mapsto [1, -1]$, then $f_A(E) = 1$ indicates that the strategy currently being used within the environment has resulted in the policy A being carried out exactly. If $f_A(E) = 0.2$ then the policy has been partly carried out and if $f_A(E) < 0$ then not only has the policy not been carried out but the opposite has (to some degree) occurred.

Now that a means of evaluating parts of a structure's performance within an environment has been described some method of choosing between alternative structures and hence strategies is required.

4.1 Selecting a Structure

Each individual $\{I_1, ..., I_n\}$ contains a set of policies $\subseteq \{A, ..., N\}$. Associated with each policy K, of individual j, is a weight $\mu_j(K)$ which determines the individuals view on, and importance of that policy. For example if $\mu_j(K) \in [1, -1]$ then 1 may represent strong agreement of K, 0 no opinion and -0.2 slight disagreement.

It is now possible to describe a selection algorithm which works within the framework described above. Each individual assigns a rating to each structure according to

$$r_{ij} = \sum_{K=A}^{N} \lambda_i(K)\mu_j(K) \ .$$

The larger the value of r_{ij} the higher the rating individual j gives to structure i. Agreements between a structure and an individual on a policy cause an increase in r_{ij}, while disagreements cause a decrease. Specifically, consider the case in which structure i, contains a policy K, which it is against then $\lambda_i(K) < 0$. Furthermore if individual j is against the policy then $\mu_j(K) < 0$ and so $\lambda_i(K)\mu_j(K) > 0$ and r_{ij} will increase. This corresponds to an agreement on policy between the individual and the structure. Table 1 provides a summary of all the cases.

Table 1. The sign of the $\lambda_i(K)\mu_j(K)$ value for a policy K.

Policy K	Individual j	
	For $\mu_j(K) > 0$	Against $\mu_j(K) < 0$
Structure i For $\lambda_i(K) > 0$	$+^{ve}$	$-^{ve}$
Against $\lambda_i(K) < 0$	$-^{ve}$	$+^{ve}$

Once the ratings of the structures have been calculated for an individual j then the individual can vote. An example of how to select between the different ratings would be to simply select the structure which results in the highest rating value. In that case the vote of individual j is given by

$$v_j = \{i : r_{ij} = \max\{r_{ij} : i = 1, ..., m\}\} \ .$$

The votes for each of the individuals are calculated and grouped together to give the poll $\mathbf{P}$. If an individual's vote is determined as above, then the poll is given by

$$\mathbf{P} = \{v_1, v_2, ..., v_n\} \ .$$

A structure is now selected by examining $\mathbf{P}$. For example, if the most 'popular' structure is to be chosen then this is simply $S_{\text{mode}(\mathbf{P})}$. Once a structure has been chosen its policies act within the environment.

There is of course one vital component missing. As it stands the above algorithm will always result in a given set of individuals selecting the same structure regardless of the structure's performance in the environment. A means of accountability needs to be introduced. Then after the environment has been allowed to develop for some interval the voting procedure can be repeated and perhaps a new structure chosen.

4.2 Accountability

A mechanism of accountability is needed to ensure that a structure which performs badly in the environment is less likely to be selected at the next vote. Two ways in which this can be achieved are now discussed.

If the set of structures remains constant and the one selected performs 'badly' in the environment (with regard to the local fitness functions) then its likelihood of reselection in a subsequent vote needs to be reduced. The values of the weights of individuals ($\mu_j(K)$s) are now the only variables in deciding which structure is chosen. By using information from the local fitness functions the weights can be altered so that the rating of a structure which is performing badly is likely to be reduced.

If the weights of each individuals remain the same throughout the interval between votes then it is the structures that must change. The structure which is currently selected will be re-selected unless others can alter their policies so that their ranking and hence chance of selection is increased. One means by which this may be carried out is for a structure to take a poll of a set of individuals and ascertain their views on and weights assigned to a particular policy. Then by adapting its own policies accordingly, a structure may increase its chances of success.

In practice some combination of the above two strategies can be expected to be required.

4.3 Generalisations

For simplicity of explanation it has been assumed that the set of policies is linearly independent. In practice this will often not be the case and much interaction can be expected. In particular effects such as pleiotropy (a single policy affecting several aspects of the environment) and polygeny (a single aspect of the environment being affected by the interaction of many policies) can be expected.

The weights assigned to each individual can not be expected to remain constant as the environment changes. As well as some random variation, more structured changes to individuals can be expected as global aspects of the environment become fixed (e.g., when all structures agree on some policy) and individuals concentrate their effort more on minor aspects of the environment. In implementing such a situation an individual would need to have its weights normalised to ensure that small concerns are not swamped.

The model discussed above does not allow for the environment to act directly on the structures. In some applications it may be necessary to allow for this in 'extreme' cases. For example, a policy may result in a crisis within the environment which if left unchecked would result in the entire environment being wrecked by the time of the next vote. In such cases the selected structure may be allowed to change policy immediately. More generally a structure's policies may be allowed to adapt between votes. This may be carried out via polls of individuals

Finally the means of voting discussed above is very simple, with the most popular structure being chosen. In practice it is possible to introduce other forms of voting. If for example, a single transferable vote system were to be used then an individual would be allowed to rank the three structures which it rates the highest. Another alternative procedure would be to split the set of individuals into smaller groups each of which selects a structure. The most popular of this set is then chosen to act within the environment.

5 Applications to Symbolic Mathematical Computation

In applying adaptive techniques to problems from symbolic computation the following need to be considered:

1. A solution encoding — Possible solutions need to be represented in terms of some (possibly subsymbolic) components.
2. A fitness function — A means of determining the performance of solutions and/or parts of solutions. This need not be quantitative. A qualitative measure, which determines the best of two given solutions, may be sufficient.
3. A means of altering the encoding structure — In order for new solutions to be considered, a means of generating them is required. New solutions are often produced as variations of those which have already been evaluated.

An example of how the adaptive techniques described in this paper may be useful to an area of symbolic mathematical computation is now briefly discussed. Consider a symbolic integration problem in which some function is given and the aim is to find a function which when differentiated is the given function. Possible solutions could be encoded in terms of combinations of primitive functions, e.g., sin, cos, tanh. To assign a fitness to a solution it may be differentiated, and the result compared in some way to that which is required. This may be numeric or symbolic differentiation and comparison. In generating new solutions from those which have been evaluated there are many possible methods which could be used, including: recombination (GA), and mutation (EP).

6 Conclusion

In attempting to solve some AI problems an approach which operates at a level below that of traditional logical-symbolic approaches is needed. Adaptive algorithms such as those inspired by natural evolution are one approach by which subsymbolic processing can be performed. This paper has reviewed several forms of adaptive algorithm, and a novel form of adaptive algorithm based on the concept of a voting system has been introduced. However, further research is needed before this algorithm can reach the level of precision given to descriptions of other forms of adaptive algorithm.

References

1. Belaw, R.K., Booker, L.B.: (eds.) Proc. Fourth Int. Conf. Genetic Algorithms Morgan Kaufmann (1991).
2. Davis, T.E., Principe, J.C.: A Markov chain framework for the simple genetic algorithm. J. Evol. Comput. 1 (1993) 269–288.
3. Fogel, D.B.: System identification through simulated evolution: A machine learning approach to modeling. Ginn Press (1991).
4. fogel, D.B.: Evolving artificial intelligence. PhD Thesis, University of California, San Diego, USA (1992).
5. Fogel, D.B.: Evolving behaviours in the iterated prisoner's dilemma. J. Evol. Comput. 1 (1993) 77–97.
6. Fogel, L.J., Owens, A.J., Walsh, M.J.: Artificial intelligence through simulated evolution. J. Wiley, New York (1966).
7. Forrest, S., Mitchell, M.: What makes a problem hard for a genetic algorithm? Some anomalous results and their explanation. Machine Learning 13 (1993) 285–319.
8. Goldberg, D.E.: Genetic Algorithms in Search, Optimization, and Machine Learning. Addison-Wesley (1989).
9. Grefenstette, J.J.: (ed.) Proc. First Int. Conf. Genetic Algorithms Lawrence Erlbaum Associates (1985).
10. Grefenstette, J.J.: (ed.) Proc. Second Int. Conf. on Genetic Algorithms lawrence Erlbaum Associates (1987).
11. Holland, J.H.: Adaption in natural and artificial systems. University of Mitchigan Press (1975).
12. Jones, T.: A model of landscapes. Santa Fe Institute, 1660 Old Pecos Trail, Suite A., Santa Fe, NM 87505, USA (1994).
13. Kirkpatrick, S., Gelatt, C.D., Vecchi, M.P.: Optimization by simulated annealing. Science 220 (1983) 671–680.
14. Lindsay, R.K.: Artificial evolution of intelligence. Contem. Psych. 13 (1968) 113–116.
15. Mason, A.J.: Crossover non-linearity ratios and the genetic algorithm: Escaping the blinkers of schema processing and intrinsic parallelism. Report No. 535b, School of Engineering, University of Auckland, Private Bag 92019, New Zealand (1993).
16. Nettleton, D.J., Garigliano, R.: Evolutionary algorithms and a fractal inverse problem. BioSystems (to appear).
17. Radcliffe, N.J.: Forma analysis and random respectful recombination. Proc. Fourth Int. Conf. on Genetic Algorithms Morgan Kaufman (1991).
18. Solomonoff, R.J.: Some recent work in artificial intelligence. Proc. IEEE 54 (1966) 1687–1697.

An Interpretation of the Propositional Boolean Algebra as a k-algebra. Effective Calculus

Luis M. Laita[1], Luis de Ledesma[1],
Eugenio Roanes-Lozano[2] and Eugenio Roanes-Macías[2]

[1] Universidad Politécnica de Madrid, Dpto. I.A. (Fac. Informática)
Campus de Montegancedo, 28660-Boadilla del Monte (Madrid), Spain
[2] Universidad Complutense de Madrid, Sec. Dptal. Algebra (Fac. Educación)
c/ Santísima Trinidad 37, 28010-Madrid, Spain

Abstract. We construct in the first part of the paper a Boolean algebra, isomorphic to a propositional Boolean algebra $(\mathcal{C}, \vee, \wedge, \neg, \rightarrow)$, that is also a k-algebra, and such that the ideals of the Boolean algebra correspond exactly to the ideals of the k-algebra.

An implementation on a Computer Algebra System is given in the second part of the paper. This implementation makes possible, for instance, to compare propositions, or to calculate a minimal base of an ideal or filter.

The use of well-known techniques from Commutative Algebra and Computer Algebra allows us to decide problems such as the ideal membership (in this particular k-algebra) with methods of lower complexity than calculating Gröbner Basis.

Besides, following our approach, results in the k-algebra can be directly translated into the Boolean algebra.

We also show how we try to apply this interpretation to verification of Knowledge Basis.

1   Introduction

The following equivalence between a Boolean algebra and a Boolean ring is well known:

i) Given a Boolean Algebra $(\mathcal{C}, \vee, \wedge)$, if the symmetric difference of two elements P,Q in $\mathcal{C}$ is: $P \triangle Q = (P \wedge \neg Q) \vee (Q \wedge \neg P)$, then $(\mathcal{C}, \triangle, \wedge)$ is a ring (called a Boolean Ring) such that every element is idempotent with respect to $\triangle$.

ii) If in the Boolean Ring $(\mathcal{C}, \triangle, \wedge)$ an operation "$\cup$" (union) is defined as: $P \cup Q = P \triangle Q \triangle (P \wedge Q)$, then $(\mathcal{C}, \cup, \wedge)$ is a Boolean Algebra.

iii) The Boolean Algebra $(\mathcal{C}, \cup, \wedge)$ is isomorphic to $(\mathcal{C}, \vee, \wedge)$.

Instead of considering this equivalence, we shall construct a k-algebra, to be denoted $(\mathcal{A}, +, \cdot)$, that will be shown to be isomorphic in a different way to $(\mathcal{C}, \triangle, \wedge)$.

For the sake of brevity, most proofs will be omitted.

2 The Ring $(\mathcal{A}, +, \cdot)$ and the Boolean Algebra $(\mathcal{A}, \tilde{+}, \cdot, \mathbf{1}+)$

Let $\mathcal{A}$ be a Boolean ring. It is known that for $a \in \mathcal{A}$, $a + a = 2a = 0$, so that $\mathcal{A}$ is a ring of characteristic 2, and the elements $a \in \mathcal{A}$ have order ≤ 2. These two conditions are verified by the k-algebra $\mathcal{A} = k[x, y, ..., z]$ over the base field $k = \mathbb{Z}/2\mathbb{Z}$ under the conditions $x^2 = x, y^2 = y, ..., z^2 = z$.

2.1 The Ring $(\mathcal{A}, +, \cdot)$

Let $R = k[x, y, ..., z]$ be a polynomial ring over the field $k = \mathbb{Z}/2\mathbb{Z}$ and I the ideal $< x^2 - x, y^2 - y, ..., z^2 - z >$ of R. We shall denote by $\mathcal{A}$ the quotient ring R/I, i.e. $\mathcal{A} = (\mathbb{Z}/2\mathbb{Z})[x, y, ..., z]/< x^2 - x, y^2 - y, ..., z^2 - z >$.

Proposition 1. *Every element of $\mathcal{A}$ is its own opposite.*

Proposition 2. *Every element in $\mathcal{A}$ is idempotent with respect to the product.*

Proof. If $a \in \mathcal{A}$, it can be expressed

$$a = \delta_0 + \delta_x x + \delta_y y + ... + \delta_z z + \delta_{xy} xy + ... + \delta_{xyz} xyz + ...$$

where all the δ belong to $\mathbb{Z}/2\mathbb{Z}$. As in the quotient ring R/I all the variables $x, y, ..., z$ are idempotent

$$a^2 = \delta_0^2 + \delta_x^2 x^2 + \delta_y^2 y^2 + ... + \delta_z^2 z^2 + \delta_{xy}^2 x^2 y^2 + ... + \delta_{xyz}^2 x^2 y^2 z^2 + ... = a \ .$$

Proposition 3. *Every element of $\mathcal{A}$ is a zero-divisor.*

Proof. If $a \in \mathcal{A}$, $a(1 + a) = 0$.

2.2 The Boolean Algebra $(\mathcal{A}, \tilde{+}, \cdot, \mathbf{1}+)$

Definition 4. We shall define a new operation $\tilde{+}$ in $\mathcal{A}$

$$\text{if } a, b \in \mathcal{A}, \text{ then: } a \tilde{+} b = a + b - ab \ .$$

Corollary 5. *If $a, b \in \mathcal{A}$: $a + b = (1 + a)b \tilde{+} a(1 + b)$.*

Remark. We shall give product a higher precedence than $\tilde{+}$.

Proposition 6. $(\mathcal{A}, \tilde{+}, \cdot, \mathbf{1}+)$ *is a Boolean algebra (1 is the infimum, 0 is the supremum and the complement of $a \in \mathcal{A}$ is $1 + a$). Consequently, the De Morgan laws are verified.*

3 Ordering the Boolean Algebra $(\mathcal{A},\tilde{+},\cdot,1+)$

3.1 Defining an Ordering from the Operations

Definition 7. Let $\leq$ be an ordering in the Boolean algebra $(\mathcal{A},\tilde{+},\cdot,1+)$ defined in the usual way: $\forall a,b \in \mathcal{A},\ a \leq b \Leftrightarrow a \cdot b = a$.

Proposition 8. *In any Boolean algebra, in particular in* $(\mathcal{A},\tilde{+},\cdot)$

$$a\tilde{+}b = b \Leftrightarrow a \cdot b = a \Leftrightarrow (1+a)\tilde{+}b = 1 \Leftrightarrow a \cdot (1+b) = 0 .$$

Proposition 9. *The ordering defined above is precisely "is a multiple".*

Corollary 10. *For any* $a,b \in \mathcal{A}$: $a,b|(a \cdot b)$; $(a\tilde{+}b)|a,b$.

Proposition 11. *In* $(\mathcal{C},\vee,\wedge,\neg)$ *the order relation defined as above corresponds to "implies".*

Proposition 12. $b|a \Leftrightarrow (1+a)|(1+b)$ *(that corresponds to* $A \rightarrow B \Leftrightarrow \neg B \rightarrow \neg A$ *in the propositional Boolean algebra).*

4 The Isomorphism φ

4.1 Constructing the Isomorphism φ

Definition 13. Let $(\mathcal{C},\vee,\wedge,\neg,\rightarrow)$ be a propositional algebra given by the propositional variables $P,Q,...,R$ for $\underline{0}$ and $\underline{1}$ respectively denoting the contradiction and the tautology. Consider the k-algebra $(\mathcal{A},\tilde{+},\cdot,1+,\text{"is a multiple"})$, where

$$\mathcal{A} = (\mathbb{Z}/2\mathbb{Z})[p,q,...,r]/<p^2-p,q^2-q,...,r^2-r> .$$

We define

$$\varphi: (\mathcal{C},\vee,\wedge,\neg,\rightarrow) \longrightarrow (\mathcal{A},\tilde{+},\cdot,1+,\text{"is a multiple"})$$

in the following way: $\varphi(P) = p,\ \varphi(Q) = q,...,\varphi(R) = r$ and for any $A,B \in \mathcal{A}$: $\varphi(A \vee B) = a\tilde{+}b$, and $\varphi(\neg A) = 1+a$. As a consequence of the De Morgan laws, $\varphi(A \wedge B) = ab$.

Theorem 14. φ *is an homomorphism.*

Corollary 15. $\varphi(\underline{1}) = 1,\ \varphi(\underline{0}) = 0$ *and* φ *preserves ordering.*

Proposition 16. φ *is surjective.*

Proposition 17. φ *is injective (it follows from the antisimmetry of the ordering).*

Remark. To be precise, we consider $\mathcal{C}/\leftrightarrow$ and $\mathcal{A}/=$ in order the relations $\rightarrow$ and $\leq$ to be antisimmetric.

4.2 Examples

Example 1. If the propositional variables of $\mathcal{C}$ are P y Q (and their negations are: $\neg P$ y $\neg Q$) then $\mathcal{C}$ has 16 elements and $(\mathcal{C}, \rightarrow)$ is the transitive closure of

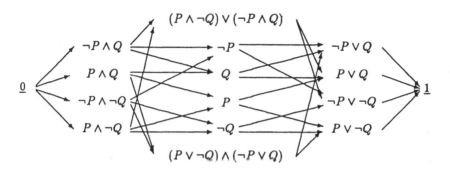

The image in φ is $(\mathcal{A},$"is a multiple"$)$, being

$$\mathcal{A} = (\mathbb{Z}/2\mathbb{Z})[p,q]/<p^2 - p, q^2 - q> \quad .$$

As the elements of $\mathcal{A}$ can be expressed this way: $\delta_0 + \delta_p p + \delta_q q + \delta_{pq} pq$ it follows that there will be $2 \cdot 2 \cdot 2 \cdot 2 = 16$ such elements. The relation "is a multiple" will be given by the transitive closure of the diagram

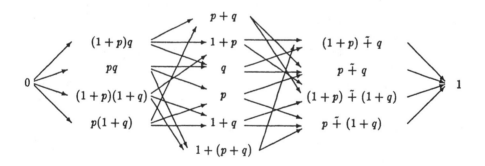

5   More About the Ordering

5.1   Results

Proposition 18. *For any* $a, b, c \in \mathcal{A}$
   $c|a \Rightarrow c|(a \cdot b) \quad ; \quad c|a$ *and* $c|b \Rightarrow c|(a \tilde{+} b) \,$.

Proposition 19. *For any* $a, b, c \in \mathcal{A}$
   $b|a \Rightarrow (b \tilde{+} c)|a \quad ; \quad b|a$ *and* $c|a \Rightarrow (b \cdot c)|a \,$.

5.2 Atoms and Co-atoms

Definition 20. Given $B \in C$, $B \neq \underline{0}$, B is said to be an atom iff B is a minimal element in the ordering.
Given $D \in C$, $D \neq \underline{1}$, D is said to be a co-atom iff D is a maximal element in the ordering.

Proposition 21. *In A the co-atoms for "is a multiple" are the irreducible.*

Proposition 22. *For any $B \in C$, B is an atom $\Leftrightarrow \neg B$ is a co-atom.*
For any $b \in A$, b is an atom $\Leftrightarrow 1 + b$ is a co-atom.

Proposition 23. *For $A = (\mathbb{Z}/2\mathbb{Z})[p, q, ..., r]/< p^2 - p, q^2 - q, ..., r^2 - r >$ the atoms of $(A, "is a multiple")$ can be written*

$$m = [\delta_p \cdot p + (1 - \delta_p)(1 + p)] \cdot [\delta_q \cdot q + (1 - \delta_q)(1 + q)] \cdot .... \cdot [\delta_r \cdot r + (1 - \delta_r)(1 + r)]$$

where $\delta_i \in (\mathbb{Z}/2\mathbb{Z})$, i.e., $\delta_i = 0$ or $\delta_i = 1$ (a dual expression could be obtained for co-atoms). Therefore, atoms of A are polynomials of maximum degree.

Proof. i) Lets suppose that m can be written that way. If $m|a \Leftrightarrow \exists l \in A : a = l \cdot m$ and, as $l \in A$, l can be written

$$l = \alpha_0 + \alpha_p \cdot p + \alpha_q \cdot q + ... + \alpha_{pq} \cdot pq + ... + \alpha_{pqr} \cdot pqr + ...$$

From idempotency and distributivity, it follows that $a = l \cdot m$ is the result of adding monomials that are m or 0 and therefore $a = m$ *or* $a = 0$.
ii) If m is an atom of $(A, "is a multiple")$ it should contain at least one monomial of the type shown in the proposition (if such was not the case, we could multiply the polynomial by one variable such that neither itself nor its negation appeared in any of the monomials of maximum degree of the polynomial).
So, m can be written $m = h \tilde{+} g$ where h is of the form shown in the proposition. But then $m|h$ and therefore $h = m$ (in order m to be minimal).

Corollary 24. *If there are n variables in the k-algebra A, there are 2^n atoms in $(A, |)$. The same will occur in C if it is generated from n propositional variables (and their negations).*

6 Ideals and Filters

6.1 Definitions

Definition 25. The (principal) ideal of the Boolean algebra $(C, \vee, \wedge, \neg, \rightarrow)$ generated by $Q \in C$ is $E_Q = \{X \in C : X \rightarrow Q\}$.
The (principal) filter generated by $P \in C$ is $E^P = \{X \in C : P \rightarrow X\}$.

Proposition 26. *The ideals and filters of C correspond by φ with the ideals and filters of the Boolean algebra A, and these are exactly the same as the ideals of the k-algebra A.*

6.2 Results

Proposition 27. *If the ideal of $\mathcal{A}$ generated by p intersects the filter generated by $1 + p$ then both are the whole ring, i.e.*

$$E_p \cap E^{(1+p)} \neq \emptyset \Leftrightarrow E_p = E^{(1+p)} = \mathcal{A} \ .$$

Proof. $\Leftarrow$) Is clear.
$\Rightarrow$) In this case there exists $a \in E_p \cap E^{(1+p)}$ and thus $(1 + p)|a|p$. Therefore $(1 + p) \in <p> \Rightarrow 1 \in <p>$ (respectively, $0 \in E^{(1+p)}$).

Theorem 28. $(\mathcal{A}, +, \cdot)$ *is a principal ideal ring. Dually, every filter is principal.*

Proof. It is easy to proof that the ideal generated by $S = \{s_1, s_2, ..., s_n\} \subseteq \mathcal{A}$, is also generated by $s_1 \tilde{+} s_2 \tilde{+} ... \tilde{+} s_n$. Dually, the filter generated by S is generated by $s_1 \cdot s_2 \cdot ... \cdot s_n$.

Remark. In this special kind of ring it is clear how to compute a Gröbner Basis of an ideal should be much more expensive than calculating the element that generates the ideal in the way suggested in Theorem 28.

Theorem 29. $\mathcal{A}$ *is an unique factorization ring (but it follows from Proposition 3 that this ring is not a UFD).*

Proof. Let a be any element in $\mathcal{A}$, and consider the filter E^a. Then a is the product of all the co-atoms (irreducible) of E^a, and this is the only way we can express a as a product of irreducible elements.

Proposition 30. $\mathcal{A}$ *is an Artinian ring ($\mathcal{A}$ is finite).*

7 Implementation in REDUCE

7.1 The Environment

Let us translate this interpretation of the Propositional Calculus as a k-algebra over the base field $\mathbb{Z}/2\mathbb{Z}$ into a Computer Algebra System (REDUCE). The prompt will be represented with "◇ :" to distinguish the inputs from the outputs.

We shall begin implementing the one-to-one correspondence between the Boolean algebra and the k-algebra. The operations in the propositional algebra will be infix operators (y,o) and we shall respectively describe tautology and contradiction as 1 and 0.

The logical operators "and" and "or" are respectively named "y" and "o" to avoid conflicts with already existing REDUCE commands.

```
◇ : operator neg,F,G;
◇ : infix y,o;
◇ : for all F,G let (F y G) = f*g;
◇ : for all F,G let (F o G) = f+g-f*g;
◇ : for all F let neg(F) = 1+f;
```

This k-algebra is a class ring over an ideal that is generated by monic univariate polynomials. That's why the canonical form of a polynomial in the k-algebra can be calculated using the remainder function. In the case of REDUCE it is enough to implement it as follows (although it is possible to do it explicitly using remainder too).

If the literals of the Boolean algebra are for instance P, Q, R, we work in $(\mathbb{Z}/2\mathbb{Z})[p, q, r]/<p^2 - p, q^2 - q, r^2 - r>$, and so, we add the rules

$\diamond$: p²-p:=0;
$\diamond$: q²-q:=0;
$\diamond$: r²-r:=0;

As we proved that in this k-algebra every element is idempotent (with respect to the product), we add

$\diamond$: for all f let f²-f = 0;

As the base field was of characteristic 2, we order REDUCE to change to a modular arithmetic (modulo 2):

$\diamond$: on modular;
$\diamond$: setmod(2);

Now an implementation of φ is already available. For example, the image of $P \vee (Q \vee R)$ will be computed just typing

$\diamond$: P o (Q o R);

and the output is (although REDUCE uses only capital letters we will preserve our notation in this paper)

r*p*q + r*p + r*q + r + p*q + p + q

7.2 Equivalence of Propositions

Propositions can be compared immediately. Two propositions will be equivalent iff there images by φ are equal. As in the final set the computation is made in modular arithmetic, a simple procedure can decide the equivalence

```
$\diamond$ : procedure equivalent(F,G);
begin
    if f-g = 0 then write("They are equivalent")
        else write("They are not equivalent");
end;
```

Lets prove automatically that $(\mathcal{C}, \vee, \wedge, \neg)$ is a Boolean algebra (as the image of $\neg A$ is $1 + a$, the complement always exists)

◇ : equivalent(A y B , B y A);
◇ : equivalent(A o B , B o A);
◇ : equivalent(A o (B o C) , (A o B) o C);
◇ : equivalent(A y (B y C) , (A y B) y C);
◇ : equivalent(A y (B o A) , A);
◇ : equivalent(A o (B y A) , A);
◇ : equivalent(A y (B o C) , (A y B) o (A y C));
◇ : equivalent(A o (B y C) , (A o B) y (A o C));

The answer is in all cases:

They are equivalent

As we could expect, the same answer is obtained if we test if the De Morgan laws holds, just by typing

◇ : equivalent(neg(A y B) , neg(A) o neg(B));
◇ : equivalent(neg(A o B) , neg(A) y neg(B));

7.3   Computation of Bases of Ideals and Filters

In the isomorphism φ, the images of propositional expressions are similar to those in the MOD2 form (an implementation can be found in the *"Logic"* package of Maple V.2). But our implementation differs from that one because we directly work in the quotient ring described in Sect. 2. The construction of the isomorphism provides a theoretical background that allows us to deal easily with structures (such as ideals and filters).

Moreover, the ideals and filters are all principal, and consequently, problems such as the ideal membership can be decided avoiding the use of tools of a high complexity (such as Gröbner Basis). The most arduous part of the work needed in our approach is to expand the polynomials completely, so the worse-case complexity of our calculations is exponential on the degree of the polynomial.

Using Theorem 28, it is easy to implement a procedure such that, given the elements that generate an ideal, obtains a single element that also generates the ideal. A possible implementation appears below (S is a list of elements that generate the ideal).

```
◇ : procedure baseideal(S);
begin
    scalar gen;
    off modular;
    gen := 0;
    for i := 1 : length(S) do gen := gen o part(S,i);
    on modular;
    setmod(2);
    return gen;
end;
```

The procedure basefilter(S) can be obtained from the former substituting gen := 0 by gen := 1 and o by y.

8   An Application to A.I.

We are trying to apply this theoretical interpretation to study verification of Knowledge Basis. The idea is that we can translate the facts and rules added by the expert into polynomial conditions and then check their compatibility.

Example 2. Lets suppose that the expert adds the rule: $A \rightarrow B$ and the fact: $A \wedge \neg B \wedge C$ simultaneously. Then we have:

$$A \wedge \neg B \leftrightarrow \underline{0} \quad ; \quad A \wedge \neg B \wedge C \leftrightarrow \underline{1} .$$

what is obviously a contradiction.

From our polynomial point of view, we have:

$$a \cdot (1 + b) = 0 \quad ; \quad a \cdot (1 + b) \cdot c = 1 \Leftrightarrow a \cdot (1 + b) \cdot c - 1 = 0 .$$

We can translate adding rules and facts as moving from our k-algebra A (that is already a class ring) into A/J, where J is an ideal generated by the polynomial translations of the rules and facts. In our example:

$$J = <a \cdot (1 + b), a \cdot (1 + b) \cdot c + 1> = <1> = A .$$

(how to calculate the element that generates J follows from Theorem 28).

The contradiction in the propositional k-algebra is translated in this case into the degeneracy of the ideal J, that has become the whole ring. Consequently, the class ring A/J has collapsed into $\{0\}$. So, $1 = 0$, what is somehow a contradiction in the polynomial k-algebra (as it is a class ring constructed from a polynomial ring whose base field is $\mathbb{Z}/2\mathbb{Z}$).

References

[Ha]   Halmos, P.R. Lectures on Boolean Algebras. Springer-Verlag (1974)

[He]   Hearn, A.C.: REDUCE User's Manual v.3.3. Rand Pub. (1987)

[Hr]   Hermes, H.: La teoría de retículos y su aplicación a la lógica mateática. Conf. Mat. VI, CSIC-Madrid (1963)

[Ch]   Char, B.W. et al.: Maple V. Library Reference Manual. Springer-Verlag (1991)

[Mo]   Monk, D.: Handbook of Boolean Algebras. North-Holland (1989)

[MW]   McCallum, M., Wright, F.: Algebraic Computing with REDUCE. Oxford Univ. Press (1991)

[Ra]   Rayna, G.: REDUCE. Software for Algebraic Computation. Springer-Verlag (1987)

Subdefinite Computations and Symbolic Transformations in the UniCalc Solver

Alexander Semenov, Alexander Babichev, Alexander Leshchenko

Novosibirsk Division of the Russian
Research Institute of Artificial Intelligence
pr. Lavrent'eva 6, Novosibirsk, Russia, 630090
e-mail: semenov@isi.itfs.nsk.su

Abstract. In the present article, we consider the method of subdefinite computations, which can be regarded as a branch of constraint propagation, and describe how the techniques of computer algebra can be applied to improve the efficiency of the method. The article also presents the UniCalc problem solver which is based on the method of subdefinite computations. In conclusion, we present some examples that demonstrate the capabilities of the solver and the results of using symbolic transformations.

1 Introduction

The method of *constraint propagation* (CP) is widely used at present, and there are numerous applications based on this method [Ku1]. Most problem-solving algorithms in this method are limited to treating a finite set of values; they usually produce a single set of values that satisfy the constraints given. This limitation is usually due to the use of the most common solution technique – search with backtracking.

However, problems often arise in practice in which the variables can assume continuous values, possibly from an infinite set, and the result should contain all feasible solutions of the system. The solve such problems, several approaches within CP exist; one of them is the method using interval labels [Da1, Hy1]. The essence of this method is to use intervals for the definition of the domains of values for the variables and apply interval mathematics to perform operations on the variables. The computation mechanism itself, in case interval extensions are used, can be different, which can produce different solutions (local, semiglobal, and global).

In the context of research on knowledge representation and processing, carried out in the Novosibirsk division of the Russian Research Institute of Artificial Intelligence, the method of so-called *subdefinite computations* was proposed [Na1, Na3]. This method can be regarded as an analog of CP with interval labels, although we believe that it is broader, since it enables one to treat simultaneously continuous and discrete domains, allows other representations of imprecise objects besides intervals, and can be extended with new "subdefinite" objects. With the use of the apparatus of subdefinite computations, several application systems were built in the Institute; these include Time-Ex [Bo1], a time scheduling system, and UniCalc [Un1], a solver of systems of algebraic equations. While experimenting with UniCalc, we have discovered some properties of the approach in question that decrease the efficiency of its use. The properties are related both to the nature of the algorithm and to the use of interval computations. To overcome these drawbacks, we applied some methods of computer algebra. This paper attempts to describe the method of subdefinite computations, the UniCalc solver, and the techniques of computer algebra that were used.

The general plan of the paper is as follows. Chapters 2 and 3 describe the method of subdefinite computations and some issues of its practical implementation. In Chap. 4, we describe the UniCalc solver, and in Chap. 5, the symbolic preprocessor that performs symbolic transformations in UniCalc. In Chap. 6, we supply some examples of applying symbolic transformations to the solution of problems with UniCalc. The more technical information and the directions for future research will be treated in Chap. 7.

2 The Algorithm

2.1 Concepts and Designations

Let a calculation model M be determined by the set of variables X and the set of relations over these variables R. We shall denote the model by $M = (X, R)$. Let A be the value domain of the model's variables, and let $^*A$ denote the set of all nonempty subsets of the set A. The values $^*a \in {}^*A$ containing one member only are called *precise*, while other values are called *subdefinite*. The value $^*a$ corresponding to the entire set A is *the fully indefinite value*. Let us map uniquely each variable x in the set X into a variable $^*x$ whose value domain is the set $^*A$. This maps the set of variables X onto a set $^*X$ of variables $^*x$. The variables $^*x$ will be called *subdefinite variables*.

The subdefinite description of the model M is a set (M, h), where $h = ({}^*a_1, \ldots, {}^*a_n)$ is a vector of subdefinite values. Narin'yani showed in [Na1, Na3] that for subdefinite descriptions it is possible to construct a finite automaton that generates finite sequences of states. The sequences are terminated either with the 'end' state, or the 'conflict' state. In the first case, the automaton produces a vector h which is an n-dimensional parallelepiped in the space A^n containing the set of the values of the variables X that satisfy the model M.

The process of inference/calculations over models usually involves interpreting each relation with a help of a set of functions allowing us to find the values of other variables from known values of some variables.

Since the variables $^*X$ rather than X are used while working with subdefinite descriptions, it is necessary to define operations over subdefinite values and subdefinite functions. Narin'yani showed [Na1] that if we map each m-ary operation s over values from A into an operation $^*s$ over undefined values from $^*A$ according to the formula

$$^*s({}^*a_1, \ldots, {}^*a_m) = \{a = s(b) \mid b \in {}^*a_1 \times {}^*a_2 \times \ldots \times {}^*a_m\}, \qquad (1)$$

and if each m-ary function f of variables in X is mapped into a function $^*f$ of variables in $^*X$ by the formula

$$^*f(^*x_1, \ldots, ^*x_m) = \{x = f(b) \mid b \in {^*x_1} \times {^*x_2} \times \ldots \times {^*x_m} \cap R\}, \qquad (2)$$

where R is an additional relation connecting the variables $x_1, \ldots, x_m$ (for uncorrelated variables $R = {^*A^m}$), then it is possible to use the inference/calculation apparatus for the initial models for inference/calculations on subdefinite models.

2.2 The Calculation Algorithm for Subdefinite Models

Suppose we have a model with a subdefinite description (in [Na3], such models are called *GCM, generalized calculation models*). According to Sect. 2.1, the inference/calculation procedure for such models may be described as a procedure of calculating the interpretation functions of subdefinite variables. If a certain relation of the initial model relating variables $x_1, \ldots, x_m$ is interpreted by a set of the following functions:

$$x_i := f_i(x_1, \ldots, x_{i-1}, x_{i+1}, \ldots, x_m), \ i = 1, \ldots, m \qquad (3)$$

then, using (2), the corresponding interpretation functions for the subdefinite model may be defined as follows:

$$^*x_i := {^*f_i}(^*x_1, \ldots, ^*x_{i-1}, ^*x_{i+1}, \ldots, ^*x_m) \cap {^*a_i}, \ i = 1, \ldots, m \qquad (4)$$

where $^*a_i$ is the current value of $^*x_i$.

The inference/calculation process formulated in [Na3] for GCM may be presented in the form of the following algorithm (let $^*X_t$ denote the set of the variables in $^*X$ whose values have become more precise at the step t, and let $F \mid {^*X_t}$ denote the set of interpretation functions having at least one argument which belongs to $^*X_t$):

Step 1. $t = 1$; $^*X_t = \{ ^*x_1, \ldots, ^*x_n \}$. (Note that all model variables have values: they are either initialized or their value is fully indefinite, i.e., the set A.)

All the model interpretation functions are placed in set $F \mid {^*X_1}$ and all the members of this set form the set of active functions.

Step 2. Combine the function set $F \mid {^*X_t}$ with the active interpretation functions.

Step 3. Choose an arbitrary subset in the produced set. This subset is joined to the working function set and removed from the active function set.

Step 4. Remove an arbitrary subset of functions from the working function set. The functions of the subset are calculated at this step. The results of these calculations are compared with the values of subdefinite variables in the common memory, and variables that have changed their values form the set $^*X_{t+1}$ defining set the $F \mid {^*X_{t+1}}$.

If a result of calculating some interpretation function yields the value of a variable that is equal to the empty set the model is incompatible. The algorithm terminates.

Step 5. If the set of active functions, the working function set and the set $*X_{t+1}$ are simultaneously empty, go to Step 7.

Step 6. $t = t + 1$ and go to step 2.

Step 7. Output the values in the set $*X = \{ *x_1, \ldots, *x_n \}$.

This algorithm has the following features:

- The model can be underdetermined or overdetermined and the parameters of the model may be imprecise or unknown;
- No distinction is made between the model's arguments and results: all variables have values that are subdefinite in various degrees; the calculations are performed for all variables of the model;
- This algorithm determines a parallel, asynchronous, undetermined process with flow data-driven control;
- According to (4), subdefiniteness of the variables never increases and the calculation is converging. For all models having only finite subdefinite values, this procedure terminates in a finite number of steps.

Consider an example. Let the model be specified by two equations:

$$8a - b = 15, \quad 6b + 4c = 230$$

and suppose that it is known that a, b, and c are integers in the range from 1 to 100. Hence, the subdefinite value of each of these variables can be represented by a set of integers with the initial state $A = \{1, \ldots, 100\}$. This relations are interpreted by the following functions:

$$f_1 : a := (b + 15)/8;$$
$$f_2 : b := 8a - 15;$$
$$f_3 : b := (230 - 4c)/6;$$
$$f_4 : c := (230 - 6b)/4.$$

Introducing the subdefinite description, we have:

$$*f_1 : *a := \{a = (b + 15)/8 \mid b \in *b\} \cap *a;$$
$$*f_2 : *b := \{b = 8a - 15 \mid a \in *a\} \cap *b;$$
$$*f_3 : *b := \{b = (230 - 4c)/6 \mid c \in *c\} \cap *b;$$
$$*f_4 : *c := \{c = (230 - 6b)/4 \mid b \in *b\} \cap *c.$$

In compliance with the above algorithm, we obtain

$$*X_1 = \{*a = \{1, \ldots, 100\}, *b = \{1, \ldots, 100\}, *c = \{1, \ldots, 100\}\};$$
$$F \mid *X_1 = \{*f_1, *f_2, *f_3, *f_4\}.$$

The Table 1 illustrates the algorithm's execution:
From this table we get the following result:

$$\{*a = \{2, 3, 4, 5, 6\}, *b = \{1, 9, 17, 25, 33\}, *c = \{8, 20, 32, 44, 56\}\}.$$

Table 1. The execution of the algorithm

Subdefinite values	Working function	Active functions		
$^*a = \{2, 3,\ldots,14\}$	$^*f_1$	$^*f_2,$	$^*f_3,$	$^*f_4$
$^*b = \{1,9,\ldots,97\}$	$^*f_2$	$^*f_3,$	$^*f_4$	
$^*b = \{1,9,17,25,33\}$	$^*f_3$	$^*f_1,$	$^*f_4$	
$^*c = \{8,20,32,44,56\}$	$^*f_4$	$^*f_1,$	$^*f_3$	
$^*a = \{2,3,4,5,6\}$	$^*f_1$	$^*f_2,$	$^*f_3$	
$^*b = \{1,9,17,25,33\}$	$^*f_3$	$^*f_2$		
$^*b = \{1,9,17,25,33\}$	$^*f_2$			

To find an every separate solution one should combine the values from different sets and verify the system. In our case such solutions are:

$$\{2, 1, 56\}, \{3, 9, 44\}, \{4, 17, 32\}, \{5, 25, 20\}, \{6, 33, 8\}$$

3 Implementation of the Algorithm

3.1 Implementation of Subdefinite Numbers

The implementation of the algorithm just described assumes the choice of a set A corresponding to the class of models; in addition acceptable operations and functions over values in A and variables in X are to be extended for objects in $^*A$ and $^*X$. The active data type apparatus [Na2] has been proposed and used to implement various types of subdefinite data (integers, reals and booleans, sets, objects of planimetry, etc.) [Te1].

We consider only models over real numbers; therefore the set A is the field of real numbers $\mathbb{R}$ or the ring of integer numbers $\mathbb{Z}$. We assume that all the arithmetic operations as well as exponentiation and root extraction are defined over values in the set A. The set of acceptable functions includes exponential, logarithmic and all trigonometric functions. Since computer number representations are finite, the set A will be represented by finite intervals in $\mathbb{R}$ or $\mathbb{Z}$ which are bounded by some numbers $MinA$ and $MaxA$ playing the role of $-\infty$ and $+\infty$, respectively (these numbers differ for real and integer sets A). The operations on elements of A are defined as follows: if $a, b \in A$ and $\odot$ is a operator, then

$$a \odot b = \max(a \odot b, MinA) \text{ if } a \odot b < 0$$
$$a \odot b = \min(a \odot b, MaxA) \text{ if } a \odot b \geq 0$$

Assuming these conventions and considering $^*A$ as the set of all intervals in the set A, we can extend all operations on elements of A to operations on $^*A$ defined by (1), using the appropriate operations of interval mathematics. Note that any nonempty subset of set A may be embedded into some interval. Therefore, assuming that $^*A$ is the set of all intervals of A, we can only increase subdefiniteness of solutions without losing any of them. It was shown in [All] that interval continuous analytic functions are monotonic by inclusion, i.e., if f is a continuous analytic function of interval variables $X_1, \ldots, X_n$ and $X_1 \subseteq Y_1, \ldots, X_n \subseteq Y_n$, then $f(X_1, \ldots, X_n) \subseteq f(Y_1, \ldots, Y_n)$. In view of this property, to extend functions over variables from A, it is possible to use their interval extensions in the corresponding continuity domains, since the interval thus obtained will always include the set determined by (2).

3.2 Representation of Models

The most appropriate method for the algorithm described above is the network representation of models. Here, a network is represented as a bipartite oriented graph with two types of vertices: objects and operators. Objects represent model variables, and operators represent functional links between objects (interpretation functions). The outgoing edges point to the operators whose arguments are the corresponding objects, and incoming edges specify objects which provide values to these operators. A network is associated with a discipline of its execution, which substantiates the inference algorithm. This substantiation depends on the computer's architecture, its computational capabilities, as well as a number of other factors. For instance, sets may be selected at step 3 according to certain priorites, from a queue or randomly. Interpretation functions are computed at step 5 sequentially or in parallel, etc. In the current implementation of the processor, the interpretation functions are selected depending on their ordinal number and are computed sequentially. To perform computations on a functional network of the structure described, a virtual dataflow processor using the apparatus of active data types was developed.

4 The UniCalc Solver

To use the algorithm described above for practical purposes, we have developed the UniCalc solver whose nucleus is the flow processor considered in Sect. 3. The UniCalc solver was designed to solve arbitrary systems of algebraic and algebraic-differential relations. Here, a relation is considered to be an equation, inequality or a logical expression. According to the algorithm used, the system to be solved can be either overdetermined or underdetermined, and the system's parameters (coefficients, variables, initial conditions for the Cauchy problem) can be imprecise and expressed as intervals. Such a system may contain only integer and real variables or combine both integer and real variables.

As a result of solving algebraic systems, we either find a parallelepiped that contains all roots of the system, or a message about the system's incompatibility is issued. If the system has a single root, then the parallelepiped will be reduced to a point (with a given accuracy). If the system has several roots, to locate each of them it is necessary to add the appropriate relations, or use the built-in tool for automatic root locating.

The solver is an integrated environment supporting input of the system to be solved, its modification, calculations, viewing results, specifying accuracy, etc. To input and modify the system UniCalc has the built-in text editor. To write the problems, a source language close to the common used mathematical notation is provided. All problems are processed with the above algorithm. To translate the source input into a network, the solver includes a translator and preprocessors, in particular, for symbolic transformations and solution of systems of algebraic-differential relations.

The UniCalc's user interface offers a number of services to support problem solving. If solving problem requires large computation times, it is possible to suspend calculations to see the intermediate results, and depending on convergence rate, either to continue computation or to stop it. A feature for locating roots is used when exact solutions are in large intervals. This process involves dichotomic division of the obtained interval for the selected variable. This tool is also useful to find global function extreme values.

5 Symbolic Preprocessor

The main purpose of the preprocessor is to optimize the network and perform some types of symbolic manipulations.

5.1 Optimizing the network

As a rule, a real-world computation model contains a large number of expressions relating a set of variables. Distinct expressions in the model often contain identical subexpressions which are duplicated in the network; this leads to growth of the network and increases of computation time. Therefore, identifying common subexpressions and using a single copy of each in the network enables one to decrease the size of the network considerably and increase computation speed. To achieve this, when constructing the network for the dataflow processor, an intermediate translation of the input expressions to the internal representation of the symbolic preprocessor is performed. In this representation, the expressions are written as dynamic n-ary Kantorovich schemes [Ka1] which are stored in a hash table. The hashing key is built from a signature of the expression, which enables one to take account of the number of operands and commutativity of operations, significantly reducing the search and insertion times. This approach ensures that only one copy of each subexpression is stored, resulting in an optimal network. The latter objective is also achieve by reduction of similar terms in sums and products when the internal representation is being constructed.

All relations in the network usually considered in constraint propagation algorithms are assumed to be binary. However, the use of the intermediate representation that we propose creates a possibility for including n-ary relations in the network; in particular, this is valid for the operations of addition and multiplication. Since these operations are the most frequently used in the construction of expressions, our approach essentially reduces the number of temporal variables and the volume of extraneous computations.

5.2 Improving interval operations

As was noted above, the present version of the dataflow processor uses algorithms of interval mathematics to implement operations on subdefinite values. It was shown in [Al1] that intervals do not obey the distributivity law; also, the inverse element with respect to addition does not exist. As a consequence, computations may yield results that are significantly wider than the true ones. This can be improved by applying symbolic transformations. In particular, in some cases the resulting intervals can be made more narrow by just reduction of similar terms. For example, if an expression contains the product $2 \times x \times x - x^2$ and $x = (-2, 3)$, then a direct computation yields $(-21, 18)$, whereas the symbolic preprocessor will transform the expression to x^2, with the result $(0, 9)$.

More precise bounds for interval expressions can be produced by applying special algorithms. UniCalc implements one of such algorithms, proposed in [As1]. This algorithm finds intervals of monotonicity; to do this, it computes the partial derivatives for the expression in the symbolic form. Additional computation time is required to find the partial derivatives, but then the algorithm produces bounds that are almost exact.

5.3 Solution of systems of linear algebraic equations

In the investigation of the algorithm of subdefinite computations, we discovered that it is bad at solving linear systems of algebraic equations. Thus, if the matrix of the system is not with diagonal domination, then the resulting intervals will be almost infinite, and with the diagonal domination, the computation time can be extremely large. To eliminate this shortcoming of UniCalc when solving systems of linear equations, a symbolic change of variable is done, with simultaneous reduction of similar terms; as a result, we obtain a triangular system of equalities, which define some variables via others. Next, a network is constructed for the system, and the final computations are performed on the network. If the system in question is underdefined, the system of equations will have a trapezoid form, and interval solutions will be produced as the result. If the system is overdefined, the triangular system will be followed by a set of equalities for numbers, and so the inconsistency of some equations will be noticed when the change of variables is done.

5.4 Using symbolic differentiation

Presently, the user can do only two types of symbolic transformations – solution of systems of linear equations and differentiation. The user can invoke the differentiation function explicitly, including it in the statement of a problem, or implicitly, by defining computations that require differentiation. In the case of explicit invocation of the function, the required partial derivatives of the given function are calculated at compile time (the differentiation variables and their orders are the parameters of the function); next, reduction of similar terms is performed, and the resulting formulas replace the call of the function. Afterwards, the functional network for computations is constructed. Before constructing the network, the user can view and print the expressions produced after differentiation and simplification. Symbolic differentiation not only enables the user to study the behavior of a function and simplify the statement of the problem; it also makes it possible to solve optimization problems (since UniCalc does not contain explicit means for the solution of problems in this class). An example of solving optimization problems with help of symbolic differentiation is given in Chap. 6. An example of the implicit call to this function is the calculation of the exact bounds for interval functions, mentioned in 5.2.

6 Examples of application of symbolic transformations

To estimate the efficiency of the solver and the range of possible applications, we have considered quite a number of various problems. We have solved linear and nonlinear systems of equations and inequalities, mixed problems, various integer problems, optimizing interval problems, systems of differential equations, etc. Below, we consider three types of problems for which you can use symbolic transformations effectively.

SYSTEMS OF LINEAR ALGEBRAIC EQUATIONS. In what follows, we demonstrate the effect of applying special means for solution of systems of linear equations (see 5.3). Consider the following system of four linear equations in four variables.

```
 x1 + 2*x2 + 3*x3 + 4*x4 = 1;
4*x1 + 3*x2 +   x3 + 2*x4 = 1.3;
2*x1 +   x2 + 4*x3 + 3*x4 = 2;
3*x1 + 4*x2 + 2*x3 +   x4 = 0.7;
```

If we attempt to solve it with UniCalc's standard algorithm, we obtain a solution that is practically undefined:

```
x1 = [-2.5e+18, 2.5e+18];
x2 = [-2.5e+18, 2.5e+18];
x3 = [-2.5e+18, 2.5e+18];
x4 = [-2.5e+18, 2.5e+18];
```

On the other hand, by applying symbolic transformations, we easily obtain the exact solution:

```
x1 =  0.4625;
x2 = -0.3375;
x3 =  0.2875;
x4 =  0.0875;
```

OPTIMIZATION PROBLEMS. For continuous functions of real variables, the following mathematical statement of an optimization problem (here we consider both the problems of unconditional optimization and of constrained optimization) is possible: the first derivative of the function at the optimum point is zero, and the sign of the second derivative at the point depends on the kind of optimum (maximum or minimum). If the function has several optimum points of desired type, an interval will be output containing all such points; to separate the solution, one can then apply the automatic root search procedure with an appropriate search direction. For example, the problem of finding the global maximum for a polynomial is stated in the UniCalc language as follows:

```
(* Find the maximum *)
f(x) := x^6 - 26 * x^3 + 45 * x^2 - 10 * x + 1;
fmax = f(x) ;
    (* The first derivative equals zero *)
dif(f(), x) = 0;
    (* The second derivative at a maximum point is negative *)
dif(f(), x:2) < 0;
```

After the computation, we have the following solution:

$$x = 1.2176, \quad fmax = 11.86.$$

FINDING EXACT BOUNDS FOR A FUNCTION. Below, we supply some examples that demonstrate the effect of applying symbolic differentiation to the computation of bounds of interval functions.

Consider the following polynomial in two variables,

$$f(x, y) = x^2 \times y - 3x \times y + y^2 - 5x \times y^2 + 6x^3,$$

defined in the domain $x = [-2, 2], y = [-3, 3]$. If we attempt to find its bounds with the ordinary interval operations, we will see that its values lie in the interval $[-168, 177]$; however, by applying the special mechanism (including symbolic differentiation and reduction of similar terms), we can find the bounds that are almost exact: $[-50.2727, 81.2774]$.

Next, we give the bounds of the intervals, computed by different methods, for one more function, which is often treated in optimization (the so-called "three hump camel function"):

$$f(x, y) = 2x^2 - 1.06x^4 + 1/6x^6 - x \times y + y^2.$$

The usual interval computations for the domain $x = [-2, 4], y = [-2, 4]$ yield the bounds

$$[-287.36, 738.667],$$

whereas the exact bounds that we can determine are

$$[-0.7458 \times 10^{-10}, 455.307].$$

Note that the possibility treated here provides another approach to determination of global extrema of functions defined analytically.

7 Conclusion

In this article, we have described the method of subdefinite computations, the solver of mathematical problems UniCalc based on this method, and the application of techniques of computer algebra to improve the efficiency of the method. To confirm the practical applicability and usefulness of this approach, we have successfully solved a large number of various problems.

At present, UniCalc is implemented for PC-compatible computers running MS Windows or MS DOS, version 3.0 or later, with at least 640K of RAM. The problems that have been solved on an IBM PC AT contained up to 300 relations.

The investigation of the method of subdefinite computations and the use of techniques of computer algebra is continuing. In particular, our objectives for the nearest future are as follows:

1. In subdefinite computations.
   - further development of the methods for representation of subdefinite data (for instance, using multi-intervals) with the corresponding modification of the computation process;
   - investigation of techniques for the implementation of the algorithm considered above on parallel computers and transputers;
   - extending the notion of subdefiniteness to other objects (for example, subdefinite relations and subdefinite functions).
2. In computer algebra:
   - investigate the dependence of the efficiency of the algorithm on the form of the input expressions and develop the means for transforming the expressions to the optimal form automatically;
   - use the methods based on using Grobner bases to solve certain types of systems of polynomial equations;
   - include the operations of polynomial arithmetic in the language of the solver.

References

[Al1] Alefeld G., Herzberger Ju.: *Introduction in Interval Computations*. Academic Press, New York, 1983.

[As1] Asaithambi N.S., Zuhe Shen, Moore R.E.: On computing the range of values. *Computing* 28 (1982) 225 – 237

[Bo1] Borde S.B., Pon'kin S.A., Salychev M.V.: Subdefinitness and calendar scheduling. *Proceedings of the East-West Conference on Artificial Intelligence EWAIC'93. Moscow, Russia* (September 1993) 315 – 318

[Da1] Davis E.: Constraint propagation with interval labels. *Artificial Intelligence* 32 (1987) 99 – 118

[Hy1] Hyvonen E.: Constraint reasoning based on interval arithmetic. *Proceedings of IJCAI – 91* (1991) 1193 – 1198

[Ka1] Kantorovich L.V.: On a system of mathematical symbols, convenient for electronic computer operations. *Proceedings of USSR Acad. of Sciences* 113 N4 (1957) 738 – 741 (In Russian)

[Ku1] Kumar V.: Algorithms for constraint-satisfaction problems: a survey. *AI Magazine* (Spring 1992) 32 - 44

[Na1] Narin'yani A.S.: Subdefinite models and operations with subdefinite values. *Preprint, USSR Acad. of Sciences, Siberian Division*, Computer Center, Novosibirsk **400** (1982) — 33 p. (In Russian)

[Na2] Narin'yani A.S.: Active data types for representing and processing of subdefinite information. *Actual Problems of the Computer Architecture Development and Computer System Software.* Novosibirsk (1983) 128 - 141 (In Russian)

[Na3] Narin'yani A.S.: Subdefiniteness in knowledge representation and processing systems. *Transactions of USSR Acad. of Sciences, Technical Cybernetics* N5 (1986) 3 - 28 (In Russian)

[Te1] Telerman V.V.: Active data types. *Preprint, USSR Acad. of Sciences, Siberian Division*, Computer Center, Novosibirsk **792** (1988) — 30 p. (In Russian)

[Un1] Babichev A.B., Kadyrova O.B., Kashevarova T.P., Leshchenko A.S., Semenov A.L.: UniCalc, A Novel Approach to Solving Systems of Algebraic Equations. Proceedings of the International Conference on Numerical Analysis with Automatic Result Verifications. Lafayette, Louisiana, USA, 1993. *Interval Computations* N2 (1993) 29 - 47

Lecture Notes in Computer Science

For information about Vols. 1–877
please contact your bookseller or Springer-Verlag